# Handbook
## of
## Transducers

# Handbook
# of
# Transducers

**HARRY N. NORTON**
*Altadena, California*

Prentice Hall, Englewood Cliffs, NJ 07632

Library of Congress Cataloging-in-Publication Data

Norton, Harry N.
  Handbook of transducers.

    Bibliography: p.
    Includes index.
      1. Transducers—Handbooks, manuals, etc.   I. Title.
TK7872.T6N67  1989          621.381          88-15246
ISBN 0-13-382599-X

Editorial/production supervision: WordCrafters Editorial Services, Inc.
Cover design: Wanda Lubelska
Manufacturing buyer: Mary Ann Gloriande

© 1989 by Prentice-Hall, Inc.
A Division of Simon & Schuster
Englewood Cliffs, New Jersey 07632

Printed in the United States of America

10  9  8  7  6  5  4  3  2  1

ISBN 0-13-382599-X

Prentice-Hall International (UK) Limited, *London*
Prentice-Hall of Australia Pty. Limited, *Sydney*
Prentice-Hall Canada Inc., *Toronto*
Prentice-Hall Hispanoamericana, S.A., *Mexico*
Prentice-Hall of India Private Limited, *New Delhi*
Prentice-Hall of Japan, Inc., *Tokyo*
Simon & Schuster Asia Pte. Ltd., *Singapore*
Editora Prentice-Hall do Brasil, Ltda., *Rio de Janeiro*

# Contents

# *Preface*

If you would have tried a few decades ago to buy almost any of the types of electrical transducers that are now available you would have gotten a blank stare as if you had asked for a pocket calculator or a digital wristwatch. Most measuring instruments in those days were meters and gages that had to be read by looking at them. Some types of electrical transducers did exist, but they were usually made laboriously by hand and were found primarily in research laboratories. The rapid growth of the electronics field brought about ever-increasing demands for measuring devices capable of producing a signal that could be read on a meter that was located not at the point of measurement, but at a more convenient and often centralized location where many such measurements could be observed at the same time. Electrical transducers then became devices manufactured on a production line.

Further growth of the transducer field was stimulated by the increasing availability of computers that could accept such signals and then manipulate them in various ways. As more electronic equipment and electrical transducers became available, engineers found more uses for them. Automatic control instrumentation changed from pneumatic, mechanical, and hydraulic systems to electronic systems. It eventually became possible to obtain measurements from sites that human observers did not want to go to or couldn't possibly go to: the insides of a nuclear reactor; a satellite in space; a weather-monitoring buoy in the Arctic Ocean; a deep oil well; and a test site for nuclear weapons, to name just a few.

Now, as we are approaching the end of the twentieth century, we see transducers used extensively for measurement as well as control instrumentation in all scientific and industrial fields. We see them more widely used in consumer products—in automobiles, appliances, alarm systems, and medical thermometers and blood-pressure monitors for home use. And we see their use mandated by governments for pollution control (e.g., noise and vibration). There is no doubt that transducer applications will keep increasing further.

Many new engineers, scientists, teachers, technologists, and managers are entering fields in which a knowledge of transducers is required. This poses the need for a new handbook that (1) shows which transducers are available for the various categories of measurements, (2) describes how these various transducers operate, (3) clarifies the nature of the different types of transducers that are used for similar measurements, and (4) explains the essential characteristics of all these transducers. This book is intended to fill that need. While it should prove very useful to those involved in transducer design, it is primarily aimed at those who use transducers in measurement and control instrumentation and is thus applications oriented.

This book is meant to be used by anyone having a high-school degree, as long as the curriculum included physics and basic algebra. A college degree is not necessary. On the other hand, even those with a college degree in science or engineering may have forgotten some of the basic concepts underlying the physical quantities that are measured by transducers. Therefore, each chapter summarizes these fundamentals. The units of measurement used throughout the book are those of the Système International (SI), sometimes called the "SI metric system"; however, conversions to "U.S. customary units" are included and, in a few instances, shown in parentheses together with the corresponding SI unit. Because of the comprehensive nature of the book, much of the material is presented in a condensed manner; a bibliography at the end of the book then points to sources of additional, more detailed information.

The book should prove useful as a reference book in trade schools, colleges, and universities. It should also be particularly helpful to anyone working with instrumentation in all industries and sciences in areas ranging from transducer design, through hardware selection, systems design, applications and marketing, to the interpretation of data generated by transducers, as well as to those working on, or responsible for, larger systems which contain instrumentation systems whose operation they need to understand.

Harry N. Norton
Altadena, California

# Handbook
## of
# Transducers

# 1

# *Instrumentation System Fundamentals*

## 1.1  INTRODUCTION

Transducers are used as *sensing devices* in measurement systems as well as control systems. *Measurement systems* are used to obtain *information*: one or more quantities are sensed and the measured values are displayed. How the displayed information is used is basically up to a human observer. *Control systems* are used to keep a quantity at a desired value. A human operator adjusts settings on electronic equipment, the sensing device feeds information to this equipment, and the equipment then sends a signal to a control device that adjusts the quantity.

## 1.2  MEASUREMENT SYSTEMS

The simplest measuring "system" would be a sensing device that also displays the measured value, such as a mercury-in-glass thermometer or a pressure gage. If the measured value needs to be recorded at certain times, a human observer with a clipboard, pencil, and wristwatch can be employed for this purpose. A more advanced method utilizes an automatic camera that takes pictures of the gage or thermometer, together with a clock, at regular intervals. In these examples, the operator, clipboard, pencil, and wristwatch, or the camera and clock, become a part of the system.

When a measured value is to be displayed some distance away from the point of measurement, a link between the two points becomes necessary.

This link can be mechanical (e.g., an automobile speedometer cable), pneumatic (a pipe filled with air whose pressure is varied by the sensing device, with a pressure indicator used for display), or electrical (an electrical cable). Electrical wiring is used in *electronic measuring systems,* in which the sensing device (transducer) has an *electrical* output and the display device accepts an *electrical* signal. Only transducers with electrical output are covered in this book, since almost all modern instrumentation systems are electronic.

### 1.2.1 Measurand

The term *measurand* is used throughout the book and needs to be clearly understood. The measurand is the quantity, property, or condition that is measured (then sensed and converted into a usable electrical output) by a transducer. Thus, if the measurand is temperature, it is measured by a temperature transducer; if it is pressure, it is measured by a pressure transducer. The chapters of this book subsequent to the introductory chapters are organized by measurand.

### 1.2.2 Basic Electronic Measuring System

A basic electronic measurement system is shown in Figure 1–1. It consists of:

1. The *transducer*, which converts the measurand into a usable electrical output.

2. The *signal conditioner*, which converts the transducer output into the type of electrical signal that the display device will accept.

3. The *display device* (or readout device), which displays the required information about the measurand.

4. The *power supply*, which feeds the required voltages to the signal conditioner, to all except "self-generating" (see Chapter 2) types of transducers, and, at times, to certain kinds of display devices.

Because of their immense variety, transducers constitute the key portion of each of the measuring systems in which they are used. Signal con-

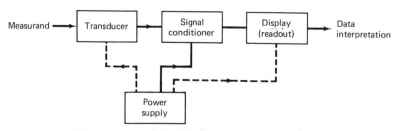

**FIGURE 1–1.** Basic electronic measuring system.

ditioners may vary in complexity from a simple resistance network or imped-
ance-matching device to multistage amplifiers with or without demodulators,
analog-to-digital converters, and other elaborate circuitry. Among commonly used display (readout) devices are circular-chart and strip-chart recorders, analog or digital meters, character printers, oscilloscopes (which
may be equipped with a camera so that permanent records can be obtained),
and discrete-level indicator lights.

A few types of measuring systems (e.g., synchro systems) operate without any signal conditioning. Most systems, however, employ signal conditioning either packaged as a separate unit, or included in the transducer or
the readout equipment. Similarly, the power supply function may be included
in the readout equipment or provided as a separate unit.

### 1.2.3 Multiple-Data Measuring Systems

Most measurement systems are designed to handle and display the outputs
of two or more transducers. The transducers feeding into such a multiple-
data measuring system can be of the same type (e.g., several thermocouples)
or of different types (e.g., temperature, pressure, and vibration transducers).
Signal conditioning in the system can be minimized by "standardizing" the
transducer output, that is, having each transducer provide the same full-
scale output to the system, regardless of type or measuring range.

Typical multiple-data measuring systems are illustrated in Figure 1–2.
Each system provides for at least some amount of signal conditioning in
addition to whatever conditioning is incorporated within the transducer.
Transducer excitation power, if any, is either connected to all transducers
simultaneously or switched to each transducer (in selectable systems) as it
is being read out, usually by a second set of contacts in the stepper or selector
switch. The latter method reduces transducer power consumption substantially.

The simplest system, shown in Figure 1–2(a), is one in which the transducer to be read is selected manually, such as by means of a rotary selector
switch. When a number of different measurements must be monitored repeatedly at relatively short intervals, a system such as the one shown in
Figure 1–2(b) can be used, wherein an automatically operating stepper or
sequencer scans the transducer outputs and the readout device also displays
an identifying number for the measurement being displayed. The temporary
("volatile") display on a digital meter can be augmented or replaced by a
permanent record, such as one obtainable from a printer. A simultaneous
display of several measurements on a multichannel strip-chart recorder, as
in Figure 1–2(c), is most frequently used when several related measurements
are expected to fluctuate rapidly. Numerous variations of the systems illustrated exist, including those in which the transducer outputs, together
with a timing signal (*clock*) and a means of identifying each measurement,

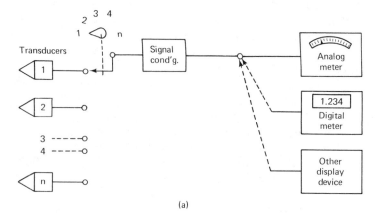

(a)

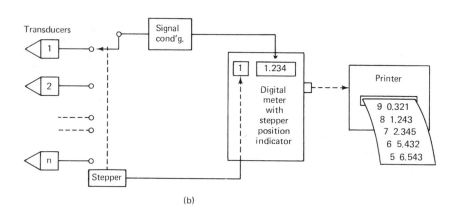

(b)

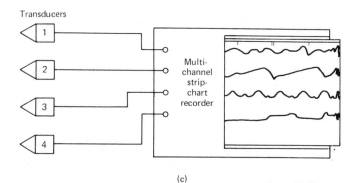

(c)

**FIGURE 1–2.** Typical multiple-data measuring systems: (a) manually selected measurements; (b) automatically selected measurements; (c) simultaneously displayed measurements.

are stored on magnetic tape which can be played back later into one or more display devices.

## 1.3   TELEMETRY SYSTEMS

Although all remote-display electronic measuring systems could be termed *telemetry* systems, this term is usually reserved for multiple-data systems using a modulated high-frequency carrier to transmit the information about the measurements from one point to another.

A generalized basic telemetry system is illustrated in Figure 1–3. The outputs from the transducers or other sensing devices, which may or may not require signal conditioning, are fed to a commutator (*multiplexer*), which combines them into a single *composite* signal. This signal is applied to the high-frequency transmitter, where it modulates the output of an oscillator. The modulated *carrier* is amplified and then fed to an antenna. The transmitting antenna, which can be highly directional, radiates the modulated carrier toward a receiving antenna. The received signal is amplified and applied to a *demodulator*, which separates the modulating information from the high-frequency carrier. This process reconstitutes the composite signal at the receiving end of the system. A *decommutator* is then employed to extract signals corresponding to the respective sensing-device outputs, so that each measurement can be displayed and evaluated individually. A *data processor* may or may not be required for the desired form of display.

In some types of telemetry systems the radio link is replaced by a conducting link. An example of this is the *carrier-current system* used by utility companies, in which the modulated carrier is coupled directly on a power transmission line and then decoupled from this line at the receiving end. Another example is *landline* (or "hardline") telemetry, where either a modulated high-frequency carrier is transmitted to a remote receiving station through a coaxial cable, or multiconductor shielded cables are used to feed a number of individual transducer outputs to a remote display center.

### 1.3.1   Carrier Modulation

The manner in which the transmitter's carrier signal is modulated—the type of *modulation*—deserves a more detailed description since it normally determines the nomenclature of the telemetry system (see Figure 1–4). The frequency of an *amplitude-modulated* (AM) carrier remains constant while its amplitude changes with the modulating signal. The frequency and amplitude of a *phase-modulated* (PM) carrier remain constant while its phase changes with the modulating signal. The amplitude of a *frequency-modulated* (FM) carrier remains constant while its frequency changes with the modulating signal.

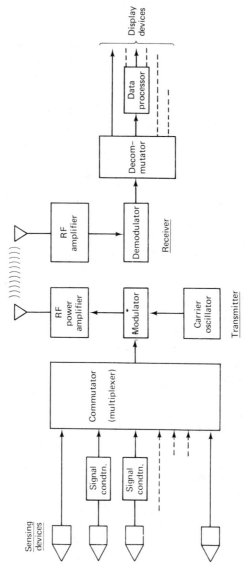

**FIGURE 1–3.** Basic radio telemetry system.

(a)

Phase modulation (PM)

Amplitude modulation (AM)　　　　　Frequency modulation (FM)

(b)

**FIGURE 1–4.** Principal types of carrier modulation: (a) unmodulated carrier; (b) modulated carrier.

### 1.3.2 Multiplexing

Two different methods are used to combine individual measurements into a composite signal for transmission over a single data link.

*Frequency-division multiplexing* (see Figure 1–5) allows the continuous display of several measurements. Each sensing-device output is fed to a different *subcarrier oscillator* (SCO). Each SCO is tuned to a different frequency in the general range 0.4 to 70 kHz. The SCO outputs are then linearly summed to form the composite signal. At the receiving station the composite signal is fed through band-pass filters, one filter for each SCO used in the transmitter, and tuned to the respective SCO frequency. The demodulated output of each band-pass filter then represents the corresponding measurement.

*Time-division multiplexing* (TDM) involves the time sharing of a number of individual measurements. This method does not permit a continuous display of each measurement. However, the measurement can be reconstituted from samples of the sensing-device output if the sampling occurs frequently enough. The sampling rate for any given measurement depends on its expected rate of fluctuation with time. The measurements are sampled by *commutating* them, that is, by switching them sequentially into a common output circuit.

Figure 1–6 shows a simple commutation scheme as well as a method for *subcommutation* which permits a number of relatively slowly varying *measurement* signals to be switched sequentially into one segment of a commutator to whose other segments the relatively rapidly varying measurement

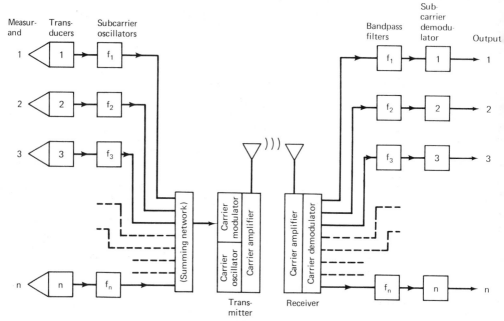

**FIGURE 1–5.** Frequency-division multiplex telemetry system.

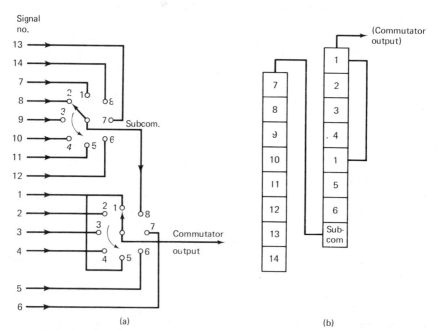

**FIGURE 1–6.** Commutation (multiplexing) and subcommutation (submultiplexing): (a) electromechanical representation; (b) solid-state representation.

signals are connected. In the example illustrated, the commutator output contains signal 1 sixteen times as often as signals 8 through 15, and it contains signals 2, 3, 4, 5, and 6 eight times as often as signals 7 through 14. This is because the subcommutator advances by only one segment for each full rotation of the commutator, and because signal 1 is *cross-strapped* from commutator segment 1 to segment 5. The example shows a representation typical for electromechanical commutators as well as an equivalent layout for a solid-state commutator that uses solid-state switching logic instead of a motor-driven rotary switch. Virtually all modern multiplexers are solid state.

The reverse process, *decommutation*, is used at the receiving end of the data transmission system to separate the time-sharing measurement signals from the commutated composite data stream. To synchronize the decommutator with the commutator, the transmitted data contain appropriate synchronization signals. They can also contain calibration reference signals, such as the zero- and full-scale levels of a measurement-circuit reference voltage, useful in evaluating the received data accurately.

### 1.3.3  Modulation Methods for TDM Data

One of the following four basic methods (see Figure 1–7) is generally used to modulate the telemetry subcarrier or carrier with time-division-multiplexed composite signals. In each method the sequential sampling of measurement signals results in a series of pulses (*pulse train*). The simplest of these methods is *pulse-amplitude modulation* (PAM), in which the amplitude (height) of each pulse is an analog of the measurement-signal value at the time it is being sampled by the commutator. The other three methods require a converter to modify the analog measurement signals into the appropriate pulse-type signals.

In *pulse-duration modulation* (PDM) the duration of each pulse (*pulse width*) represents the value of the measurement-signal sample. In *pulse-position modulation* (PPM) this value is represented by the position, in time, of a pulse. A reference pulse train can be transmitted together with the signal pulse train to serve as a repetitive time reference for the positions of the signal pulse. Both PDM and PPM are forms of *pulse-time modulation*.

*Pulse-code modulation* (PCM) is the most efficient of the four modulation methods because the transmitter power needed to send a given amount of information is less than for the other three methods. The analog signals are converted into a pulse code, usually a series of binary digits, by an analog-to-digital converter (ADC). The value of the sampled measurement signal is thus represented as a discrete amplitude increment, by a digital word. The number of bits used to form each word dictates the resolution obtainable for the data; for example, 127 discrete increments are available when seven-bit words are used ($2^7 - 1$), whereas 511 discrete increments are available when nine-bit words are used in a PCM system (see Table 1–1).

**Table 1–1.   RESOLUTION OF DIGITIZED
ANALOG MEASUREMENT AS
FUNCTION OF LENGTH OF
DIGITAL WORD**

| *Number of Discrete Increments* | *Word Length (bits)* |
|:---:|:---:|
| 1 | 1 |
| 3 | 2 |
| 7 | 3 |
| 15 | 4 |
| 31 | 5 |
| 63 | 6 |
| 127 | 7 |
| 255 | 8 |
| 511 | 9 |
| 1023 | 10 |
| 2047 | 11 |
| 4095 | 12 |
| 8191 | 13 |
| 16 383 | 14 |
| 32 767 | 15 |
| 65 535 | 16 |
| 131 071 | 17 |
| 263 143 | 18 |
| 524 287 | 19 |
| 1 048 575 | 20 |
| 2 097 151 | 21 |
| 4 194 303 | 22 |
| 8 388 607 | 23 |
| 16 777 215 | 24 |
| 33 554 431 | 25 |

Alternative methods of digital encoding are *frequency-shift keying* (FSK), used in PCM/FM systems, and *phase-shift keying* (PSK), where the transducer output is converted into fixed-step changes of the phase of the modulating signal.

In some time-division multiplex systems a "zero" reference pulse, of the same type as the signal pulse but of a fixed low level, is inserted between consecutive signal pulses. The resulting *return-to-zero* (RZ) waveform affords only a 50% duty cycle but facilitates signal separation after decommutation. The PAM illustration (Figure 1–7(a)) shows such a waveform. When no such signal separation is required, the 100% duty cycle *non-return-to-zero* (NRZ) waveform is used.

Combined multiplexing (time division as well as frequency division) is often used in FM systems in which groups of measurements are commutated into subcarrier oscillators.

The nomenclature of a telemetry system is given by the type of modulation used; for example, in a PAM/FM system a PAM pulse train frequency-modulates the radio-frequency (RF) carrier.

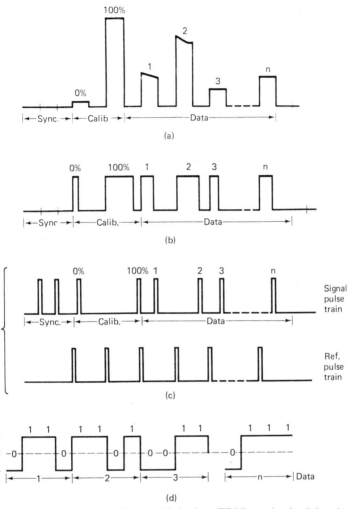

**FIGURE 1–7.** Time-division multiplexing (TDM) methods: (a) pulse-amplitude modulation (PAM); (b) pulse-duration modulation (PDM); (c) pulse-position modulation (PPM); (d) pulse-code modulation (PCM) (NRZ waveform, four-bit binary encoding).

## 1.4 DATA CONDITIONING, PROCESSING, AND DISPLAY

### 1.4.1 Analog Data

The "raw" data provided by a measuring or analyzing system often require a number of different operations to be performed on them to facilitate the determination of the required information (the *data reduction*).

Amplifiers can be used to make the full-scale amplitude of the system output data compatible with the capabilities of a given display unit. Filters can be used to remove noise from data signals or to eliminate high-frequency components (*low-pass filter*), low-frequency components (*high-pass filter*), or frequency components above and below a given frequency band (*band-pass filter*). *Amplitude discriminators* can be employed to create an "on-off" signal as a function of the difference in amplitude between the data signal and a reference signal. *Frequency discriminators* (frequency-to-dc converters) convert frequency variations into amplitude variations. *Rectifiers* convert ac amplitude variations into dc amplitude variations. Bias networks can be used to display only that portion of a data signal that is above or below a preset level. A composite system containing such devices is shown in Figure 1–8.

Data can be recorded on magnetic tape or stored in a computer memory for later playback or readout (*data storage*). Analog data can be digitized (*analog-to-digital converter*), and digital data can be converted into analog form (*digital-to-analog converter*).

### 1.4.2 Digital Data

The processing of digital data, such as PCM telemetry data or digitized analog data, is usually handled by computer systems. The availability of digital computer systems, varying widely in cost, capability, and complexity, has facilitated data reduction to such an extent that such systems have become attractive to designers and users of even relatively small measuring or analysis systems.

A typical system for digital data processing and display is illustrated

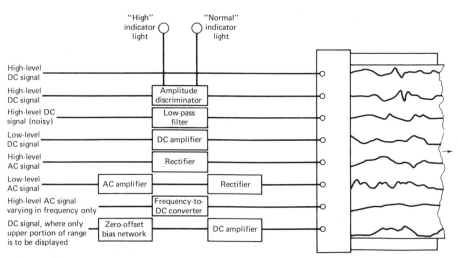

**FIGURE 1–8.** Analog data conditioning (with displays on multi-channel oscillograph).

in Figure 1–9. The incoming composite data stream (from which any RF carrier has been removed) usually consists of a number of sequential data frames. Each data frame starts with a frame synchronization word (typically between 7 and 31 bits in length) and one or more additional identifier words (often including a word indicating at what time the data were acquired); these are followed by the data words. Each word is a group of bits representing either one digitized analog measurement, one event count, or one group of state or mode indications. It is desirable to keep the length of all data words (the number of bits in each word) the same.

The data stream is usually stored on magnetic tape. In some systems it is necessary to delay any further processing at that time, such as when the computer is at a different location or when the computer is not available to the user at the time data are being received. The tape is then played back through the computer at a later time (*recorded data processing*). In the system illustrated, the data stream is simultaneously fed directly to the computer (*real-time data processing*). If any errors occur during processing, or if the data must be processed again for other reasons, the recorded data are then still available on a reel of magnetic tape.

The data are interfaced to the computer. The essential portions of the

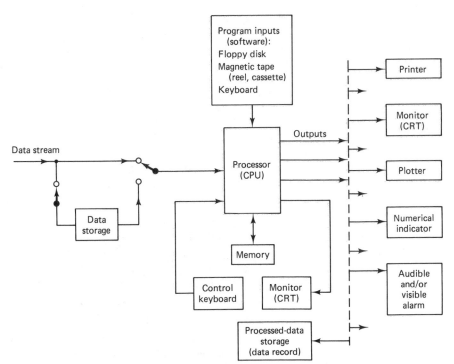

**FIGURE 1–9.** Digital data processing and display system, block diagram.

data computer are the *processor* (central processing unit or CPU) and the *memory* (typically a magnetic core). Programs required to process the data and to display them in the desired form on various display devices are stored on magnetic tape or disks, or generated from a keyboard. Reader units then feed the appropriate programs to the CPU and its memory. At least one operator's console is a part of the computer system. A keyboard control unit allows the operator to modify or override stored data-processing programs, to alter data display, and to perform diagnostic operations when computer functions are improper.

Among the variety of display units available for computer systems, the most popular are printers and cathode-ray-tube (CRT) monitors, both of which can display data in alphanumeric form (numbers and characters of the alphabet). Printers provide a permanent record of data on paper strips or sheets. Strip printers print one or two relatively short groups of characters per line of an approximately two-inch-wide paper tape (similar to cash-register receipt strips). Line printers (page printers) print a relatively large number of characters on one line of a wide prefolded paper sheet, and then reset to the next line so that sequential pages of data printouts can be obtained. CRT units can similarly show page-type displays of data in alphanumeric form, but do not provide a permanent record. They can also be programmed for display in graphical form. When permanent data records are required in graphical form, X-Y plotters are employed to provide plots of the variation of a given measurand with time or another reference or measurand.

### 1.4.3  Data Processing

Computer systems allow many different types of special data processing in addition to a sequential display of decommutated data. A few examples are:

1. Converting the decimal equivalent of a digital data word into a decimal number representative of the measured value, expressed in engineering units on the basis of a calibration record (*engineering-unit conversion*).

2. Limiting the display of each measurement to those times when the change of the measured value, compared to its previous value, is significant (i.e., exceeds a specified tolerance). Such *data suppression* facilitates the evaluation of data from a multimeasurement system by a single observer and results in shorter data records.

3. Comparing each data value to predetermined upper and/or lower limits and providing an alarm when the limit is exceeded (*alarm limit test*). The alarm can be in the form of a special character (e.g., an asterisk) next to the display of the data word. It can also be in the form of a warning light or audible tone.

4. Accumulating successive values of the same measurement over a

specified period of time, averaging those values, and then displaying the average value (*data averaging*).

5. Accumulating successive values of the same measurement over a specified period of time (or a specified number of data words), determining the largest of these values, and displaying the largest value (*peak search*).

6. Performing mathematical operations on data for one or more measurements and displaying the results (*computer-derived data*), such as multiplying a current measurement by a voltage measurement to display electrical power.

7. Comparing variations of a measurement, over a specified interval, with computer-stored data representative of a "model" of such variations, and displaying data resulting from such a comparison.

## 1.5  CONTROL SYSTEMS

The purpose of measuring systems is to provide the user with information (data). In control systems employing a human operator as part of the control loop, this information can then be used by the operator to effect a control function manually (e.g., increase a temperature, reduce a pressure, stop a flow, fill a tank, or change a speed). In automatic control systems the output of the sensing or analyzing device is used to effect a control function without the use of a human operator. The former are known as *open-loop* control systems, the latter as *closed-loop* control systems.

The most commonly used automatic control systems are closed-loop systems employing feedback. A feedback loop includes a forward signal path, a feedback signal path, and a signal summing point, which together form a closed circuit. A typical basic closed-loop control system is illustrated in Figure 1–10. It operates in the following manner (equivalent terms commonly used in process control are shown in brackets):

A specific quantity within a *controlled system* [*process*] is to be maintained at a specified magnitude. This *controlled quantity* [*controlled variable*] is measured by a sensing device, usually a transducer [*transmitter*]. The output of the sensing device, which may or may not have to be conditioned in some manner, is fed to a *comparing element*, or *summing point* [*set point*], in a *regulating device* [*controller*]. At this point, the signal fed back from the *sensing device* [*feedback signal*] is compared with a *reference signal* [*set-point signal*]. If the two signals are of the same magnitude, or within a relatively narrow tolerance from each other [*dead band*], no further action occurs. If the two signals differ from each other by an amount larger than that tolerance, a regulating signal is sent to a *control device* [*final controlling element*]. This signal causes the control device to change a quantity or condition (*manipulated quantity*) [*manipulated variable*] in the controlled

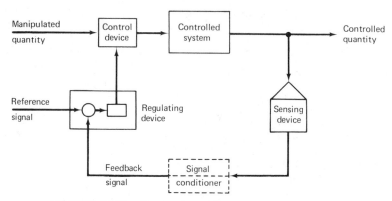

**FIGURE 1–10.**   Basic closed-loop control system.

system. The control action remains in effect until the controlled magnitude is at its proper level, as indicated by the feedback signal equaling the setpoint signal.

One example of a simple feedback control system is speed control of an internal combustion engine. If the angular speed (rate of rotation) of the engine's output shaft is to be the controlled quantity, a tachometer can be connected to the shaft. The output of the tachometer is compared with a preset reference signal. Differences between the two signals then cause a valve to operate so as to admit more or less fuel to the engine and, hence, to maintain the angular speed of the output shaft at the desired value.

There are many types of control action. The simplest is *on-off control,* as exemplified by the thermostatic furnace control used in most homes. When the room temperature drops below the set point, the furnace is turned on. When the temperature rises and reaches the set point again, the furnace is turned off. This control action usually results in a noticeable temperature change, particularly if the dead band of the controller is too wide. When the temperature of a heating device must be kept at a more constant level, as in many industrial applications, *proportional control* can be applied. This control action provides a continuous linear relation between the output and the input of the controller. A small deviation in the temperature from the desired value, as sensed by a temperature transducer, causes a small regulating action to restore the temperature to that level. A large deviation causes a large regulating action. In this manner the temperature is controlled more closely than by on-off control.

Among additional types of control action are *derivative control,* in which the controller output is proportional to the rate of change of the output, and *integral control,* in which the rate of change of the output is proportional to the input. The set point may be set manually, automatically, or in accordance with a program.

# 2

# *Transducer Fundamentals*

## 2.1 TRANSDUCER TERMINOLOGY

*A transducer is a sensor, but a sensor is not necessarily a transducer!* What does this statement mean? It means that transducers are really a subcategory of *sensing devices,* or *sensors,* the devices that furnish the input signals to instrumentation systems. A sensor can be something very simple like a shunt, for current measurement, or a voltage divider, for voltage measurement. Or it can be something very complex like a mass spectrometer or a multichannel infrared imaging spectrometer. The last two instruments belong to another subcategory of sensors, called *analyzers,* since their main application is in *chemical* analysis. A transducer, then, is a sensing device (or sensing instrument) that provides a usable electrical output in response to a *specific* measurand. The *measurand,* as explained in Section 1.2.1, is the *physical* quantity (or property or condition) that is to be measured.

Transducers have been, are being, and will probably always be called by different names in different technical disciplines. In the process industries, especially the petrochemical field, they are still very often referred to as transmitters (e.g., pressure transmitter instead of pressure transducer). This usage tends to be confusing to people involved with telemetry, where *transmitter* refers to a radio transmitter (see Section 1.3). In some facilities they are called *sensors* (e.g., temperature sensor); this is perfectly acceptable since, as just explained, all transducers are sensors (e.g., pressure sensors, force sensors, and temperature sensors). In some fields, notably in the area of electro-optical devices, they are called *detectors*, and it would be very difficult to convince workers in this area to call an IR detector an infrared light-intensity transducer. At one time, the word "cell" was popular for certain transducers. The term "load cell," meaning force transducer, is still very popular. Most of us think of a gage as a dial-type indicator; however, at various times transducers have been referred to as gages. Some transducers, particularly when they are small in size, are still occasionally called pickups (e.g., a vibration pickup). Transducers that have configurations enabling them to be immersed into a fluid are often called probes (e.g., probe-type temperature transducers are often called temperature probes). Many contractions ending in "-meter" are still used: for example, "accelerometer" for acceleration transducer, "flowmeter" for flow-rate transducer, and "tachometer" for angular-speed transducer. Although efforts at standardizing on the use of the term "transducer," described shortly, have been reasonably successful, many of the alternative terms have remained so popular that there should be no serious compunctions about using them in a handbook such as this.

During the aerospace boom of the 1960s a great many new types of transducers were specified, designed, manufactured, and used. The chaotic nomenclature situation became a nuisance, especially to users. An effort to produce a usable standard for transducer nomenclature and terminology was undertaken by the Instrument Society of America, initially to benefit primarily the aerospace industry, and later to facilitate communication with regard to transducers in all industries and sciences. Under the chairmanship, and later the directorship, of the author of this book, a standard was drafted, concurrently with the author's preparation of the *Handbook of Transducers for Electronic Measuring Systems* (Prentice-Hall, Inc., 1969, now out of print). ISA S37.1, "Electrical Transducer Nomenclature and Terminology," was published in 1969 and was adopted as American National Standard ANSI MC6.1-1975 in 1975. The adherence to such standards, particularly in the United States is, of course, voluntary. Many users and a number of manufacturers have been using the nomenclature and the terminology of this standard to varying degrees. Its perusal is recommended to readers of this handbook, wherein practices set forth in the standard are adhered to the extent deemed practical.

The description of a transducer is generally based on the following considerations:

1. *Measurand:* What quantity is intended to be measured?

2. *Transduction Principle:* What is the operating principle of the electrical portion of the transducer in which the output originates?

3. *Significant Feature:* This can be the type of *sensing element* employed (the element of the transducer that responds directly to the measurand), or a particular type of construction, or the function of

internal circuitry, or a descriptor used to differentiate between transducers using the same transduction principle.

4. *Range:* What are the upper and lower limits of the measurand values the transducer is intended to measure?

Transduction principles are explained in Section 2.2, and sensing elements are described in the chapters following Chapter 4. The "Basic Concepts" section of each chapter explains *measurand modifiers.* These are often necessary in transducer descriptions; e.g., a pressure transducer can be designed to measure *absolute, gage, or differential* pressure, and an accelerometer can be *uniaxial, biaxial, or triaxial.* Table 2–1 lists descriptors used for transducer nomenclature. The "Designation Letter" is not really a part of this listing; however, if properly used, it allows virtually any type of transducer to be designated by a four-letter code. For example, a semiconductor-strain-gage absolute-pressure transducer would be identified by the code "PAKK," since (column 1) P = pressure, (column 2) A = absolute, (column 3) K = strain-gage transduction, and (column 4) K = semiconductor (-element). The letter assignment shown is not intended to be a "standard."

The sequence in which descriptors are used in transducer nomenclature depends on how and where the nomenclature is used. In drawing and specification titles, and in indexing, cataloging, and inventorying, the sequence follows the one just shown, i.e., "transducer, displacement, potentiometric, 0 to 10 cm." In text and conversation, however, the sequence is reversed, e.g., "(Actuator motion was measured by a) 0-to-10-cm potentiometric dis-

**Table 2–1. DESCRIPTORS USED IN TRANSDUCER NOMENCLATURE[1]**

| Designation Letter | Measurand | Measurand Modifier | Transduction Principle | Significant Features |
|---|---|---|---|---|
| A | Acceleration | Absolute | Photovoltaic | Absolute |
| B | Chemical property[2] | Biaxial, bidirectional | Encoder type | Incremental |
| C | Current (electr.) | Compression | Capacitive | With integral dc-dc converter or signal amplifier or impedance converter |
| D | Displacement (position) | Differential | Photoconductive | Differential-pressure type |
| E | Voltage (potential) | Triaxial, tridirectional | Electro-optical, photoelectric (general) | With integral signal-to-dc voltage converter |
| F | Flow (flow rate) | Gamma-ray | Microwave (type) | Conductor (metal wire, foil, or film) element |
| G | Attitude | Gage (= relative to ambient) | Gyro type | Immersion probe |

Table 2–1. (*continued*)

| Designation Letter | Measurand | Measurand Modifier | Transduction Principle | Significant Features |
|---|---|---|---|---|
| H | Humidity (& moisture) | Visible (+ near UV, + near IR) | Hall effect | Hygrometric |
| I | Light intensity | Infrared | Ionization | With integral analog-to-digital converter |
| J | Torque | Ultraviolet | Pyroelectric | Damped-oscillation type |
| K | Strain | Continuous | Strain gage | Semiconductor (element) |
| L | Level (of "liquid") | Linear (rectilinear) | Reluctive (or inductive) | Dew-point sensing |
| M | Magnetic flux density | Medical, biomedical | Magnetic, electromagnetic | Mechanical-element (general) type |
| N | Nuclear radiation | Angular | Nucleonic (type) | Narrow-band type |
| O | — | (nonspecific or no modifier) | (nonspecific or no modifier) | (nonspecific or no modifier) |
| P | Pressure | Point | Potentiometric | Phase-displacement type |
| Q | Heat flux | Mass | Resistive (semiconductor) | Thermal or heat-transfer type |
| R | Density | Remotely sensed | Resistive (conductor) | Oscillating-fluid type |
| S | Speed or velocity | Sound (acoustic) | Servo type | Conductivity-sensing (type) |
| T | Temperature | Tension | Thermoelectric | Toothed-rotor or turbine type |
| U | Viscosity | Uniaxial, unidirectional | Ultrasonic (type) | For underwater use |
| V | Vacuum | Volumetric | Vibrating element | in "bare" form, e.g., in wire or cable form (thermocouples), or as "bare bead" (thermistors) |
| W | Force (and mass) | Neutron detection | Photoconductive | Wide-band type |
| X | — | X-ray | Switch type (or discrete-increment type) | Surface sensor (or nonintrusive, e.g., clamp-on) |
| Y | Attitude rate | Special purpose | (other) | Scintillation counting |
| Z | Power (electr.) | Multidirectional (more than three) or multifunction | Piezoelectric | Psychrometric |

[1] From H. N. Norton, ed., *Sensor Selection Guide* (Lausanne, Switzerland: Elsevier Sequoia S. A., 1984 (by permission)).
[2] Will involve a different set of modifiers.

placement transducer.'' As a general rule, transducer nomenclature should be as complete as possible and practical; it is better to describe too much than too little.

## 2.2 TRANSDUCTION PRINCIPLES

The principles of transduction, i.e., the operating principles of the transduction elements of typical transducers used for specific measurands, are covered in detail in the sections devoted to design and operation in the main body of this book. The most commonly used of these principles are defined and explained here. Most transduction elements are *passive* in nature: they require excitation power from some external source. Some types of transduction elements, however, notably electromagnetic, piezoelectric, thermoelectric, and photovoltaic elements, do not require external excitation. Such transduction elements are *self-generating.*

*Capacitive transduction elements convert a change in measurand into a change in capacitance* (Figure 2–1). Since a capacitor consists, basically, of two electrodes separated by a dielectric, the change in capacitance can be caused either by the motion of one of the electrodes to and from the other electrode, or by changes in the dielectric between two fixed electrodes.

*Inductive transduction elements convert a change in measurand into a change in the self-inductance of a single coil* (Figure 2–2). The inductance changes can be effected by the motion of a ferromagnetic core within a coil or by externally introduced flux changes in a coil having a fixed core.

*Reluctive transduction elements* (Figure 2–3) *convert a change in measurand into an ac voltage change due to a change in the reluctance path between two or more coils (or separated portions of one or more coils), with ac excitation applied to the coil system.* This category includes ''variable-reluctance,'' ''differential-transformer,'' and ''inductance-bridge'' elements. The change in reluctance path is usually effected by the motion of a magnetic core within the coil system.

*Electromagnetic transduction elements convert a change in measurand into an electromotive force (output voltage) induced in a conductor by a change in magnetic flux, in the absence of excitation* (Figure 2–4). The

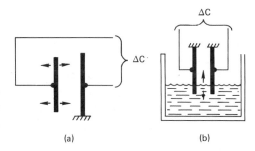

**FIGURE 2–1.** Capacitive transduction: (a) moving plate, constant dielectric; (b) fixed plates, changing dielectric.

(a)          (b)

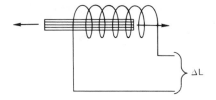

**FIGURE 2–2.** Inductive transduction.

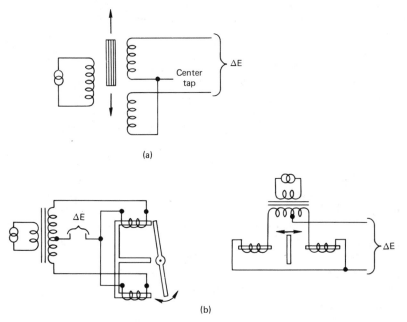

(a)

(b)

**FIGURE 2–3.** Reluctive transduction: (a) differential transformer; (b) inductance bridges (variable reluctance).

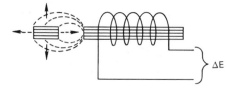

**FIGURE 2–4.** Electromagnetic transduction.

change in flux is usually effected by relative motion between an electromagnet and a magnet or portion of magnetic material.

*Piezoelectric transduction elements* (Figure 2–5) *convert a change in measurand into a change in the electrostatic charge (Q) or voltage (E) generated by certain materials when they are mechanically stressed.* The stress is typically developed by compression or tension forces, or by bending forces exerted upon the material (the *crystal*) directly by a sensing element or by a mechanical member linked to a sensing element.

*Resistive transduction elements convert a change in measurand into a change in resistance* (Figure 2–6). Resistance changes can be effected in conductors as well as semiconductors by such means as heating or cooling, applying mechanical stresses (so as to utilize the *piezoresistive effect*), wetting or drying of certain electrolytic salts, or moving the wiper arm of a rheostat.

*Potentiometric transduction elements convert a change in measurand into a voltage-ratio change by a change in the position of a movable contact (wiper) on a resistance element across which excitation is applied* (Figure 2–7). The ratio given by the wiper position is basically a *resistance ratio.*

*Strain-gage transduction elements convert a change in measurand into a change in resistance due to strain, in either two or four arms of a Wheat-*

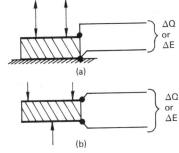

**FIGURE 2–5.** Piezoelectric transduction: (a) compression, tension; (b) bending.

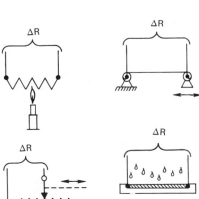

**FIGURE 2–6.** Resistive transduction.

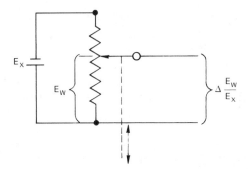

**FIGURE 2-7.** Potentiometric transduction.

*stone bridge.* This transduction principle is a special version of resistive transduction; however, it involves either two or four resistive strain transducers (*strain gages*) connected into a Wheatstone-bridge circuit across which excitation is applied, so that the output is a voltage change (Figure 2-8). Upward arrows in the illustration indicate increasing resistances, and downward arrows indicate decreasing resistances, in the arms of the bridge, that are simultaneously effected (in a *four-active-element bridge*) by a change in measurand due to the placement and connection of the individual resistive elements; in the example illustrated (an unbonded-strain-gage element) the indicated directions of the resistance changes would occur as the sensing link moves toward the left.

*Photoconductive transduction elements convert a change in measurand into a change in the resistance (or conductance) of a semiconductor material due to a change in the amount of illumination incident upon the material* (Figure 2-9).

*Photovoltaic transduction elements convert a change in measurand into a change in the voltage generated when the illumination incident upon a junction between certain dissimilar materials changes* (Figure 2-10).

*Thermoelectric transduction elements convert a change in measurand into a change in the electromotive force (emf) generated by a temperature difference between the junctions of two selected dissimilar materials* (due

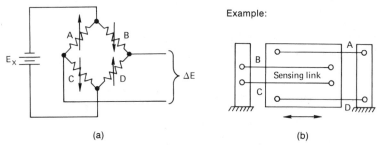

(a)                    (b)

**FIGURE 2-8.** Strain-gage transduction: (a) basic circuit; (b) example.

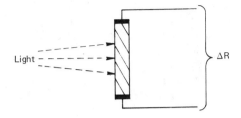

**FIGURE 2–9.** Photoconductive transduction.

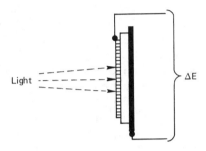

**FIGURE 2–10.** Photovoltaic transduction.

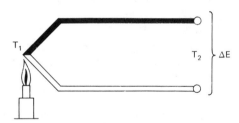

**FIGURE 2–11.** Thermoelectric transduction.

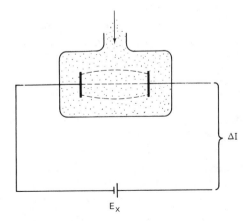

**FIGURE 2–12.** Ionizing transduction.

to the *Seebeck effect*). In the basic thermoelectric element shown in Figure 2–11, a junction between the output terminals (at which temperature $T_2$ prevails) would be formed, for example, by a voltmeter connected across the terminals.

*Ionizing transduction elements convert a change in measurand into a change in ionization current,* such as through a gas between two electrodes (Figure 2–12).

## 2.3  GENERAL TRANSDUCER CHARACTERISTICS

In general, transducer characteristics can be categorized into design, performance, and reliability characteristics. Sections 2.4, 2.5, and 2.6 define and explain a fairly complete set of these characteristics. All of these should be considered when specifying a transducer or selecting one from manufacturers' catalogs and bulletins; however, not all of them may be important or applicable to a given type of transducer. On the other hand, a few specialized characteristics are only important to one category of transducers; these are not included here, but are explained in the chapter dealing with this category.

Throughout this book, when two or more terms are shown for some of the transducer characteristics, as well as for some other expressions, the preferred term (which is usually a term defined in ANSI Standard MC6.1, "Electrical Transducer Nomenclature and Terminology") is printed in italics. Alternative terms whose usage is generally acceptable in a proper context are shown in parentheses. Terms whose usage is not recommended are shown in quotation marks.

## 2.4  DESIGN CHARACTERISTICS

The design characteristics of a transducer describe or specify how the transducer is (or should be, in a user-written specification) designed and constructed and what its "rating" is in terms of measuring range. As shown in Table 2–2, they pertain to the range, the electrical design, and the mechanical design of a transducer.

### 2.4.1  Measurand Characteristics

A transducer is normally designed to sense a specific measurand and to respond only to this measurand. For example, pressure transducers provide an output indicative of pressure, and acceleration transducers (accelerometers) provide an output representative of acceleration. Other measurands can often be calculated by their known relationship to the measurand sensed by the transducer. Thus, velocity can be calculated from measurements of displacement and time, and altitude (above sea level) as well as water depth

**Table 2-2. TRANSDUCER DESIGN CHARACTERISTICS**

| *Measurand* | *Electrical* | *Mechanical* |
|---|---|---|
| Range | Excitation | Configuration |
| Overrange | Isolation | Dimensions |
| Recovery | Grounding | Mountings |
|    time | Source impedance | Connections |
| | Load impedance | Case material |
| | Input impedance | Materials in contact with |
| | Output impedance |    measured fluids |
| | Insulation resistance | Case sealing |
| | Breakdown voltage rating | Identification |
| | Gain instability | |
| | Output | |
| | End points | |
| | Ripple | |
| | Harmonic content | |
| | Noise | |
| | Loading error | |

(below sea level) can be derived from pressure measurements. Various determinations can be inferred from one or more transducer output signals. For example, bearing wear can be inferred from accelerometer outputs, and electrical power can be calculated from the outputs of a current transducer and a voltage transducer. Each transducer, however, is specified by its basic measurand and its measuring range.

The *range* of a transducer is given by the upper and lower limits of measurand values it is intended to respond to within specified performance tolerances. A range can be *unidirectional* (e.g., "0 to 5 cm"), or *bidirectional* (either symmetrically, e.g., "±20 N", or asymmetrically, e.g., "−10 to +30 N"), or *expanded* (*zero-suppressed*) (e.g., "90 to 120 L/s"). The algebraic difference between the two range limits is the *span* of the transducer. The span of a 90-to-120 L/s flowmeter is 30 L/s; the span of a −10-to-+30 N force transducer is 40 N.

The *overrange* (overload, maximum measurand; *proof pressure* for a pressure transducer) is the maximum magnitude of measurand that can be applied to a transducer without causing a change in performance beyond specified tolerances. The *recovery time* is the amount of time allowed to elapse after removal of an overrange condition before the transducer again performs within the specified tolerances.

### 2.4.2 *Electrical Design Characteristics*

The basic electrical design characteristics of a transducer are illustrated in Figure 2-13, in which a transducer is viewed as a "black box," that is, without regard to its internal workings and primarily as a device with which other electrical or electronic equipment must interface electrically.

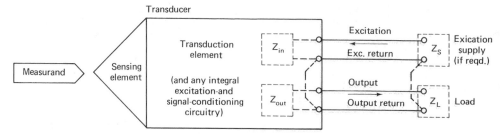

**FIGURE 2–13.** Basic "black-box" electrical characteristics of a transducer.

With the exception of the self-generating types (unless they include integrally packaged conditioning circuitry), transducers require *excitation,* i.e., externally supplied electrical power (expressed in terms of voltage or current, or both) applied to them for their proper operation. The impedance of the excitation supply presented to the transducer is the *source impedance* $(Z_s)$, of which the impedance of the excitation cabling is considered a part. In the opposite direction, the impedance of the transducer presented to the excitation supply is the *input impedance* $(Z_{in})$. The impedance measured across the output terminals of the transducer is the *output impedance* $(Z_{out})$. In the opposite direction, the impedance presented to these output terminals by the external circuitry they connect into is the *load impedance* $(Z_L)$. The impedance of the cabling between the transducer and the external circuitry (the *load*) is included in $Z_L$.

The diagram in Figure 2–13 is intended to be general. The excitation return and output return are separate for many types of transducers, such as the strain-gage types (see Figure 2–8), as well as for the many designs incorporating excitation conditioning or signal conditioning (or both) inside the transducer housing. Some types of transducers (e.g., the potentiometric versions; see Figure 2–7) use only one return, and this is indicated by the dashed jumpers at both ends of the return lines. Other types, e.g, resistive transducers (see Figure 2–6), may be simple two-terminal devices in which the passive transduction element is essentially connected between "excitation" and "excitation return."

Shielding and grounding considerations are very important in instrumentation systems, in which signal levels are typically quite low and can be affected by noise pickup, ground loops, and common-mode voltages. When a transducer is enclosed in a housing (case) the return lines (and internal grounds) are usually not electrically connected to the case; instead, a single-point ground is established at the excitation supply or at the load. The excitation and output lines may stay entirely disconnected from an external ground (earth ground), and only the shielding of the wires is eventually connected to such a ground. At the transducer end, the shield around the wires is kept unconnected, unless it is connected to a terminal leading to an internal shield (*a guard*) which itself is not connected to the case or any internal grounds (*floating grounds*). In some transducers, special provisions, such as transformer coupling, are included to isolate the excitation side completely from the output side.

When two or more portions of a transducer are electrically insulated from each other, the resistance between them, as measured while a specified dc voltage is applied, is the *insulation resistance*. The degree of insulation can also be expressed in terms of the *breakdown voltage rating,* the magnitude of ac or dc voltage that can be applied across specified insulated portions without causing arcing and without causing conduction between the portions, above a specified value of current. The *breakdown voltage,* which is established by a test generally considered destructive, is the magnitude of voltage at which arcing or excessive conduction occurs.

*Output* is the electrical quantity produced by a transducer and is a function of the applied measurand. The output is usually a continuous function of the measurand (*analog output*) in the form of voltage amplitude, voltage ratio, or current, or sometimes just as changes in capacitance, inductance, and so on. *Frequency output,* where the number of cycles or pulses per second are a function of the measurand, and *frequency-modulated output,* i.e., frequency deviations from a "center" frequency (e.g., "3000 ± 200 Hz"), are also forms of analog output. *Digital output* represents the measurand in the form of discrete quantities coded in some system of notation (e.g., binary code). Output that represents the measurand in the form of discrete or quantized values not coded in any system of notation is sometimes referred to as *discrete-increment output*; such an output is exemplified by that of a switch-type transducer.

*End points* are the output values at the lower and upper limits of the range of a transducer. They can be the mean of end-point readings determined over two or more consecutive calibration cycles. When end points are specified, a tolerance is usually applied to them (e.g., "0.00 ± 0.02 and 10.00 ± 0.01 V dc"). No such tolerances are allowed for *theoretical end points*, the points between which the theoretical curve is established (see Section 2.5.1). Theoretical end points are not necessarily established at 0% measurand, 0% output (lower end point) and 100% measurand, or 100% output (upper end point); however, when they are set at these values, they are referred to as *terminal end points*.

As for any other electronic device, it is important for transducers to be appropriately matched to, and interfaced with, the associated measurement system. One of the areas that tends to be overlooked is matching output impedance to load impedance. A mismatch in these impedances can cause *loading error,* which increases with the ratio of output impedance to load impedance. Careful attention must also be paid to manufacturer-recommended excitation supply characteristics.

Certain electrical characteristics need to be watched (and probably need tolerances assigned to them in a transducer specification) when excitation-conditioning or output-conditioning circuitry is incorporated within a transducer. The output of an ac-to-dc converter (demodulator) in a transducer, for example, may contain a measurable ac component (*ripple*). Similarly, the output from an integrally packaged amplifier may contain random disturbances (*noise*) and may be subject to changes in the characteristics of amplifier components that result in *gain instability*. When the output of a transducer is sinusoidal ac, it may contain distortions due to the presence of harmonics (frequencies other than the fundamental frequency); this *harmonic content* is usually expressed as a percentage of the rms output of the transducer.

### 2.4.3 *Mechanical Design Characteristics*

Mechanical design characteristics complement the electrical design characteristics in that they define primarily the *physical* interfaces of the transducer.

The mass ("weight") typically starts the list of these characteristics. Next is the configuration (best shown in the form of a drawing), which shows all pertinent dimensions and the location and orientation of all external mechanical, electrical, and fluid connections, including any mounting holes. The locations of any special provisions, such as those used for zero and gain adjustment, are included on the drawing. All external connections should be identified by type and (unless they are defined by an industrial or government standard) by their dimensions and materials. Case (housing) materials (and finish) and the type of case sealing should be specified. The materials that can come in contact with a measured fluid should also be stated; alternatively, the categories of measured fluids that may come in contact with those portions of a transducer that can be exposed to them can be specified (e.g., "noncorrosive liquids and gases" or "nonconductive fluids").

Some applications require that certain industrial or governmental standards or codes be adhered to by the transducer. This is often the case when, for example, a transducer must operate as part of a sealed system, must be explosion proof, must be waterproof, or must operate in a hazardous environment such as might contain nuclear radiation. The codes or standards that the transducer can meet are then stated.

Identification (nameplate information) is another important mechanical characteristic. Such information either is shown on a separate nameplate attached to the transducer case, or is directly etched or engraved on the case. Except for those rare instances when such information is limited by a very small transducer size, it is possible to include fairly complete information. It should really be possible for a user to install and connect a transducer without looking up drawings and specifications. Thus, the nameplate information should include proper nomenclature and the most pertinent characteristics, such as range, excitation, output, identification of electrical connections, and, of course, the name and location of the manufacturer and the part number and serial number of the transducer. Additional information may be required by applicable standards or codes and (when user specified) by special needs of the user.

## 2.5 PERFORMANCE CHARACTERISTICS

Transducer performance characteristics, summarized in Table 2–3, are generally categorized as follows:

**Table 2–3. TRANSDUCER PERFORMANCE CHARACTERISTICS**

| *Static* | *Dynamic* | *Environmental* |
|---|---|---|
| Resolution | Frequency response | Operating environmental |
| Threshold | Transient response: | effects |
| Creep | Response time | Operating temperature |
| Hysteresis | Rise time | range |
| Friction error | Time constant | Thermal zero shift |
| Repeatability | Natural frequency | Thermal sensitivity shift |
| Linearity | Damping | or |
| (+ reference line) | Damping ratio | Temperature error |
| Sensitivity | Overshoot | or |
| Zero-measured output | Ringing frequency | Temperature error band |
| Sensitivity shift | | Temperature gradient |
| Zero shift | | error |
| Conformance | | Acceleration error |
| (+ reference curve) | | or |
| or | | Acceleration error band |
| Static error band | | Attitude error |
| (+ reference line or | | Vibration error |
| curve) | | or |
| Reference lines: | | Vibration error band |
| Theoretical slope | | Ambient-pressure error |
| Terminal line | | or |
| End-point line | | Ambient-pressure error |
| Best straight line | | band |
| Least-squares line | | Mounting error |
| Reference curves: | | Nonoperating |
| Theoretical curve | | environmental effects |
| Mean-output curve | | Type-limited environmental |
| | | effects |
| | | Conduction error |
| | | Strain error |
| | | Transverse sensitivity |
| | | Reference-pressure error |

1. *Static characteristics,* which describe performance at room conditions, with very slow changes in the measurand, and in the absence of any shock, vibration, or acceleration (unless one of these is the measurand); although there is some disagreement as to what conditions constitute *room conditions,* they have generally been established as the following (unless specifically stated otherwise): a temperature of 25 ± 10°C, a relative humidity of 90% or less, and a barometric pressure of 880 to 1080 mbar (88 to 108 kPa).

2. *Dynamic characteristics,* which relate to the response of a transducer to variations of the measurand with time.

3. *Environmental characteristics,* which relate to the performance of a transducer after exposure (*nonoperating* environmental characteristics) or during exposure (*operating* environmental characteristics) to specified external conditions (such as temperatures, shock, or vibration).

### 2.5.1  Static Characteristics

An ideal or theoretical output-measurand relationship exists for every transducer. If the transducer were ideally designed by ideal designers, and if it were made from ideal materials by using ideal methods and workmanship, the output of this ideal transducer would always indicate the true value of the measurand. The output would follow exactly the prescribed or known *theoretical curve* which specifies the relationship of the output to the applied measurand over the transducer's range. Such a relationship can be stated in the form of a table of values, a graph, or a mathematical equation. Figure 2–14 illustrates a theoretical curve both in general terms (percent of full-scale output, or *% FSO*, vs. measurand expressed in percent of range) and for the example of a pressure transducer whose range is 0 to 1000 psia (0 to 6895 kPa) and whose output is 0 to 5 V dc, for the case of a *linear* output–measurand relationship, which causes the curve to be a straight line.

The output of an actual transducer, however, is affected by the nonideal behavior of the transducer, which causes the indicated measurand value to deviate from the true value. The algebraic difference between the indicated value and the true (or theoretical) value of the measurand is the transducer's *error.* Error is usually expressed in % FSO, sometimes in percent of the output reading of the transducer ("% of reading"), or in terms of units of the measurand. *Accuracy* is defined as the ratio of error to full-scale output, usually expressed in the form "within ± ____% FSO," sometimes in terms of units of measurand or in percent of the error/output ratio.

Although the simplest way to consider transducer errors is in terms of maximum deviations from a specified reference line or curve which defines the output-measurand relationship over the transducer's range (*error band*),

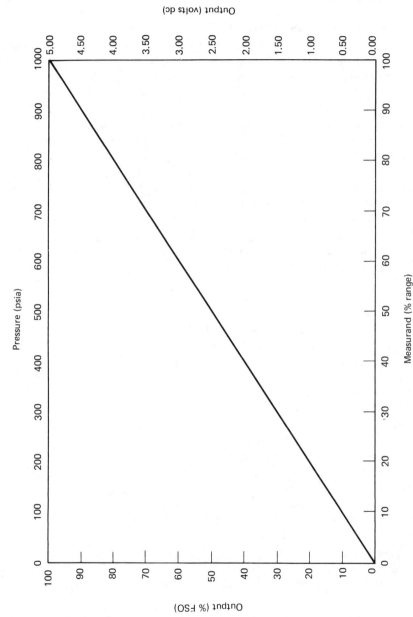

**FIGURE 2–14.** Output-measurand relationship of ideal linear-output transducer (including application of general example to a typical dc output pressure transducer).

the existence of individual errors such as nonlinearity, nonrepeatability, hysteresis, zero shift, and sensitivity shift must be recognized, and the nature of these errors must be understood. The effects of these errors on transducer behavior and data obtained should be known; such knowledge can often be used to correct final data and increase data accuracy.

Error characteristics are determined by *calibration*. This term usually implies a *static calibration,* performed for the purpose of determining static characteristics, unless the term *dynamic calibration* (see Section 2.5.2) is specifically used. A (static) calibration is a test during which known values of measurand are applied to a transducer and corresponding output readings are recorded. The resulting test record, when in the customary tabular form, is the *calibration record.* When it is in graphical form, it is referred to as *calibration curve.* Note that a calibration curve can also be plotted on the basis of a calibration record, manually or by the use of a computer. A single performance of this test over the entire range of the transducer (unless a *partial-range calibration* is specified), once with increasing and once with decreasing measurand, is called a *calibration cycle.* A complete calibration usually comprises two or more calibration cycles, which are commonly referred to as "Run 1," "Run 2," and so on.[1] Individual errors, as determined by calibration, are explained subsequently.

*Hysteresis* (see Figure 2–15) is the maximum difference in output, at any measurand value within the (specified) range, when the value is approached first with increasing and then with decreasing measurand. Many types of transducers exhibit hysteresis, which is typically caused by a lag in the action of the sensing element. Hysteresis is expressed in % FSO. The hysteresis seen when only a portion of the range is traversed (e.g., 0 to 30%, as shown in the illustration) is always less than the total hysteresis. Some type of transducers, notably potentiometric transducers, exhibit an error that looks like hysteresis but should not be confused with it. This error is typically caused by sliding friction between the wiper arm and the potentiometric element and is called *friction error.* Such friction effects can be minimized by *dithering* the transducer, i.e., applying intermittent or oscillatory acceleration forces to it (sometimes called "tapping"). When such a transducer is dithered during a calibration, its true hysteresis can be established. However, unless dithering is specified, friction error is included with hysteresis. *Friction error* is often determined as the maximum change in output at any measurand value within the range before and after minimizing friction within the transducer. A calibration during which dithering is employed (e.g., by mounting a small buzzer against the transducer) is called a *friction-free calibration.* Such a calibration should be used only when the intended application of the transducer can reasonably be expected to minimize friction error equally.

---

[1] Calibrations of certain categories of transducers, e.g., temperature transducers, are performed quite differently (see the subsequent chapters dealing with these categories).

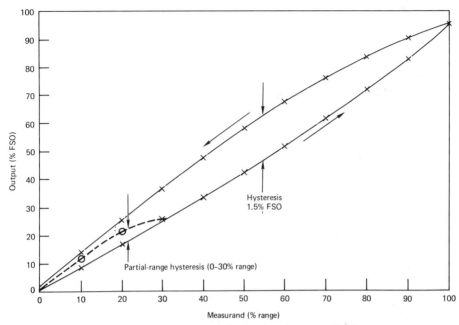

**FIGURE 2–15.** Hysteresis (scale of errors 10:1).

*Repeatability* (sometimes called "reproducibility") is the ability of a transducer to reproduce output readings when the same measurand value is applied to it consecutively under the same conditions and in the same direction. It is expressed as the maximum difference between output readings as determined by two calibration cycles (see Figure 2–16), unless otherwise specified, and is usually stated as "within ____% FSO." If the sampling is increased by increasing the number of calibration cycles, a better statistical measure of repeatability can be obtained.

*Linearity* is the closeness of a transducer's calibration curve to a specified straight line. It is expressed as "within ±____% FSO" (in a specification) or as "within +____, −____% FSO" (as a result of a calibration)—in effect, the maximum deviation of any calibration point from the corresponding point on the specified straight line during any one calibration cycle. When more than one calibration run is made, the worst linearity seen during any one calibration cycle is stated. "Linearity," when not accompanied by a statement explaining what sort of straight line it is referring to, is meaningless. A transducer may have an independent linearity within ±0.5% FSO while its terminal linearity is within ± 3.5% FSO. The specific type of reference line is stated either by adding a modifier to the word "linearity" (e.g., "terminal linearity") or by adding a statement such as "referred to the best straight line." The various types of linearity are as follows.

*Theoretical-slope linearity* is referenced to the *theoretical slope,* the

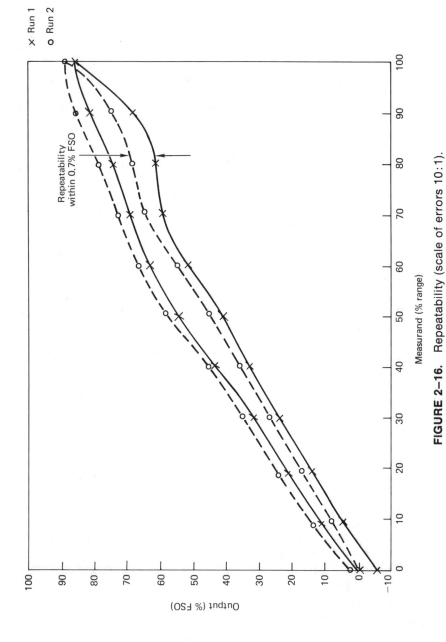

**FIGURE 2–16.** Repeatability (scale of errors 10:1).

straight line between the *theoretical end points*. These are usually close to 0% FSO (for 0% range) and close to 100% FSO (for 100% range), but can be purposely offset (e.g., 5% FSO at 0% range and 95% at 100% range). Since no tolerances apply to theoretical end points, the theoretical slope can always be drawn without referring to any measured values.

*Terminal linearity* is referenced to the *terminal line* (see Figure 2–17), a special form of theoretical slope for which the theoretical end points are exactly 0% and 100% of both the range and the full-scale output.

*End-point linearity* is referenced to the *end-point line,* the straight line between the *end points*, i.e., the outputs at the upper and lower range limits obtained and averaged (unless otherwise specified) during any one calibration. End-point tolerances should be specified.

*Independent linearity* is referenced to the *"best straight line"* (see Figure 2–18), a line midway between the two parallel straight lines closest together and enveloping all output values on a calibration curve. The best straight line can be drawn only after a calibration has been completed.

*Least-squares linearity* is referenced to the *least-squares line,* that straight line for which the sum of the squares of the residuals is minimized. The term "residual" refers to the deviations of output readings from their corresponding values on the straight line calculated. The calculation is usually performed with the aid of a computer.

Some additional types of linearity have been used at times, such as

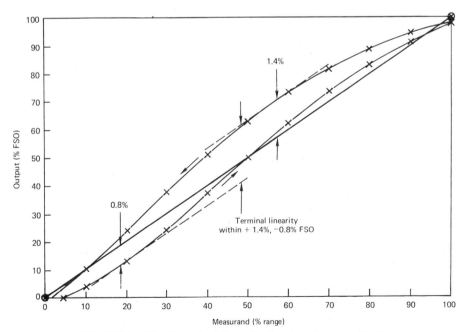

**FIGURE 2–17.** Terminal linearity (scale of errors 10:1).

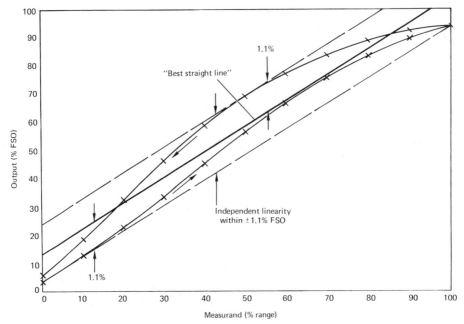

**FIGURE 2–18.**   Independent linearity (scale of errors 10:1).

"independent linearity with forced zero" or "with fixed *Y*-intercept," which require that the best straight line also pass through a specified point (*point-based linearity*). It has been argued that specifications (and calibration results) for linearity actually include hysteresis, since linearity is determined by means of a bidirectional calibration cycle; some commercial literature, therefore, refers to "combined linearity and hysteresis," referred to a specific straight line, typically the "best straight line."

*Conformance* (or "conformity") is a term that has sometimes been applied to the closeness of a calibration curve to a specified curve for an inherently nonlinear transducer. It is typically referred to a theoretical curve, although least-squares and other fits have also been used.

*Resolution* and *threshold* are both descriptive of "smallest increment," but are quite different characteristics. When the measurand is continuously varied over the range, the output of certain transducers will not be perfectly smooth, but instead will change in small (but measurable) steps. This is typical for potentiometric transducers with wire-wound elements: small step changes in the output occur as a result of the wiper sliding from wire turn to wire turn. The magnitude of the output step changes as the measurand is continuously varied over the range is the *resolution* of the transducer. It is expressed in % FSO and is not equal for potentiometric transducers with wire-wound elements. Some steps will be larger and some smaller, varying with minute variations in turn spacing and wire thickness. The term *average resolution,* then, is applied to the reciprocal of the total number of output steps over the range, multiplied by 100 and expressed in % FSO (or percent of voltage ratio, i.e., % *VR*). The magnitude of the largest of all observed output steps, also expressed in % FSO or % VR, is the *maximum resolution.* The resolution of digital-output transducers (for which step changes are inherent and equal) is given by the number of bits in the data words or, in the case of incremental digital-output transducers, by the number of "on" indications obtained per unit length or angle or revolution. When there are no measurable step changes in the output of a transducer, it is said to have *continuous resolution* (sometimes erroneously referred to as "infinite resolution," since "infinitesimal" would be more appropriate).

A change in measurand of finite magnitude is required to cause a change in the output of any transducer. In some types of transducers, those minimal measurand changes are not measurable. In others they may be measurable but are negligible for a given application, or they may be significant only at the lower limit of the range. The smallest change in measurand that will result in a measurable change in output is the *threshold* of the transducer. It is usually stated in terms of measurand and may have different values in different portions of the range.

*Sensitivity* (which has at times been confused with threshold) is simply the ratio of the change in output to the change in the value of the measurand. It establishes the slope of the calibration curve.

There are three characteristics which are time dependent: creep, zero shift, and sensitivity shift. *Creep* is a change in output occurring over a specific time period during which the measurand is held constant (at a value other than zero) and while all other conditions that may influence the reading (e.g., environmental conditions) are held constant. The determination of creep cannot be done by means of a static calibration; it requires a separate test. *Zero shift* is a change in the zero-measurand output over a specific period of time at room conditions (in a specification it is the maximum allowable change). The *zero-measurand output* (the "zero") is the output of the transducer under room conditions with nominal excitation and zero measurand applied. Zero shift is characterized by a parallel displacement of the entire calibration curve. *Sensitivity shift* is a change in the slope of the calibration curve due to a change in sensitivity over a specific period of time at room conditions (in a specification it is the maximum allowable change). Zero shift can be determined by a simple test, equivalent to the beginning of a calibration cycle. Sensitivity shift can be determined by performing one calibration cycle (or more than one if a statistical refinement of the result is needed) at the end of the specified time period. There is a difference between *short-term* and *long-term* zero and sensitivity shifts: the former tend to be specified in terms of hours, the latter (which could be considered reliability characteristics) in terms of months or years.

The concept of *error band* was originally developed by the author, assisted and inspired by his colleagues at a large aerospace facility, to simplify the specification and determination of transducer errors. An *error band* is the band of maximum deviations of output values from a specified reference line or curve due to causes attributable to the transducer. Since such deviations may be due to nonlinearity, nonrepeatability, hysteresis, zero shift, sensitivity shift, and so on, it can be seen that transducer characteristics are easier to specify and determine when individual characteristics need no longer be specified and determined. An error band is specified in terms of " ± _____% FSO," and is determined on the basis of maximum deviations observed over at least two consecutive calibration cycles (so as to include repeatability) and then expressed as " +____%, −____% FSO." A specific reference line or curve must be stated for an error band, and the term "error band" is modified by a term denoting applicable environmental conditions and, when required, other special conditions. The types of straight lines an error band can be referred to are the same as those used for linearity.

The *static error band* is the error band applicable at room conditions and in the absence of any shock, vibration, or acceleration (unless one of these is the measurand). Figure 2–19 illustrates a static error band referred to the terminal line, as it may have been specified (as " ±2.0% FSO") and as it may have been determined over two consecutive calibration cycles. It can be seen that the actual error band, " +1.5%, −1.1% FSO," indicates that the transducer is well within specifications. Further scrutiny of the calibration curves (which, however, is not needed for making an accept/reject decision) shows enough information for determination of individual characteristics when such knowledge is required.

When a number of transducers of the same design, range, and output ("of the same part number and dash number") are used in a given measurement system, and when errors within the error band specified for this transducer are acceptable to the system, the calibrations for each of these transducers (as long as each unit was accepted) can be considered *interchangeable* (within the error-band tolerances). This means that it would not be necessary to use individual calibration records for data reduction; the individual calibrations then serve merely as "acceptance records."

Static error bands can be referred to any of the lines explained for *linearity*. They can also be referred to any curve that can be specified by means of a graph, a table of values, or a mathematical equation. Figure 2–20 shows a static error band referred to a theoretical curve (i.e., one that can be drawn before any measured values are obtained); the shape of the curve was selected arbitrarily for the purpose of illustration. Narrower static error bands can be obtained when linearity, or conformance to a prescribed curve, is not required. Such requirements can be waived when final data are to be reduced on the basis of a complex curve that represents the actual transfer function of the transducer over its measuring range. The curve is

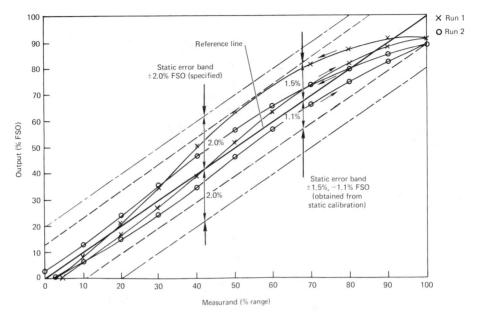

**FIGURE 2–19.** Static error band referred to the terminal line (error scale 10:1).

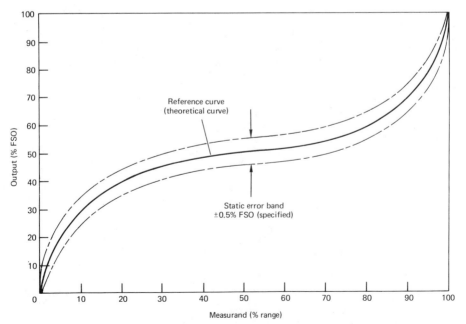

**FIGURE 2–20.** Static error band referred to a theoretical curve (error scale 10:1).

usually different for every transducer of the same part number. Data reduction then requires that a calibration record be generated for each transducer on the basis of multiorder polynomials. (This may sound difficult, but it is really quite a simple programming task.) A hard copy can then be obtained for each calibration record by using a plotter. More importantly, the conversion file is stored in the computer that is used for data reduction, so that incoming data are automatically reduced on the basis of the respective calibration records. An example of such an error band is shown in Figure 2–21. This static error band is referred to the *mean-output curve*, the curve plotted through the mean of output readings obtained during a specified number (three in the example shown) of consecutive calibration cycles. It can be seen that the deviations of output readings are due to nonrepeatability and hysteresis only.

In some applications transducer accuracy is of prime importance in only a limited portion of the range; accuracy in the other portions can be sacrificed. For such cases a *stepped static error band* can be used (see Figure 2–22 for an example; the theoretical end points were arbitrarily offset). Specifying this type of error band relieves the transducer manufacturer from providing very low error over the entire range of the transducer, while of-

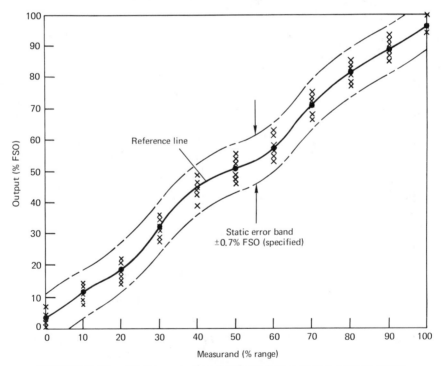

**FIGURE 2–21.** Static error band referred to a mean-output curve (error scale 10:1).

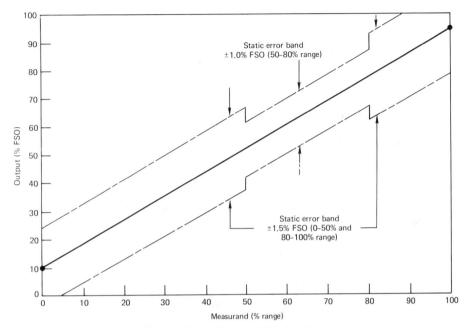

**FIGURE 2–22.** "Stepped" static error band referred to the straight line between theoretical end points 10% FSO at 0% range, 95% FSO at 100% range (error scale 10:1).

fering the user very good accuracy in the portion of the range of primary interest.

The selection of a reference line for a static error band of a bidirectional-range transducer involves the consideration of the intercept of the line with the zero-measurand output. When a theoretical slope or a terminal line is used, such a consideration is usually not necessary. However, when an end-point line is used and end-point tolerances are established, additional tolerances may have to be applied to the zero-measurand output. A stepped static error band can be useful in such cases. The band can be narrower near the zero measurand, near the end points, or in portions of the range along each direction, depending on the user's requirements.

### 2.5.2  Dynamic Characteristics

When a transducer is used for a measurement where rapid measurand variations occur, or where step changes in measurand have to be monitored with good fidelity, the transducer's dynamic characteristics must be established. These can be stated and determined in terms of frequency response, response time, or damping and natural frequency, depending on the type of transducer involved and its application. Note that some of the determinations require test methods and equipment of fairly high complexity as well as considerable expertise in test personnel. Tests have been standardized for many types of transducers by professional societies and by government laboratories such as, in the United States, the National Bureau of Standards. Unless otherwise stated in a specification, specified dynamic characteristics are applicable at room conditions.

*Frequency response* is the change with frequency of the output-measurand amplitude ratio within a stated range of frequencies of a sinusoidally varying measurand applied to a transducer. It is also the change with frequency of the phase difference between this measurand and the output. Frequency response is usually specified as "within ±____% (or ±____ dB) from ____to ____Hz" and should be referred to a frequency within the specified frequency range and to a specific measurand value. Figure 2–23 shows two typical response curves considering only amplitude ratio, not phase difference (the output will lag behind the measurand). Curve A shows the response of a transducer that can be used for static as well as dynamic measurements; in this example the frequency response is within ±5% from 0 to 300 Hz, referred to 10 Hz. Curve B shows the response of a transducer usable only for dynamic measurements; the response is within ±5% from 10 to 3500 Hz, referred to 100 Hz. A reference amplitude was not stated for these generalized examples. A commonly found colloquial expression for the response of curve A is that the response is "from *dc* to

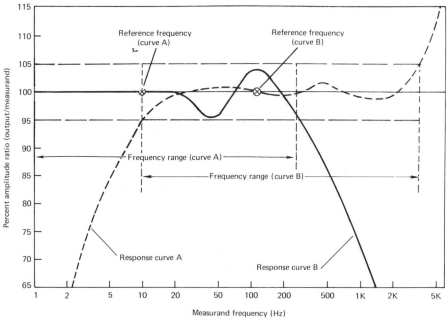

**FIGURE 2–23.**  Frequency response.

300 Hz (within ±5%)." Similarly, the response of curve B could be called "*flat* (within ±5%) between 10 and 3500 Hz."

Response time, rise time, and time constant characterize the response of a transducer (that is not underdamped) to a step change in measurand (see Figure 2–24). When such a step change is applied, the output will change nonlinearly toward the final value (100% output change) over a period of time. The length of time required for the output to rise to a specified percentage of its final value (as a result of a step change in measurand) is the *response time*. The percentage is typically stated in the form of a modifier of "response time," e.g., "95% response time" or "98% response time." A special term and symbol have been assigned to 63% (actually 63.2%) response time: the *time constant* τ. Another term, *rise time,* is used to state the length of time for the output to rise from a small specified percentage to a large specified percentage of its final value. Unless otherwise specified (e.g., "5 to 90% rise time" in Figure 2–24), the percentages should be assumed to be 10% and 90% of the final value. The general term for a transducer's response to a step change in measurand is *transient response*. In some cases it is possible to calculate frequency response from a transducer's transient response, its mechanical properties, or its geometry; if so, it should be referred to as *calculated frequency response* and the basis for calculation should be identified.

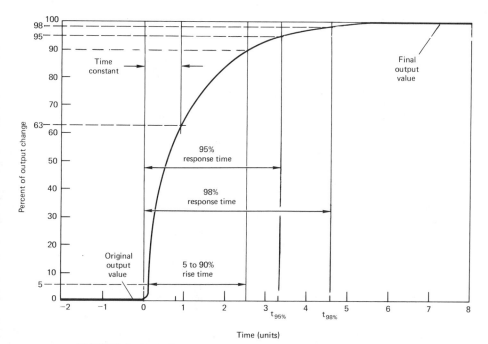

**FIGURE 2–24.**  Response time, rise time, and time constant.

*Damping* is the energy-dissipating characteristic which, together with natural frequency (see shortly), determines the upper limit of frequency response as well as the transient-response characteristics of a transducer. In response to a step change in measurand, an *underdamped* system oscillates about its final steady value before coming to rest at that value (see Figure 2–25), an *overdamped* system comes to rest without overshoot, and a *critically damped* system is at the point of change between the underdamped and overdamped conditions. When a transducer's sensing element is set into free oscillation, the frequency of this oscillation is the *natural frequency*. It is important that this oscillation is free, not forced. Natural frequency has also been defined as that frequency of a sinusoidally applied measurand at which the output lags the measurand by 90°. Figure 2–25 shows that time constant is still a valid characteristic of an underdamped transducer, but that such characteristics as 95% response time could not be stated due to the *overshoot* and oscillations about the final value; the frequency of this oscillatory transient is the *ringing frequency* of the transducer.

The ratio of the actual damping to the degree of damping required for critical damping is the *damping ratio* (damping factor). A damping ratio of 1.0 indicates critical damping, damping ratios larger than 1.0 signify overdamping, and damping ratios less than 1.0 represent underdamping. Figure 2–26 shows responses of a typical mechanical sensing element (spring-mass

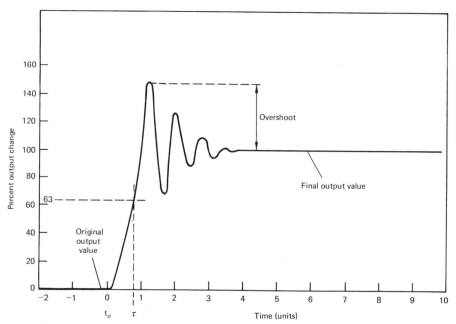

**FIGURE 2–25.** Response of underdamped transducer to step change in measurand.

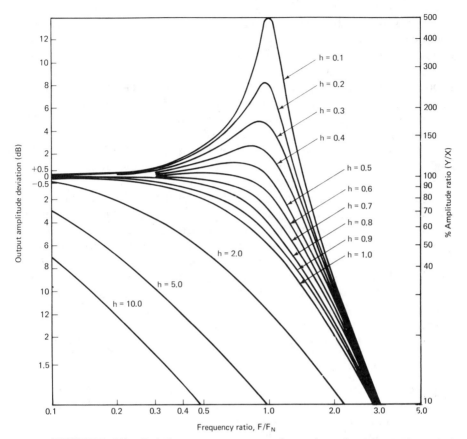

**FIGURE 2–26.** Relative response curves for various damping ratios *h* as a function of the ratio of frequency *F* to natural frequency $F_N$.

system) to a sinusoidally varying measurand for various damping ratios. Some types of transducers are inherently overdamped, others achieve some degree of damping by incorporating a fluid (*viscous damping*) or electromagnetic action (*magnetic damping*), and still other types are purposely left underdamped.

### 2.5.3 Environmental Characteristics

The static performance characteristics as well as the dynamic performance characteristics of transducers are specified, and verified, as those which the transducer exhibits at room conditions and in the absence of any external conditions (*environmental conditions*) that may affect the transducer's performance. When a transducer can reasonably be expected to operate under conditions (*operating environmental conditions*) other than those under which it was calibrated, the *environmental effects* must be known and the resulting deviations from static performance (*environmental errors*) must be limited by tolerances (in a specification) and determined by tests. Such additional environmental tests (temperature tests, vibration tests, ambient-pressure tests, etc.) may have to be performed on each transducer used; more commonly they are performed on a *sampling* basis (test one of every *n* transducers of each model and range), but sometimes only on a *qualification* basis (test one representative transducer). Environmental testing of transducers requires considerable skill and expertise from test personnel as well as appropriate test equipment and test setups whose behavior in the course of a test is well understood.

Besides operating environmental conditions, there are other environmental conditions to which a transducer may be exposed, but the transducer is not expected to operate within specified tolerances (or operate at all) while exposed to them. However, the transducer is expected to perform within specified tolerances *after exposure* to such *nonoperating environmental conditions*. When nonoperating environmental conditions, including those encountered during storage (e.g., while the device is in a warehouse or is installed in its application awaiting activation of the system in which it is to operate), shipping, and handling, are known or suspected to alter the behavior of a transducer, they should be included in a specification and the absence of out-of-tolerance nonoperating environmental effects should be verified by testing.

*Temperature effects* must be known and accounted for, for essentially all types of transducers. The *operating temperature range* is the range of ambient temperatures, given by their lower and upper extremes (e.g., ''−50 to +250 °C'') within which the transducer is intended to operate and within which all specifications related to temperature effects apply (unless they specifically relate to nonoperating temperatures). Some manufacturers of transducers incorporating elements intended to compensate for temperature effects call this the ''compensated temperature range.'' When the temperature of a *measured fluid* can cause significant temperature effects in the transducer, the *fluid temperature range* is specified, sometimes instead of the operating temperature range, and then governs temperature-related specifications.

For some transducers, temperature effects (thermal effects) are stated only in terms of the zero shift (*thermal zero shift*) and sensitivity shift (*thermal sensitivity shift*), which cause a parallel displacement and a slope change, respectively, of the calibration curve. Knowledge of these individual errors is useful when the temperature prevailing while a measurement is made is known and appropriate corrections to final data are to be made. However, thermal effects on hysteresis and repeatability are not included in such specifications.

A more general and inclusive way of specifying thermal effects on performance characteristics is given by the term *temperature error,* the maximum change in output (at any measurand value within the transducer's range) when the (operating or fluid, as applicable) temperature is changed from room temperature to specified temperature extremes. The simplest way of specifying tolerances on thermal effects is provided by the error-band concept, whose use also facilitates verification. The *temperature error band* is simply the error band (see Section 2.5.1) that is applicable over the operating (or fluid) temperature range. This form of specification is particularly useful when the error band is referenced to a theoretical slope, such as the terminal line (see Figure 2–27).

It is also important to specify the *maximum* (*ambient* or *fluid*) *temperature*, the highest (or lowest) temperature that a transducer can be exposed to without being damaged or subsequently showing a performance degradation beyond specified tolerances.

When a transducer is exposed to a step change in (ambient or fluid) temperature, a transient output deviation (*temperature gradient error*) can appear in its output (see Figure 2–28). Tolerances on this error should be specified for a stated rate of change of temperature, for the two temperatures between which the step change occurs, and for a specific measurand value.

Temperatures will also affect dynamic characteristics, particularly when they employ viscous damping. Specifications should then cover thermal effects appropriately (e.g., "Damping ratio: 0.7 ± 0.2, between 0 and

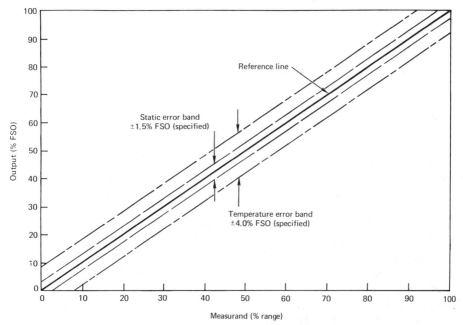

**FIGURE 2–27.** Temperature error band referred to terminal line (error scale 2:1).

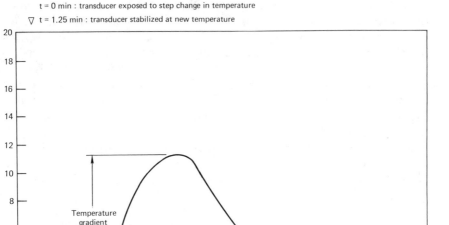

**FIGURE 2–28.** Temperature gradient error (typical example shown for output at zero measurand).

100°C,'' and ''Frequency response: within ±5% from 30 to 5000 Hz, referred to 100 Hz, between −20 and +120°C'').

*Acceleration effects* are the effects of quasi-steady-state accelerations on internal elements of a transducer, causing errors in its output. The acceleration error is typically more severe when acceleration is applied along one axis of a transducer than when it is applied along other axes. Error-causing acceleration may act directly on a mechanical sensing element or its linkage and cause spurious deflections; upon structural supports, causing distortion which may even result in failure; upon bearing-supported rotating members, causing eccentric loading and increased friction; and in a number of other ways, causing mass shifts, deformations, and distortions.

When a transducer is to be used in an application where it will experience acceleration (e.g., on moving vehicles, or on moving mechanical members), the possibility of acceleration errors must be considered and tolerances must be established for such errors. In order to reach an understanding about these with the transducer's manufacturer, one begins by agreeing on a labeling of the axes of the transducer (see Figure 2–29). The manufacturer can then inform the user that this particular transducer design is more sensitive to accelerations along, say, the X-axis than along the

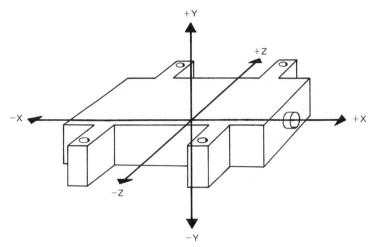

**FIGURE 2–29.** Typical labelling of acceleration axes for a transducer.

$Y$- or $Z$-axes. With this information in hand, the user can assure that the transducer gets installed in such an orientation that acceleration along its $X$-axis is minimized.

*Acceleration error* is the maximum difference (at any measurand value within the transducer's range) between output readings taken without and with the application of specified constant acceleration along specified axes. In a specification the maximum expected acceleration in each axis and, if necessary, in each of the two directions, is specified in conjunction with the maximum allowable acceleration error [e.g., "Acceleration error: $\pm 2.5\%$ FSO for: $10g(+X)$, $2g(-X)$, $5g(\pm Y)$, $3g(\pm Z)$." This error can also be specified or determined in terms of *acceleration sensitivity* within a stated range of acceleration values acting along specified axes (e.g., "$0.2\%/g$, over 0 to $10g$, $X$-axis, and $0.15\%/g$, over 0 to $10g$, $Y$- and $Z$-axes"). For acceleration transducers, sensitivity to acceleration acting along other than the measured axis is known as *transverse sensitivity.*

When the error-band concept is used, the *acceleration error band* will include acceleration errors as well as static errors; the latter may actually be reduced by accelerations in some designs.

Some types of transducers are so sensitive to acceleration forces that even the acceleration due to the earth's gravity can cause undesirable effects. The error due to the orientation of a transducer relative to the direction in which gravity acts upon it is called *attitude error*. When this error in the different transducer axes is known, it is often possible to install the transducer in such a position that attitude errors are minimized.

*Vibration effects,* the effects of vibratory acceleration, can affect transducers in the same manner as steady-state acceleration. More severe effects, however, are connected with the frequencies of vibration. As the vibration frequency is varied over a stated range and along a specific axis, amplified vibrations (*resonances*) of internal elements can occur at one or more frequencies. Figure 2–30 illustrates typical vibration effects; a potentiometric transducer was chosen for this example, which includes evidence of a reduction in friction error due to vibration; it also shows vibration error at several resonances as well as the equivalent *vibration error band,* at the largest resonance, identified as one-half the vibration error band since error bands always have bipolar tolerances.

*Vibration error* is the maximum change in output (at any measurand value within the transducer's range) when vibration levels of specified amplitudes and ranges of frequency are applied to the transducer along specified axes (at room conditions). Note that different resonances and, hence, different vibration errors may be observed for different measurand values, particularly when the transducer incorporates a mechanical sensing element. It may be necessary, therefore, to predict the measurand value most likely seen by the transducer while it is exposed to the most severe vibration environment and then to specify and verify vibration errors at that value.

*Ambient-pressure effects* can be observed in some transducer designs when the transducers are calibrated at room barometric pressure and then

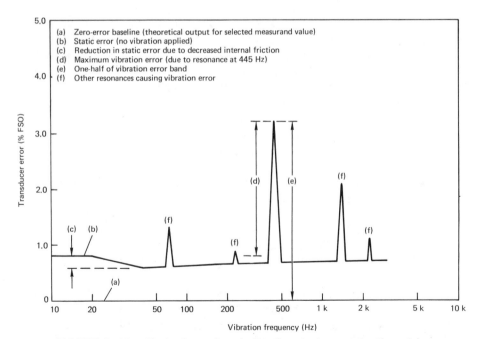

**FIGURE 2–30.** Typical results of vibration test on potentiometric transducer at a given program of vibration amplitudes, from 10 to 3000 Hz, at one measurand value, along one transducer axis.

used either at very high altitudes or on aircraft or spacecraft, where ambient pressures approach a vacuum, or when they are used far underground (e.g., in deep mines or wells) or at deep submersion underwater, where pressures are very high. Transducer performance can be affected by resulting case deformations and changes to the internal geometry of a transducer. It can also be degraded to the point of failure when a poorly sealed transducer is exposed to a vacuum environment and internal materials outgas, wires self-heat in the absence of air as a heat-transfer medium, internal sealed cavities start to bulge or leak, or corona-arcing occurs across high-voltage terminals.

*Ambient-pressure error* is the maximum change in output (at any measurand value within the transducer's range) when the ambient pressure is changed between specified values, usually between room pressure and a lower or higher ambient pressure. Such errors can also be stated in terms of an *ambient-pressure error band*. Ambient-pressure error has at times been referred to as "altitude error," with ranges of (low) pressures stated in terms of altitude above sea level.

*Mounting effects* can occur during the installation of a transducer and may result in subsequent performance changes when, for example, the mounting surface of the transducer is not evenly machined so that the case gets deformed when all mounting hardware is tightened, or when the torque applied to the coupling nut on a pressure fitting causes sensing-element deformations. *Mounting error* is the error resulting from mechanical deformation of the transducer caused by mounting the transducer and making all electrical and measurand connections. Mounting error is not commonly included in specifications; however, it may be necessary to verify its absence (e.g., by performing an in-situ calibration check).

*Other operating environmental effects* on the behavior of a transducer during its normal operation which should be known and included in specifications include humidity or immersion in liquid (affects poorly sealed transducers, starting with a reduction of insulation resistance); corrosive (and insulation-reducing) effects of high salt concentrations (or concentrations of other corrosive substances) in the ambient atmosphere; various effects of measured fluids on the sensing elements of transducers; the influence of ambient electromagnetic fields on transduction elements and integral circuitry; and the effects of radiation (nuclear, ionizing) on various internal transducer elements.

*Nonoperating environmental effects* on the performance of a transducer subsequent to its exposure to such environmental conditions should be understood and, if necessary, limited by inclusion in a specification. These environmental conditions do not only include temperatures, vibration, shock, moisture, etc., encountered during storage, shipping, and handling; they may also include environmental conditions in excess of limits of operating conditions specified for a transducer that can be encountered while the transducer is installed in its intended application. All overrange, or overload, characteristics can be classified as nonoperating conditions. Some examples of nonoperating environmental conditions that tend to be overlooked are room temperature, for a light sensor or radiation sensor which normally operates while cooled to low temperatures; any significant vibration, for any transducer used on a satellite, where it need not operate until after separation of the satellite from its booster stages; and low temperature, together with a liquid state, of a fluid that is intended to be measured only when the fluid is gaseous and at a significantly higher temperature.

*Type-limited environmental effects* are significant to certain types of transducers, or to transducers for certain measurands, in addition to the generally applicable (to varying degrees) environmental effects previously described. These effects include:

1. *Conduction error,* the error in a temperature transducer due to heat conduction between the sensing element and the mounting of the transducer.

2. *Strain error,* the error resulting from a strain imposed on a surface to which a transducer is mounted; this error is significant primarily for surface-temperature transducers and is not intended to relate to strain gages.

3. *Reference-pressure error,* the error resulting from changes of the reference pressure, the pressure relative to which a *differential-pressure transducer* measures pressure, within a specified reference-pressure range.

4. *Transverse sensitivity,* the response of acceleration transducers to acceleration forces in axes transverse to the sensing axis.

## 2.6 RELIABILITY CHARACTERISTICS

Although some operating and most nonoperating environmental characteristics relate to a transducer's reliability, the characteristics considered here are only those relating to the useful life of a transducer as well as to any characteristics that may have adverse effects on the system in which the transducer is installed when the transducer fails in a particular mode.

The useful life of a transducer can be expressed in one of two ways: *operating life* is the (specified) minimum length of time over which the transducer will operate, either continuously or over a number of on-off cycles whose duration is specified, without changing its performance characteristics beyond specified tolerances; *cycling life* is the (specified) minimum number of full-range excursions (or specified partial-range excursions) over which a transducer will operate (as specified) without changing its performance beyond specified tolerances.

In some cases it may also be necessary to specify or be concerned about a transducer's *storage life,* the length of time over which it can be exposed to specified storage conditions without changing its performance beyond specified tolerances.

Reliability characteristics relating to adverse effects on the system in which a transducer is installed are generally application dependent. They include any characteristics controlled by applicable health and safety codes. They further include any characteristics deemed important after a determination of the effects of failure modes has been made. An example is the results of internal short-circuiting (protection may be included in the associated electrical system or may have to be included within the transducer). Another example is the *burst-pressure rating* of a transducer, i.e., the pressure that may be applied to the sensing element or the case, as specified, of a transducer without rupture of sensing element or case, respectively. When a sensing element ruptures, the measured fluid may come in contact with materials it is not compatible with and the results may range from measured-fluid contamination to an explosion. When a case ruptures, portions of it may hit surrounding personnel or equipment and cause damage.

Because of this hazard, rare as its occurence may be, it has been necessary to specify a separate *case-burst pressure rating* for transducers used in certain critical installations. The test to verify this is necessarily destructive: a hole is drilled into the sensing element so that the pressurized fluid can enter the case. The pressure is then raised to the specified pressure rating, and any significant case deformation, cracking, or actual rupture is observed. The pressure is then bled off. This test must be performed within an enclosure that offers protection from any resulting explosion to the operator.

Some potential failures and their effects simply cannot be provided for by writing a specification. An example is an in-line flowmeter in a pipeline, such as a turbine flowmeter, whose turbine blades may separate, move downstream, and then cause a pump to fail. It is, therefore, important to assess the impact of any possible failure on the system in which a transducer is installed and then take steps to build appropriate protection into the system.

# 3

# *Criteria for Transducer Selection*

The selection of a particular transducer design is based primarily on a measurement requirement. This requirement is sometimes established by the person responsible for the selection; much more frequently, however, it is established by someone else, typically a project engineer or the cognizant engineer of a subsystem. Typically, too, the data system (including power supplies) with which the transducer must interface and operate either already exists or has already been designed. There can also be other constraints on transducer selection: there may be a project policy to use only parts (including transducers or electronic parts within a transducer) that are on an "approved list" or that have previously passed a qualification test for specific applications; or the purchasing department may have a list of "approved vendors," based on prior procurement and quality control experiences; or regulations may prohibit buying transducers from another country unless it can be proven that a manufacturer there is the only possible source. There are almost always sound reasons for such policies. It should be evident, then, that transducer selection is often an iterative process, and that there can be times when measurement requirements have to be negotiated with the originator.

The following listing of guidelines is meant to be fairly complete. For some applications a number of these considerations can be omitted. For others, additional factors may have to be considered. Some of the entries tend to imply possible incompatibilities. For example, if the measurement requirement calls for a frequency response that is flat up to 3000 Hz, and the data system can only handle up to 500 Hz, the measurement requirement will most likely have to be negotiated downward. Cost and availability can affect measurement requirements similarly.

## 3.1 MEASUREMENT CONSIDERATIONS

1. What is the real purpose of the measurement?
2. What is the measurand?
3. What range of measurand values will be displayed in final data?
4. Will the measurand only increase, or only decrease, or both?
5. What overrange conditions may occur before or during the time data are required?
6. With what accuracy must the measurement be presented in the final data?
7. What are the dynamic characteristics (e.g., fluctuation frequency range, step changes) of the measurand?
8. What frequency response or transient response must be visible in the final data?
9. If a fluid is being measured, what are its physical and chemical characteristics?
10. Where and how will the transducer be installed?
11. In what manner, and to what extent, is it permissible for the transducer to modify the measurand while it is being measured?
12. What ambient environmental conditions will the transducer be exposed to?

## 3.2 DATA SYSTEM CONSIDERATIONS

1. What is the general nature of the data system (e.g., radio telemetry, hard-wired telemetry, individual direct display)?
2. Is the data system inherently analog or digital?
3. What is the nature of the major elements in the data system:
   a. Signal conditioning, multiplexing, analog-to-digital conversion, pretransmission buffering?
   b. Data transmission link?
   c. Data processing, data storage?
   d. Data display?
4. What are the accuracy and frequency response characteristics of the end-to-end data system, exclusive of those of the transducer?

5. What form of transducer output will the data system accept with minimum additional signal conditioning?

6. What load impedance will be seen by the transducer?

7. Is frequency filtering or amplitude limiting of transducer output required, and can the data system handle this?

8. What transducer excitation voltage is most readily available?

9. How much current may the transducer draw from the excitation supply?

10. Are special transducer-related checking functions (e.g., "ready" check, electrical calibration check) required by the data system, and does the data system provide circuitry for these?

## 3.3   TRANSDUCER DESIGN CRITERIA

In the step-by-step design process presented in this chapter, criteria for transducer design and, hence, selection are based primarily on the preceding lists of considerations. However, as mentioned earlier, cost and availability factors, as well as policies governing procurement, may influence design decisions and can translate into measurement requirement changes.

1. What constraints are imposed on transducer mass, configuration, excitation, and power consumption?

2. What are the transducer output requirements?

3. Which transduction principle is most suitable?

4. What accuracy and other performance characteristics must the transducer provide: static? dynamic? environmental (operating)? environmental (nonoperating)?

5. What operating or cycling life is required?

6. If a fluid is to be measured, what will be the effects of the measured fluid on the transducer?

7. Will the transducer affect the measurand to the extent that erroneous data will be obtained?

8. What constraints are imposed on the transducer by any applicable governmental standards or industrial codes?

9. What are the failure modes of the transducer? What hazards would a failure present to the system in which it is installed? to adjacent components or systems? to the data system, especially the power supply provided by it? to the area in which the transducer operates? to personnel working in that area?

10. What is the lowest level of technical competence to be possessed by any and all personnel expected to handle, install, and use the transducer? What human-engineering requirements should be incorporated in the transducer design?

11. What testing methods (including calibration) will be used to verify performance? What tests will be performed by the manufacturer, and what tests will be run by the user? Are those tests adequate? Are the test methods correct? Is the test equipment appropriate? Are the test methods simple and well established?

## 3.4  AVAILABILITY FACTORS

1. Is a transducer that fulfills all the requirements available "off the shelf"?

2. If the answer to the preceding question is "no," the following should be considered:
   a. Will minor redesign of an existing transducer be sufficient, or will a major development effort be required?
   b. How many transducers of identical design will be procured at this time and in the future?
   c. What manufacturer has demonstrated the ability to produce a transducer similar to the required item?
   d. What has past experience been like in dealing with a proposed manufacturer?
   e. Can the transducer(s) be delivered in time to meet installation schedules?

## 3.5  COST FACTORS

1. Is the quoted cost of the transducer compatible with the measurement function it will provide?

2. What additional costs will be incurred by required transducer testing, periodic recalibration, handling, and installation?

3. Which requirement imposed on the transducer is the major cost driver?

4. What relatively minor compromises in requirements could lead to substantial savings?

5. What modifications to the data system (including power supplies) could lead toward reduced costs of a number of different transducers used in that system, and what cost trade-offs would be involved?

## 3.6   LOCATION OF PROCESSOR ("SMART SENSORS")

An additional consideration in selecting a transducer is the location of the processor that performs at least some of the data processing as well as some transducer control functions. The continuing development of microprocessors has made it possible to package such a processor integrally with the transducer, i.e., contained within the transducer housing. The capabilities of such built-in processors vary over a wide range. Some fairly simple transducers are available with a built-in processor that performs such operations as analog-to-digital conversion, storage of the calibration curve and conversion factors, and output of data in engineering units (in digital form). Additional capabilities can be internal buffering (storage) of data to provide a "data dump" when "polled" by the data system; decoding of command sequences and their conversion into transducer operating-mode changes; responding to a transducer's own observation and changing an operating mode (e.g., amplifier gain) and sampling rate; and combining data words with time tags obtained from a clock signal, with error correction codes, with data about internal operational states, and with headers that the data system accepts as the beginning of a data sequence ("frame, packet").

Transducers that incorporate such microprocessors are often called "smart" or "intelligent" sensors (or transducers). Making a choice between a "smart" and a "normal" transducer involves primarily trade-offs of cost vs. data system capability vs. convenience. For example, multiplexing standardized transducer outputs into a single analog-to-digital converter (ADC) is probably cheaper than paying for an ADC for each transducer. On the other hand, multiplexing (formatting) digital data may facilitate programmability. Or, if the existing data system already provides for programmable decalibration and engineering-unit conversion, there is really no need to have these functions also residing in a microprocessor in the transducer. On the other hand, if the data system is new and was specifically designed for having these functions performed by each transducer, the overall cost may be the same, and the cost of expanding the data system (for additional transducers) may even be less.

# 4

## *Transducer Performance Tests*

Transducer specifications, selection criteria, and tests for performance determination are strongly interrelated. Every specified characteristic, and their tolerances when also specified, should be able to be verified by a test. Some of these tests are relatively simple, and test equipment is readily available; such tests can be performed quickly and at no significant additional cost. Other tests require elaborate setups, equipment that is expensive and not readily available, and highly skilled test engineers and technicians; these tests are expensive and time consuming. Yet, for certain critical applications, a complex, lengthy, and costly test program is not only justifiable but necessary. An example of such an application is an instrumentation system that operates unattended for extended periods of time and under a variety of harsh environmental conditions, especially when an instrumentation failure can cause a system failure that carries a high penalty in terms of replacement cost, loss of revenue, critical time, or endangering human life. Such combinations of conditions can be found in some industrial installations, but are more typical for nuclear power stations, military vehicles and systems, aircraft, and space vehicles.

Because of the wide range of test complexity, it is important for the transducer user to have some knowledge of the tests needed to determine or verify transducer characteristics. The user will then be in a better position to judge what specifications to impose or not to impose, and be better equipped to understand a test report and its implications.

It is poor practice to specify anything that has never been or is never intended to be verified by a test, or to specify something that can only be verified by a test whose costs cannot be justified by the application. It is meaningless to specify error tolerances that are tighter than those offered by the best available test setup. On the other hand, it is quite acceptable for a user to ask a manufacturer for a copy of a test report that substantiates a specification in his or her catalog or bulletin. It is also acceptable for a manufacturer's sales or marketing engineer to ask a potential user what the real application is, and whether he or she really needs to specify certain characteristics or certain tolerances on them.

This chapter describes generally applicable transducer tests and typical equipment and setups used for these tests. Some considerations applicable to calibration and test methods are described under "Transducer Characteristics" in subsequent chapters.

The general categories that call for one or more of the individual performance tests described subsequently to be performed are stated in various ways. The nomenclature that is typically used for such test categories assumes that a quantity of transducers bearing the same part number (the same design, with the same optional features) are being purchased. These tests are all in the overall category of *performance verification tests*, meaning that the objective of the test is to determine whether or not the actual performance complies with applicable specifications. The results show primarily a "pass" or "fail" status, although other information about the behavior of the transducer can usually be extracted from the test records. By contrast, a *performance determination test* is performed primarily to characterize the behavior of the transducer; sometimes specifications and tolerances are established on the basis of such a test.

The performance verification test that each transducer of a lot is subjected to is the *acceptance test*, which always includes a calibration and often includes additional tests. A *qualification test* ("type test") is usually performed on only one transducer that is then considered representative of the entire lot purchased. This type of test consists of a comprehensive series of subtests, including environmental tests performed at the extremes of specified environmental levels. Some users will accept the results of a qualification test performed for another user in lieu of having such a test performed for themselves (and paying for it). In some cases, for example, when a transducer of the same model number as those being purchased but varying from it in some relatively minor way such as range has passed a qualification test (complete enough to satisfy the user's needs), the units being purchased can be declared as "qualified by similarity" to the unit that passed the qualification test.

Some applications (or procurement policies) call for a test that is intermediate between the relatively simple acceptance test and the usually quite complex qualification test; viz., a *sampling test*, which is performed on one out of a specified number of transducers (of the same part number). A sampling test includes electrical tests additional to those performed during the acceptance test, as well as some dynamic and environmental tests, often at environmental levels less than the maximum levels specified. Sampling tests are normally *nondestructive*, which means that the transducer is fully usable after the test. A qualification test, however, is usually *destructive*. Even if the transducer is still functional after completion of all tests, it should not be used, if only because the qualification test includes a life test, in which all the useful life has been "tested out" of the transducer.

## 4.1 ELECTRICAL TESTS

Different types of electrical tests are routinely performed on transducers. The most important group of electrical tests comprises output tests, i.e., measurements of the transducer's output (especially during a calibration) in whatever form it may be. Most frequently, the output is in the form of a dc or (less often) ac voltage. Sometimes, notably in potentiometric transducers, it is a voltage ratio. It can also be an ac output varying in frequency, or pulses whose repetition rate (pulses per second) varies (a pulse count). Some types of transducers have an inherently digital output (e.g., angular and linear encoders), or provide a digital output by virtue of a built-in analog-to-digital encoder. Other categories of transducers don't provide an output signal of any of the foregoing types; instead, their output is in terms of change in resistance, capacitance, or inductance. (These changes in a passive element are subsequently signal-conditioned into an active signal.)

Other electrical tests, usually performed apart from acceptance tests, are used to measure output impedance, noise (of built-in conversion electronics), loading error, threshold, and resolution; to characterize excitation parameters; and to check the degree of insulation of active circuitry from an external ground such as the transducer case (housing).

It is very important that the test equipment be properly calibrated and that the calibration be traceable to a standard maintained by the government agency responsible for providing and maintaining very accurate standards. (In the U.S.A. the responsible agency is the National Bureau of Standards (*NBS*).) Some of the equipment mentioned in what follows is of the type used as secondary or transfer standards. It is good practice for any company or facility to have their transfer standards certified periodically by an agency such as NBS. (The certification will show by how much a secondary standard deviates from a primary one, or a tertiary standard deviates from a secondary one.) Such a local standard then becomes a *transfer standard*. It is equally good practice to establish periodic recalibration intervals for all measurement equiment used for production inspection and testing. (These recalibrations are usually performed by the company or facility's own standards laboratory.)

### 4.1.1 Voltage Measurement

Electronic voltmeters with a high input impedance and now mostly with a digital display are generally used for dc transducer output voltage measurements, from millivolts to volts; electronic voltmeters of special design can be used to obtain readings in the microvolt range. Such voltmeters provide adequate accuracy as long as they are periodically recalibrated against a transfer standard such as a *voltage potentiometer*. This instrument measures dc voltages from microvolts to about 1.5 V by balancing the voltages against a known voltage. The known voltage is obtained between the wiper arm and one end of a very precisely wound potentiometer across which a battery is connected (see Figure 4–1). The wiper is moved across the potentiometer element, usually by turning a knob, until a null balance is observed on the null indicator. The wiper is mechanically linked to a calibrated dial on which the voltage necessary to balance the unknown voltage can be read.

Voltage potentiometers are calibrated by connecting a known voltage across the measurement terminals. An accurate voltage source used for this purpose is the *standard cell*, which produces slightly over one volt. Many voltage potentiometers have a built-in standard cell as well as other useful features, such as range-multiplier resistors, optical aids to facilitate the null reading, and devices that minimize thermoelectric potentials generated by frictional heating at the contacting point of the wiper arm. A voltage potentiometer always reads the open-circuit output voltage of a transducer because the potentiometer presents a theoretically infinite impedance to the transducer at the precise point of null balance.

For voltages over 1.5 V the choice of readout equipment is practically

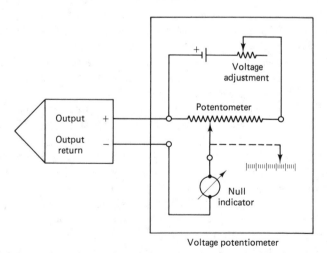

Voltage potentiometer

**FIGURE 4–1.** DC output voltage measurement by a voltage potentiometer.

limited to the digital electronic voltmeter, which also operates on the voltage-balance principle. Instead of using a continuously variable resistance (a potentiometer), this device employs precision resistors that divide the voltage into discrete increments. The unknown voltage is balanced in the voltmeter in decade steps, using transistor switching circuits to select the internal balancing voltage by successive approximation so that an increasingly finer null is obtained. The state of each decade switching network (equivalent to the position of a selector switch) is displayed by a numeral on the display. For example, on a five-digit voltmeter, a voltage of 21.453 V would be successively approximated to, first, 20 V on the "tens switch," then 1 V on the "units switch," 0.4 V on the "tenths switch," 0.05 V on the "hundredths switch," and, finally, 0.003 V on the "thousandths switch." Most digital voltmeters now provide for automatic range switching and polarity indication.

Ac voltages can also be measured by a digital voltmeter if it is equipped with an ac-to-dc converter that responds either to the average, peak, or rms value of the ac voltage. On some meters the type of value is selectable. Very accurate measurements can be obtained by using a ratio transformer in conjunction with a transfer-standard ac voltmeter or voltammeter.

### 4.1.2   Current Measurement

Precision electronic multimeters are normally used for production-type measurements. For current calibrations of such instruments, as well as for more accurate current measurements, a transfer-standard ac voltammeter is used for ac currents; dc currents are passed through a *shunt*, in this case a stable precision resistor of accurately known value (standard resistor), and the resulting voltage drop across the shunt is read on a voltage potentiometer or digital voltmeter.

### 4.1.3   Frequency and Pulse-Count Measurement

A considerable number of different types of transducers have frequency or pulse-count outputs. Such outputs are usually measured by means of an electronic counter or an events-per-unit-time (*EPUT*) meter. These indicators display the periodic or aperiodic transducer output in digital form, using special logic circuitry to count the "events" (cycles or pulses) repetitively over a preselected time interval (e.g., 0.1 s, 1 s, 10 s, etc.). Since time is used as reference, and since it is possible to determine time with close accuracy, electronic counters and EPUT meters have inherently very small errors.

Where poorer accuracy is permissible, two other types of indicators can be used. The "integrating" frequency or pulse-count meters display a voltage representing the number of "events" as integrated by suitable electronic circuitry over a fixed time interval. Somewhat better accuracy is obtained with the discriminator type of frequency meter, which converts the deviation from a given center frequency into a dc voltage over a limited range of frequencies. Discriminators are useful in measuring any rapidly fluctuating frequency-modulated transducer outputs, such as those encountered during dynamic tests of FM output transducers.

### 4.1.4  Digital Output Measurement

Digital output is displayed by an indicator containing a converter that changes the digitally coded number into a decimal number and then displays this number. Some indicators are designed to convert the output of an angular encoder directly into radians, or into degrees, minutes, and seconds.

### 4.1.5  Voltage Ratio Measurement

The output of potentiometric transducers is a voltage ratio—the ratio of the voltage between the wiper and the signal-return side of the potentiometer (resistance) element to the excitation voltage applied across the entire potentiometer element. This makes it possible to measure this form of output as a ratio, independently of excitation-supply variations.

The two types of indicators used most commonly for voltage ratio measurements are the digital electronic voltmeter with ratio-measurement provisions and the manually balanced or servo-balanced resistance-bridge ratiometer. The former is a digital voltmeter modified to measure and display the ratio between two voltages, one of which can be the excitation voltage and the other the output of the transducer. The latter (see Figure 4–2) is essentially a precision potentiometer of either the continuous or the decade type or a combination of these two, with digital indicator dials connected mechanically to the potentiometer shaft or (concentric) shafts so as to indicate the relative position of the potentiometer's wiper arm. A null indicator

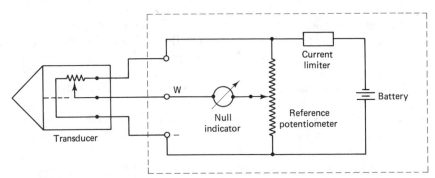

**FIGURE 4–2.**  Basic measuring circuit of a resistance-bridge ratiometer.

(or null-indicating galvanometer) is connected between the wiper and the external wiper terminal, while a battery is connected across the potentiometer as well as across the external (transducer) excitation terminals. The transducer output is read as "percent of voltage ratio (%VR)" directly off the mechanical indicating dials after a fine null reading has been obtained on the null indicator.

A current-limiting device, which can be a simple resistor, should always be connected in series with the battery in order to avoid burning out the transducer whenever it is incorrectly connected so as to apply the excitation between its wiper and "ground" terminal.

### 4.1.6 Resistance Measurement

The resistance-change output of resistive transducers is almost invariably measured by means of a resistance bridge. Various commercial resistance bridges are available, all modifications of the basic bridge circuit first used by Wheatstone in 1843 and, hence, known as the *Wheatstone bridge* [see Figure 4–3(a)]. The bridge is balanced when the current through the galvanometer (null indicator) is zero. This occurs when the ratio between resistors $A$ and $B$ equals the ratio between the unknown resistor $X$ and the "standard" (reference) resistor $S$, or

$$\frac{R_A}{R_B} = \frac{R_X}{R_S}$$

The unknown resistance is then obtained by multiplying the value of the known resistor $S$ by the ratio of resistances $A$ to $B$, or

$$R_X = R_S \frac{R_A}{R_B}$$

Interchanging the galvanometer and the battery in their circuit position does not affect bridge-balancing conditions.

The basic bridge circuit is made usable for resistance measurement by connecting a rheostat into the circuit as resistance $S$ and connecting the unknown resistance across terminals $X_1$ and $X_2$. The electrical position of the wiper of the rheostat, which can be continuously variable or vary in predetermined discrete intervals (by decade switches), is mechanically indicated by a numerical display dial, or by a number of dials indicating decade switch positions. The basic measuring circuit is illustrated in Figure 4–3(b). One important refinement of this circuit is the replacement of fixed resistor $A$ or $B$, or both, by adjustable resistors connected to a selector switch so as to allow ratios of $R_A/R_B$ other than unity, e.g., 10:1, 100:1, etc., and 1:10, 1:100, etc. Further refinements and modifications are included in such commercially available resistance bridges as the "guarded Wheatstone

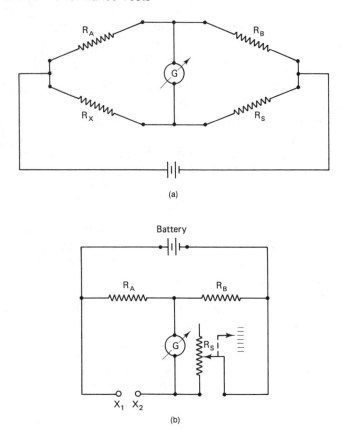

(a)

(b)

**FIGURE 4–3.**   Wheatstone bridge: (a) basic bridge circuit; (b) basic resistance bridge for resistance determination.

bridge" and the Mueller bridge, which is used exclusively for the accurate determination of the resistance of platinum-wire resistive temperature transducers.

### 4.1.7   *Capacitance and Inductance Measurement*

Although the output of a capacitive or inductive transducer is rarely in the form of a capacitance or inductance change, it is sometimes necessary to measure such outputs.

AC-excited reactance bridges (*impedance bridges*) of various types are most commonly used for measuring inductance as well as capacitance. These are similar in basic design to the Wheatstone bridge, except that no more than two legs of the bridge can be purely resistive. The "standard" or "reference" leg is reactive and is frequently a series or parallel combination of an adjustable resistor and an adjustable capacitor. The reactance (capacitive or inductive) across the transducer output terminals is then balanced against the "standard" leg of the bridge. The bridge dials are so arranged and labeled as to permit a direct reading of capacitance or inductance.

### 4.1.8 Output Impedance Tests

The output impedance of capacitive, inductive, resistive, reluctive, and strain-gage transducers whose output is not modified by active circuitry within the transducer is simply measured with an impedance or resistance bridge.

A somewhat more elaborate method is necessary for output-impedance tests on transducers, such as dc output transducers, which contain active circuitry between transduction element and output terminals. The substitution method (Figure 4–4) is frequently used for this purpose. In this method the measurand—between 50% and 100% of the transducer's range—is applied and maintained at a constant level. The output voltage is measured with a high-impedance voltmeter. A resistance decade box is then connected across the transducer's output terminals, and the resistance dials are adjusted until the output voltage is reduced to 90% of its open-circuit (no-load) value. This resistance, $R_{90}$, is used to calculate the output impedence, which is simply $\frac{1}{9} \times R_{90}$. If saturation effects are suspected within the transducer, which may yield an incorrect output-impedance reading, the resistance can be adjusted, instead, to the value $R_{99}$ required to reduce the output voltage to 99% of the open-circuit value. In this case the output impedance is calculated as $\frac{1}{99} \times R_{99}$.

### 4.1.9 Output Noise Tests

Noise in the output of potentiometric transducers frequently occurs during the stroking action of the wiper on the resistance element and is related to variations in instantaneous contact resistance. It is customary to measure

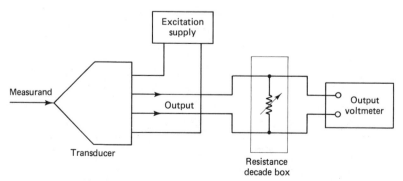

**FIGURE 4–4.** Output impedance determination by substitution method.

such noise in terms of noise resistance using the test setup illustrated in Figure 4–5. Actuation of the "push-to-calibrate" button results in a calibration deflection on the display device for a noise resistance of 100 ohms. The transducer is then cycled over its full range by suitable variation of the measurand. The noise resistance is monitored on the display device and determined by reference to the 100-ohm calibration deflection. The usual purpose of this test is to determine the maximum noise (resistance). This value can be affected by the transducer cycling rate as well as by environmental conditions such as temperature and (unless the transducer is sealed) humidity. Transducers which have been stored for a long time frequently exhibit abnormally high noise during the first few cycles of such a test. The existence of these various effects point to a need for careful definition of test conditions before starting a noise test on a potentiometric transducer.

Output noise tests on dc output transducers and other types containing active circuitry which can modify the output are considerably simpler. Using the test setup shown in Figure 4–6, the peak-to-peak output noise is read off a calibrated oscilloscope while the measurand, first at 0% and then at 100% of its range, is applied to the transducer. In some cases the use of a low-pass filter can be specified so that noise at higher frequencies is excluded from the measurement. This is specified only when noise at such higher frequencies is expected in the transducer output and the associated measuring system will in no way be adversely affected by it.

### 4.1.10   Loading-Error Tests

Loading-error tests are performed to determine the effect on the transducer's output of variations in the load impedance either between specified limits or between "infinity" (open circuit) and a specified value. The loading error, when expressed in percent of full-scale output (% FSO), is usually largest at the upper end point, except for potentiometric transducers, where it is maximum at 66% FSO. Figure 4–7 illustrates the loading error, with various

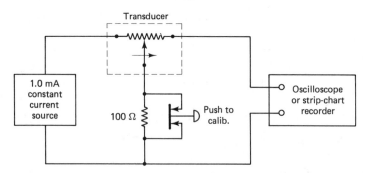

**FIGURE 4–5.** Output-noise test setup for potentiometric transducers.

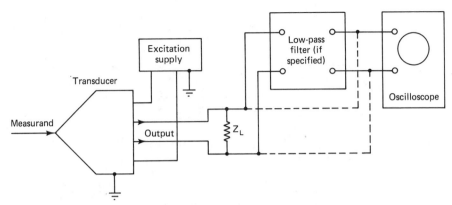

**FIGURE 4–6.** Output-noise test setup for dc output transducers.

load resistances $R_L$, of a potentiometric transducer having an element resistance (input impedance) of 7500 ohms. During this test the measurand is applied to the transducer and its level is adjusted so that the transducer's output is at the value where maximum loading error occurs. While the measurand is held constant, the transducer's output is measured first across the maximum, and then across the minimum, specified load impedance.

Since the input impedance of the readout device forms a portion of the

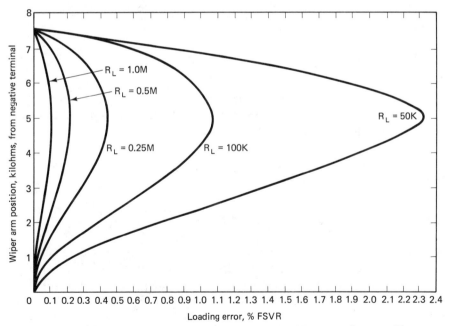

**FIGURE 4–7.** Loading error for potentiometric transducer with 7500-ohm element resistance.

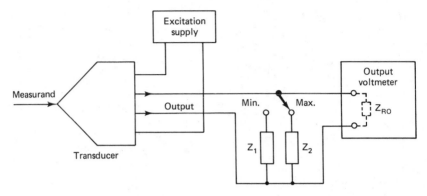

**FIGURE 4–8.**   Loading-error test setup.

load impedance, it must be considered in the determination of test impedance (usually resistance) values. As illustrated in Figure 4–8, the impedance $Z_1$ necessary to provide the maximum specified load impedance $Z_{max}$ in parallel with the input impedance $Z_{RO}$ of the readout device is

$$Z_1 = \frac{Z_{RO}Z_{max}}{Z_{RO} - Z_{max}}.$$

The impedance $Z_2$ necessary to provide the minimum specified load impedance $Z_{min}$ is

$$Z_2 = \frac{Z_{RO}Z_{min}}{Z_{RO} - Z_{min}}.$$

From this test, the loading error is determined as the output reading $E_1$ (for $Z_{max}$) minus the output reading $E_2$ (for $Z_{min}$). Expressed in percent of full-scale output, the loading error $\Delta L$ is then

$$\Delta L = \frac{E_1 - E_2}{E_1} \times 100.$$

### 4.1.11   *Resolution and Threshold Tests*

*Resolution tests* are normally performed only on potentiometric transducers utilizing wire-wound transduction elements. The purpose of this test is to verify the use of a suitable number of turns on the potentiometric winding, as well as the general quality of the winding, but primarily it verifies the absence of grossly uneven spacing between turns, of two or more turns shorting together, and of turns protruding or recessed relative to the nominal outside diameter of the wire-wound resistance element.

The resolution test (Figure 4–9) is usually performed by applying the measurand, varied from the lower to the upper transducer range limit, simultaneously to the test transducer and to a reference transducer which has a continuous-resolution transduction element (e.g., a strain-gage type of transducer). The output of the transducer under test is connected to the "y-axis" input of an x-y plotter, that of the reference transducer to the "x-axis" input. The resulting x-y plot shows the magnitude and number of all steps in the output of the test transducer, provided that the zero and gain setting of the x-y plotter's amplifiers were properly adjusted. The best results are obtained when the zero setting is shifted several times during the test so that several full-scale plots are obtained, each showing only a selected portion of the test transducer's output but with steps magnified accordingly. The maximum magnitude of any of the output steps (maximum resolution) as well as the average magnitude of all output steps (average resolution) are then determined from examination of the x-y plot.

*Threshold tests* are sometimes performed on transducers having continuous-resolution transduction elements. A threshold test is simply the determination of the smallest change in the measurand that will result in a measurable change in transducer output, by use of suitable measurand and output-monitoring equipment. This test may have to be performed at several measurand levels within the transducer's range.

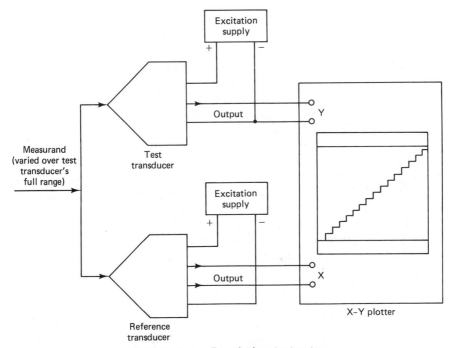

**FIGURE 4–9.**  Resolution test setup.

### 4.1.12 Excitation Tests

Excitation tests are performed to determine specification compliance of the transducer's excitation ratings and to verify the transducer's susceptibility to field hazards related to excitation.

Tests on transducer designs in which excitation is applied directly to the transduction element are usually limited to power-rating verification. Included here are input-impedance or transduction-element resistance measurements, determination of current drain at rated and maximum excitation voltages, measurements of power dissipation at rated and maximum excitation currents, and verification of proper transducer operation after applications of high-transient pulses of excessive excitation voltage or current.

Additional tests are performed on transducers incorporating excitation-regulation, modification, isolation, or conversion circuitry. Depending on the severity of applicable specification requirements, these may include performance determination at minimum and maximum values of excitation and after excitation polarity reversal, misapplication of excitation to output terminals, and output short-circuiting.

### 4.1.13 Voltage-Breakdown Test

The voltage-breakdown test, also called the high-potential, dielectric-strength, or dielectric-withstanding-voltage test, is used to verify the adequacy of electrically insulating materials and of the spacing between certain mutually insulated conducting surfaces.

The voltage-breakdown test is performed by applying a specified test voltage between mutually insulated electrical transducer connections or between ungrounded connections and the (grounded) transducer case. The duration of the application (typically 60 s) should be specified, as well as the magnitude and other characteristics of the test voltage (e.g., 500 V ac, rms, 60 Hz). Test failure is evidenced by surface discharge (*flashover*), air discharge (*sparkover*), or puncture discharge (*breakdown*) within the transducer. When limits are also placed on the surge current (measured at initial test voltage application) and on the maximum leakage current (over the entire duration of the test), these currents are measured and recorded.

Most specifications state either the breakdown voltage rating or the insulation resistance (see next). Hence, only one of these two tests is normally performed on a transducer.

### 4.1.14 Insulation-Resistance Test

The insulation-resistance test is used to measure the resistance offered by the insulating members of a transducer to an impressed dc voltage tending to produce a current leakage either through or on the surface of these members. The insulation-resistance test is commonly used to determine characteristics of the insulation between the transduction element, together with any integral signal-conditioning circuitry, and the external case of the transducer. In some transducers, where excitation-output isolation is required, the test is additionally used to measure the insulation between excitation and output connections.

The importance of this test is given largely by the high impedance circuits frequently used in electronic measuring systems. The operation of the system can be affected severely by "ground loops"—undesirable leakage currents through structural ground planes between components of the system or between excitation-supply and signal-transmission grounds. Excessive leakage currents may also lead to further deterioration of the insulation by heating or electrolysis.

The test is performed by connecting the two voltage-carrying leads from a megohmmeter, megohm-bridge, or insulation-resistance test set to the specified connection points on the transducer (e.g., all insulated receptacle pins, in parallel, and the transducer's case). The test voltage is raised to the specified level (usually 50, 100, or 500 V dc), and the resistance is read on the test-set meter. The reading is taken either when the resistance has stabilized at the value above the minimum specified resistance or after an electrification time of two minutes.

Insulation-resistance measurements are often repeated during and after various environmental tests in order to determine the effects of heat, moisture, dirt, oxidation, and loss of volatile materials on the required insulation characteristics of a transducer.

## 4.2 CALIBRATIONS

The term "calibration," in its broader sense, includes making adjustments to instruments and making marks on indicating dials and on control knobs. In its narrower sense, however, when applied to transducers (with electrical output), a calibration is purely a test during which known values of measurand are applied to the transducer and the corresponding output readings are recorded. All adjustments, stabilization, and compensation, as well as any interim checks of the operation of the transducer, are considered manufacturing processes, and these all have to be completed before calibration. Only those operations that would normally be performed during use, or just prior to use, of a transducer can be included in the calibration. An example of such a normally performed operation is the short-time, high-temperature "bakeout" of certain vacuum transducers.

Calibration methods differ widely, depending on the measurand. The key issue in a calibration is that the value of the measurand applied to the transducer has to be *known*, which means that the possible errors in that value have to be known. Some calibrations use a reference instrument for this that is of the same general type as the transducer; or a different type can be used, as long as its errors are known and are substantially smaller than those allowed for the transducer. Usually, these reference instruments are periodically compared with a transfer standard. Other calibrations may use a well-understood natural phenomenon as reference; for example, temperature transducers are calibrated by exposing them to a "fixed-point" temperature such as the freezing point of pure water, the boiling point of water, or the freezing or boiling points of other pure materials. Experts in national standards laboratories are usually available for advice on up-to-date acceptable calibration methods. The manner in which output readings are obtained is covered in Section 4.1.

Most transducers are subject to a *static calibration*, which is performed under room conditions and includes letting the measurand stabilize at various values before an output reading is recorded. These values are generally selected at equal increments, with the measurand first increasing and then decreasing. Certain types of transducers that either cannot be calibrated statically or are used primarily for dynamic measurements are subjected to a *dynamic calibration* (see Section 4.3). A few types of transducers whose behavior can be predicted accurately by measuring their output at only one measurand value may be subjected only to a *one-point calibration*.


## 4.3 DYNAMIC TESTS

Dynamic tests are performed on transducers to verify or determine their dynamic characteristics (see Section 2.5.2). These tests indicate how well the transducer output indicates fluctuations in the amplitude of the measurand. (Fidelity tests on loudspeakers or phonograph cartridges are comparable to this.) The fluctuations can be continuous (e.g., sound or vibration) or occasional (e.g., sudden increases or decreases in temperature, pressure, or flow). Certain categories of transducers are routinely subjected to a *dynamic calibration*, a thorough dynamic test, instead of a static calibration. More categories, however, receive a static calibration as well as a dynamic test; in these cases the dynamic test is generally performed only on transducers subjected to qualification tests, sampling tests, and other non-production tests.

Two types of tests exist for dynamic tests: the sinusoidal response test and the step-function response test. Both tests often (but not always) employ a reference transducer that has a known and faster response (i.e., a wider flat-response frequency range or a shorter time constant) than the transducer to be tested. Connecting both transducers to the same measurand then allows a comparison between the behavior of the test transducer and that of the reference transducer. When such a test is used as a dynamic calibration, it is called a *comparison calibration*. The amplitude changes that are typically used in the two methods are illustrated in Figure 4–10.

The *sinusoidal response test* can be performed on only a limited number of different types of transducers, primarily because of test equipment design problems. During this test, the measurand applied to the transducer is caused to undergo precisely controlled sinusoidal variations in its level. The frequency of the sinusoidal variations is increased, either continuously (*frequency sweep*) or in steps, and the transducer output is recorded continuously or for each step, respectively. The amplitude of the sinusoidal measurand fluctuation is held constant throughout the test. Typical test amplitudes are 20, 10, and 1% (the latter two are preferred) of the transducer range, peak to peak, within the lower or middle portion of the range.

The *step-function response test* can be performed on most types of transducers. It consists of recording the transducer output while the measurand applied to the transducer is caused to undergo a step change in its level. Step changes from 45% to 55% or from 10% to 90% of the transducer range are commonly used. The equipment used to create this step change must be capable of minimizing the rise time of the step, as seen by the

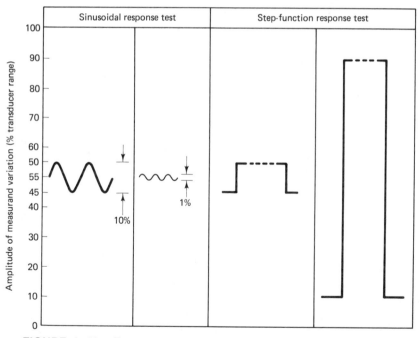

**FIGURE 4–10.** Types and amplitudes of measurand variations in dynamic response tests.

transducer, so that a true step function and not a ramp function is applied to it. The equipment must also be capable of maintaining the final level of the measurand for a time long enough to yield valid output data.

Frequency response is best determined by use of the sinusoidal response test, but equipment limitations may make this test difficult or impossible to administer. In this case the frequency response may be inferred from calculations based on results obtained from a step-function response test in combination with a knowledge of the transducer's mechanical properties or geometry; it should then be identified as a *calculated frequency response*. The step-function response test can be used for determining not only a transducer's time constant, rise time, or response time, but also its natural frequency, ringing period, damping ratio, and overshoot.

## 4.4  ENVIRONMENTAL TESTS

Most transducers are used in applications where one or more environmental conditions exceed those defined as room conditions, the ambient conditions prevailing during calibration. Unless otherwise specified, *room conditions* are generally a temperature of 25 ± 10°C, a relative humidity of 90% or less, a barometric pressure of 88 and 108 kPa (26 to 32 in Hg), and an absence of vibration and acceleration (unless one of these is the measurand). When transducers are to be exposed to more severe environments during their intended use (see Section 2.5.3), they can be expected to have performance errors greater than those indicated on their calibration record. Under these conditions they can also take a shift in their calibration curve, and they can even be damaged and stop operating entirely. When environmental conditions more severe than room conditions are specified for a transducer (usually together with widened performance tolerances under such conditions), environmental tests must be performed to verify that the transducer performs as specified. Such environmental tests are normally performed only as part of a qualification or sampling test; however, some critical applications may require at least some partial environmental testing on a 100% basis.

There are two categories of environmental tests: *operating tests*, during and after which the transducer must perform within specified tolerances; and *nonoperating tests*, during which no performance tolerances apply but after which the transducer must again perform as specified.

It rarely happens that the person responsible for transducer selection and use is also responsible for running the environmental tests. Most commonly the tests are performed by a different department in a company or facility, or by the transducer manufacturer, or by an outside laboratory. Therefore, it is best to have a written test procedure that both parties agree to before any tests are begun. This procedure should establish the test sequence, test methods, and test setups and equipment to be used, and it should show when and how test data, test results, and any additional observations will be recorded. A set of general ground rules should also be followed for all environmental tests. These can only be reflected partly in the written procedure, but can be verified by inspection and observation as well as by discussions with test personnel:

1. The transducer, and especially its electrical, mechanical, hydraulic, and pneumatic connections, must not affect the environment.
2. The transducer must be exposed to no other environments than that specified for the test.
3. The calibration equipment, especially the measurand-level sensor or indicator, must be isolated from the environment.
4. The environment should not affect the measurand seen by the transducer.
5. The level of the environment actually seen by the transducer should be monitored throughout the test.
6. At least a partial calibration, at room conditions, should always be performed right after an environmental test has been completed, to determine any latent or permanent effects of the environment on the transducer.

### 4.4.1 Temperature Tests

The primary environmental equipment used for these tests is the *temperature chamber*, typically a double-walled, well-insulated metal box with a removable door and with a number of small holes or ports, either in the door or the box, or both, as feed-throughs for electrical, mechanical, and hydraulic or pneumatic connections to the transducer. A mounting bracket for the test specimen is often attached to the inside of the door. The transducer to be tested can then be mounted to that bracket, all connections to it made through holes in the door, and the door, carrying the entire setup, reinstalled on the chamber.

Heating as well as cooling elements are installed between the inner and outer walls. The heating elements are electrical resistance heaters, and the cooling elements are made of tubing that will circulate an externally supplied coolant between the chamber walls. Many chambers also provide a compartment that can be filled with dry ice. A small blower in the chamber improves convective heat transfer and prevents thermal stratification. The internal chamber temperature can be preset, whereupon it will remain thermostatically controlled. Some chambers provide for fine control, using a continuous-type temperature sensor and a controller (which can be proportional) in a closed control loop. A temperature sensor is attached to the transducer case, its leads are connected to an external temperature indicator, and the actual temperature of the transducer can then be monitored and stabilization at the desired temperature can be determined. Performance verification tests are carried out either after stabilization at the specified temperature or after an additional specified period ("temperature soak").

### 4.4.2 Temperature Shock and Temperature Gradient Tests

Even though both the temperature shock and the temperature gradient test involve a rapid change in the temperature seen by the transducer, the first is usually a nonoperating test and the second is usually an operating test.

The temperature shock test is used to determine whether a change in ambient temperature between two specified levels and at a specified rapid rate (e.g., 5° per min) causes subsequently detectable damage or intolerable performance changes. This test usually simulates a shipping or other pre-use condition such as transportation in an unpressurized aircraft cargo compartment. Some temperature chambers can provide the desired rate of temperature change, while others require the use of a simpler but more severe test method. The transducer, with all required connections, is first stabilized in a chamber maintained at one temperature and then removed and inserted in a second chamber that is precooled or preheated to a different temperature.

A performance test (including at least a partial calibration) after subsequent stabilization of the transducer at room conditions demonstrates any permanent effects of temperature shock on the transducer.

A temperature gradient test is performed on some types of transducers to determine their temperature gradient error (see Section 2.5.3). The test is performed on a bench rather than in a temperature chamber, and its objective is to create a sudden difference in temperature between the sensing element and the case while the transducer output is being recorded continuously. One of these two temperatures is usually room temperature.

Some methods involve keeping the sensing element at room temperature while the case of the transducer is rapidly immersed into a hot or cold liquid bath. If the test specimen is a pressure transducer, for example, a short piece of tubing can be connected to its pressure port, and the assembly can be held by the tubing while the transducer case is immersed. When the sensing element is exposed, as in a flush-diaphragm pressure transducer, the sensing element can be immersed in a hot or cold bath without the transducer case being immersed in it, or the sensing element can be exposed to a step increase in temperature by setting off a flashbulb in front of it or suddenly directing hot air from a blower at it.

### 4.4.3 Acceleration Tests

Acceleration tests are performed on transducers to determine their performance while they are subjected to quasi-steady-state or slowly varying acceleration (as opposed to vibratory accelerations; see Section 4.4.4). A rotary accelerator (*centrifuge*) is most commonly used to simulate an environment in which acceleration is present. The acceleration acting on a transducer mounted to a turntable near its perimeter can be calculated very accurately from a knowledge of the radius arm (distance between the centerline of the transducer and the center of the turntable) and the angular speed (in r/min) of the turntable.

When the axes of the transducer have been properly identified (see Figure 2–29), the transducer can be mounted in various orientations so that the acceleration will act along its *x*-, *y*- or *z*-axis, in a positive or negative direction. Most acceleration specifications call for acceleration tests along each of these three axes in both directions, although not necessarily at the same acceleration levels. This means mounting the transducer on the centrifuge six times and performing an acceleration test each time. For each acceleration test, the measurand is applied to the transducer and usually held at a fixed level. The difference in transducer output readings taken just before starting the centrifuge and while it is running is the acceleration error. Acceleration tests are quite short; 30 to 60 s is usually long enough to obtain a valid output reading with acceleration applied. Remounting the transducer for each test makes the tests more time consuming.

Special fittings and fixtures are often required to apply a given measurand to the transducer while the turntable is in motion. Pressure can be applied through a rotary-seal swivel fitting in the turntable shaft. Displacement transducers can have their shaft clamped in a fixed position, while temperature transducers can be immersed in a specially designed constant-temperature bath mounted on the centrifuge turntable, with care taken to keep the acceleration from affecting the temperature of the bath during each of the short tests. Electrical connections are always made by use of slip-ring contacts in the centrifuge shaft.

### 4.4.4 Vibration Tests

Vibration tests are performed to determine the effects of (linear) vibratory acceleration on the transducer's performance. A *vibration exciter* (*shaker*) is used to apply vibration to a mounting fixture on which the transducer is installed. The axis along which the vibration is applied is determined by the orientation of the transducer's axes, as explained in the previous section for acceleration tests. Since vibration is bidirectional, it is only necessary to install the transducer in three different orientations, whereas six such orientations and associated remountings are necessary for performing a complete triaxial acceleration test. Electromagnetic shakers are normally used for vibration testing; their operating principle is similar to that of dynamic loudspeakers such as are used in many automobile radios.

The mounting fixture for the transducer requires careful design, followed by a "dry run" on the shaker while it is instrumented, so as to assure that there are no mechanical resonances and no "cross talk" in the fixture; "cross talk" refers to vibration induced in axes other than the test axis. As illustrated in the typical set-up diagram (Figure 4–11), three monitoring accelerometers are installed on the fixture in close proximity to the transducer to be tested, in order to monitor vibration along three mutually orthogonal axes—the test axis and the two transverse axes. An additional accelerometer, the *drive* (control) accelerometer, is mounted so that it measures the applied vibration. The drive is connected in a feedback circuit to the shaker control console so as to maintain closed-loop control of vibration at preselected levels. As during acceleration tests, the measurand can be applied to the transducer from external sources (as illustrated), or it can be generated and maintained locally (on the transducer, on the mounting fixture, or in the environment immediately ambient to the mounting fixture).

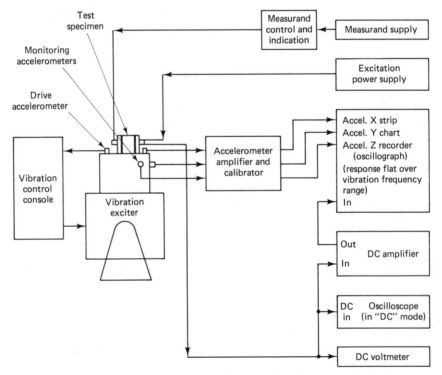

**FIGURE 4–11.** Typical transducer vibration test setup.

By use of appropriate control, filtering, and power-amplifying equipment, vibration is applied to the transducer in accordance with a predetermined vibration program stated in tabular or graphical form. Such a program usually describes vibration levels in terms of vertical displacement (for low frequencies) or acceleration (for higher frequencies) vs. vibration frequencies, and vibration frequency vs. elapsed test time.

Two types of vibration can be generated and applied to a transducer, either individually or simultaneously (combined): sinusoidal vibration, whose source is usually an audio oscillator; and random vibration, whose signal emanates from a noise source and passes through a number of narrow band-pass filters (equalization controls), allowing the power spectral density of the applied vibration to be adjusted for each narrow band of vibration frequencies. Sinusoidal vibration is periodic in nature. Random vibration is nonperiodic, described only in statistical terms, and characterized by an amplitude distribution which essentially follows a Gaussian distribution ("normal error curve"). The test duration time must be specified, unless the random vibration is combined with a sinusoidal vibration program including a frequency-vs.-time program.

Vibration effects on transducers, as determined during such tests, can usually be classified into three categories: permanent damage (mechanical failure), output variations corresponding to applied vibration levels, and output variation due to *resonances*—amplified vibrations of internal components, within narrow frequency bands, related to the resonant frequencies of these components and excited by the applied vibration when at those frequencies.

More than most other tests, vibration tests call for the talents of an experienced test engineer, one who is capable of determining the validity and possible origins of an apparent test failure as soon as it occurs. Failures ascribed to a transducer are frequently caused by loose or improper mechanical, pneumatic, or electrical connections to the transducer, by resonances or cross talk within the fixture, by insufficient tightening of the fixture or transducer mounting hardware, by inadvertent variations in the applied measurand, by incorrect settings on the control console or on an accelerometer calibrator, by improper grounding of the various interconnected pieces of test equipment, or even by unfiltered transients in the power-line voltage.

### 4.4.5 Shock Tests

Shock tests are normally performed on transducers to determine their performance after (rather than during) exposure to mechanical shocks. Various types of shock-test machines are available for conducting such tests. They are all capable of applying shocks having acceleration amplitudes over 10 000$g$ with time durations from less than 200 $\mu$s to over 100 ms. The shape of the shock pulse can be half-sine (upper or lower half of a sine wave), sawtooth (terminal peak sawtooth), or trapezoidal. Shock pulses can be defined either by shape, amplitude (of acceleration), and time duration, or by their frequency spectrum. Some shock-test machines are simple mechanical devices, such as a pendulous hammer swinging against a vertical surface or a weight dropping in free fall from a given height guided by vertical rails or rods. Other, more complex machines are hydraulic; these can often be programmed more accurately for a large variety of shock pulses.

Most transducer test specifications call for a total of six equal shocks to be applied to the test specimen while it is installed and connected as intended during its actual use. By mounting the transducer on different surfaces of a test fixture, shock is applied sequentially in both directions along each of the transducer's three ($x$-, $y$-, and $z$-) axes.

In rare cases, the performance of a transducer is also monitored during the shock applications, usually by means of an oscilloscope or oscillograph connected to the transducer's output.

### 4.4.6  Atmospheric Environmental Tests

The category of atmospheric environmental tests includes all those tests required to verify that the transducer can withstand a variety of chemical and physical effects in the atmosphere (or vacuum, or liquid) it is intended to be used in. Some such tests may also verify that there are no harmful interactions between the transducer and its ambient environment. Most of these tests apply to the housing (case) of the transducer, its material and finish, and its sealing qualities.

The tests are typically nonoperating tests, although excitation power may need to be applied to the transducer for some of the tests. Requirements for these tests usually stem from industrial codes and civilian or military government specifications, which also contain detailed test procedures.

Included among these tests are ambient-pressure tests (for effects of low pressures encountered at high altitudes, vacuum encountered in space, and high pressures acting on an underwater transducer); sunshine and ozone tests; salt-spray and salt-atmosphere tests; fungus, rain, and humidity tests; liquid-immersion tests; explosive-atmosphere tests; and sand and dust tests. Some of these tests may be waived if other evidence exists that the transducer case is hermetically sealed and corrosion resistant.

### 4.4.7  Special Environmental Tests

Certain transducer applications require that maximum levels and spectra of additional environments be established, and that analyses of the transducer's ability to operate in such environments be supplemented by tests. Some such typical special environments contain high sound-pressure levels (for which "acoustic tests" are performed), nuclear radiation, or strong electromagnetic or electrostatic fields.

### 4.4.8 Combined Environmental Tests

In most transducer applications two or more different types of environments act upon the transducer simultaneously, each at various levels at various times. When prediction of the transducer's behavior under such conditions, based on individual environmental tests, is not deemed adequate, combined environmental tests may have to be performed. Examples of these are temperature-vibration, thermal-vacuum, and temperature-altitude-humidity tests, as well as shock tests performed at cryogenic temperatures.

Since such tests typically require complex test setups, their cost effectiveness should be carefully considered before they are called for. Thorough analyses, combined with expert knowledge of transducer design, construction, materials, piece parts, and processes, may lessen the need for having these tests performed.

## 4.5 LIFE TESTS

Operating life tests are hardly ever performed, because specifications for operating life tend to be in terms of years, typically between three and ten years. Sometimes, field experience can be substituted for performing an operating life test, when it can be verified and documented that at least one transducer of the same design, taken randomly from a production lot, has operated maintenance free and within specified tolerances in its end-use application for a certain number of years. Another substitute is the accelerated life test, which calls for continuous operation of the transducer at conditions more severe than normal, but only for periods on the order of days or weeks.

Cycling life tests are usually performed as part of a qualification test when a transducer specification calls for full-range or partial-range cycling life. Equipment has been designed for rapid and automatic cycling of many categories of transducers. A calibration is performed at the beginning and end of such a test, and sometimes also at one or more intermediate intervals. When a cycling test is performed automatically and unattended, provisions should be made to stop the test (and the cycle counter) when a catastrophic transducer failure occurs.

# 5

# Displacement Transducers and Motion Sensors

## 5.1 BASIC DEFINITIONS

*Length* is a fundamental unit in mechanics. The length of a straight line is the number of times that a specified measuring rod must be applied successively until the line has been covered completely. The length of a curve is defined in the same manner except that the measuring rod must be flexible so that it always conforms exactly with the portion of the curve to which it is applied. The length of a straight line between two points is the *distance* between the two points, the spatial separation between the two points or objects. *Proximity* is the spatial closeness between two points or objects.

An *angle* is the figure obtained by drawing two straight lines from one point or by the figure formed by two surfaces diverging from the same line. The angle represents the amount of rotation of one line about a fixed point on the other line or the amount of rotation of one surface from the other surface if both diverge from the same line. If the line on which the fixed point is located is considered the *initial* line and the line rotating about the point is the *terminal* line, the angle is positive if the terminal line rotates counterclockwise and negative if the terminal line rotates clockwise.

The concepts of position, motion, and displacement are so closely interrelated in the measurement field that the terms "position transducer" or "motion transducer" have been used instead of the more correct term "displacement transducer" for the identical sensing device. *Position* is the spatial location of a body or point with respect to a reference point. *Motion* is the change in position of a body or point with respect to a reference system.

*Displacement* is the vector representing a change in position of a body or point with respect to a reference point. *Linear displacement* is a displacement whose instantaneous direction remains fixed. *Angular displacement* is the angle between the two coplanar vectors determining a displacement.

## 5.2 UNITS OF MEASUREMENT

### 5.2.1 The International System of Units (SI)

The SI (Système International d'Unités), which was derived from the *metric absolute system*, has been adopted as a standard by virtually all countries. It is based on seven fundamental units and two supplementary units.

The fundamental units are the *meter* $(m)$* as unit of length, defined as exactly 1 650 763.73 wavelengths in vacuum of the radiation corresponding to the transition between the energy levels $2p_{10}$ and $5p_5$ of the krypton-86 atom; the *kilogram* $(kg)$ as unit of mass, defined as the mass of the international prototype kilogram kept in the custody of the Bureau International des Poids et Mesures (BIPM), Sèvres, France; the *second* $(s)$ as unit of time interval, defined as the interval occupied by exactly 9 192 631 770 cycles of the radiation corresponding to the $(F = 4, M_F = 0)$ to $(F = 3, M_F = 0)$ transition of the cesium-133 atom when unperturbed by exterior fields; the *ampere* $(A)$ as unit of electric current; the *kelvin* $(K)$ as unit of temperature; the *candela* $(cd)$ as unit of luminance; and the *mole* $(mol)$ as unit for amount of substance. The latter four units are discussed in more detail in their appropriate chapters.

The two supplementary units are the *radian* $(rad)$ as unit of plane angle and the *steradian* $(sr)$ as unit of solid angle.

All other units of measurement in the SI are derived from these fundamental units and, to a limited extent, from the supplementary units.

### 5.2.2 SI Units for Displacement

Linear displacement is given by the *meter* $(m)$ and such commonly used multiples and submultiples of it as the *kilometer* $(km)$, *centimeter* $(cm)$, *millimeter* $(mm)$, and *micrometer* $(\mu m)$. The micrometer has also been referred to as the "micron."

### 5.2.3 Other Units for Displacement Compatible with the SI

Alternative units for angular displacement are the *degree* $(°)$, the *minute* of arc or "arc-minute" $(')$, and the *second* of arc or "arc-second" $('')$. These

---

*Unless specific government-sanctioned rules apply, the spelling "meter" should be considered interchangeable with "metre."

will all probably coexist with the radian for many more years. One radian equals 57.296 (roughly 57.3) degrees. One degree equals 0.01745 radian. Equivalently, $1° = 60' = 3600''$. Note that there is no space between the number and the symbol, an exception to normal SI practice. Note also that a previously used symbol for degree of arc, ''deg,'' should no longer be used.

### 5.2.4  Conversion Factors

The following table lists conversion factors from U.S. customary to corresponding SI units:

| To Convert From | to | Multiply by |
|---|---|---|
| acre | m² | $4.047 \times 10^3$ |
| astronomical unit (AU) | m | $1.49598 \times 10^{11}$ |
| circular mil | m² | $5.067 \times 10^{-10}$ |
| degree (angle) | rad | $1.745 \times 10^{-2}$ |
| degree (angle) | mrad | 17.45 |
| fathom | m | 1.829 |
| foot | m | 0.3048 |
| inch | m | $2.540 \times 10^{-2}$ |
| inch | cm | 2.54 |
| light year | m | $9.46055 \times 10^{15}$ |
| mil | m | $2.540 \times 10^{-5}$ |
| mile (U.S. statute) | m | $1.609 \times 10^3$ |
| mile (U.S., Internat'l nautical) | m | $1.852 \times 10^3$ |
| minute (angle) | rad | $2.909 \times 10^{-4}$ |
| minute (angle) | mrad | 0.291 |
| parsec | m | $3.08374 \times 10^{16}$ |
| second (angle) | rad | $4.848 \times 10^{-6}$ |
| second (angle) | μrad | 4.848 |
| yard | m | 0.9144 |

## 5.3  SENSING METHODS

Displacement transducers sense displacement, detect motion, or measure position by either contacting or noncontacting methods. *Contacting* sensing methods rely on connecting the *sensing shaft* of the transducer mechanically to the point of measurement, the point or object whose displacement is being measured. The connection can be made by attachment, using one or more fasteners, or it can be made by spring loading the transducer shaft against the point of measurement. *Noncontacting* methods employ inductive, electromagnetic, or capacitive coupling (for relatively small displacements and close-up motions) or electromagnetic waves such as visible-light beams, infrared beams, or microwave beams (for relatively large displacements and more distant motions).

To understand the importance of sensing shafts and their coupling means, one must simply realize that the output of the transducer indicates the position of the sensing shaft, not of the driving point. To make the two equal requires a shaft of the proper shape and strength as well as a suitable coupling device. The latter must be designed so that it will be free of unwanted play and backlash and so that there is no slippage in it after it is fastened. Some displacement transducer designs incorporate provisions, with known tolerances, to allow for minor misalignments between the point of measurement and the sensing shaft. Typical shaft ends and couplings for linear and angular displacement transducers are shown in Figure 5–1. Spring-loaded shafts (Figure 5–2) are required for certain applications, such as those in which only a relatively small portion of the measured object's total displacement is to be measured (the portion closest to the transducer). Spring-type sensing shafts are also used widely on switch-type position-sensing devices.

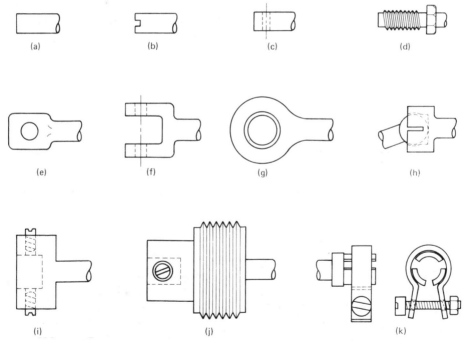

**FIGURE 5–1.** Displacement-transducer shaft ends and couplings: (a) plain; (b) slotted; (c) through-hole; (d) threaded; (e) lug; (f) clevis; (g) bearing; (h) ball joint; (i) coupler; (j) bellows-coupler; (k) collet-and-clamp.

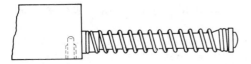

**FIGURE 5–2.**   Spring-loaded shaft.

## 5.4   DESIGN AND OPERATION

### 5.4.1   *Capacitive Displacement Transducers*

Figure 5–3 illustrates the four basic designs of transducers in which a change in capacitance is proportional to a change in displacement. In the *moving-dielectric* design both capacitor electrodes are fixed in position. A sleeve made of an insulating material with a dielectric constant different from that of air slides in and out of the electrode assembly. As the sleeve is pulled out of the electrode assembly, an increasing amount of electrode surface sees air as the dielectric and a decreasing amount of surface area sees the sleeve material as the dielectric. This results in a change of capacitance proportional to the axial motion of the sleeve. A similar principle has been used in transducers that are used to determine the thickness of a film of one liquid on top of another liquid, where the two liquids have different dielectric properties (e.g., oil on water).

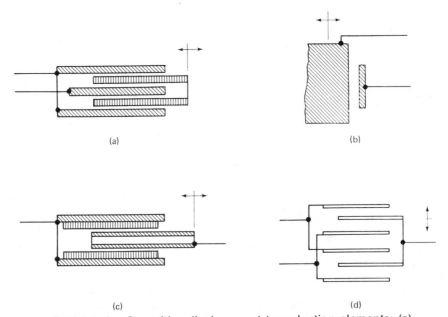

**FIGURE 5–3.**   Capacitive displacement transduction elements: (a) moving dielectric; (b) moving rotor, noncontacting; (c) moving rotor, single stator, coupled; (d) moving rotor, split stator, coupled.

Two versions of the *moving-rotor* design are illustrated in the figure. In the *single-stator* transducer a conductive cylindrical electrode slides in and out of a stationary conductive electrode whose inside surface is usually coated with a dielectric material. Rotor displacement thus causes changes in capacitance between the two electrodes. In the *split-stator* (balanced) transducer the rotor consists of a number of ganged, electrically interconnected electrodes, each of which moves between two stationary (stator) plates. The upper stator plates, interconnected, form one section of the split stator. The interconnected lower plates form the other section. As the rotor plates move, their capacitance to one section increases while their capacitance to the other section decreases. The electrodes can be connected as two active arms of an ac bridge circuit.

In the *noncontacting* capacitive displacement transducer the measured object (which must be partly or wholly electrically conductive) acts as the rotor which moves relative to the transducer, which is simply a stator insulated from its mounting. This mounting, as well as the measured object, should be grounded.

A number of known sources of error should be avoided in the design and use of capacitive transducers. The capacitance of the interconnecting cable between transducer and electronics must be very low and must remain fixed in its value. It is usually advantageous to keep this cable very short and package the electronics close to, or integral with, the transduction element. The transducer must be so constructed that no unwanted motion such as end play can occur. The electrode design should be such that the electrode is free from the effects of fringing and protected from stray fields. Compensation for temperature effects should be included in the design, or the transducer and associated circuitry should be so designed that no significant temperature errors occur. Linearization of transducer output can be provided either electrically or by shaping the electrodes appropriately.

### 5.4.2  Inductive Displacement Transducers

Transducers converting (usually linear) displacement into a change in the self-inductance of a single coil can be grouped into coupled and noncontacting versions (see Figure 5–4). The coupled designs employ a sliding, magnetically permeable core which moves within a coil (bobbin). The sensing shaft is attached to the moving core. As the core moves, the coil changes its self-inductance. The coil can be connected into an *LC* oscillator circuit so that inductance changes result in output frequency changes. More frequently, the coil acts as one arm of an impedance bridge whose ac output (when the bridge is excited with ac current) reflects the inductance changes. A second coil (reference or balancing coil) is usually connected into the adjacent arm of the impedance bridge. This coil, whose inductance is not influenced by displacement changes, is often packaged integrally with the

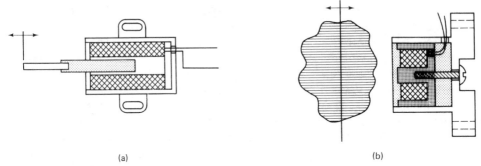

(a)                                           (b)

**FIGURE 5-4.** Basic types of inductive displacement transducers:
(a) coupled; (b) noncontacting.

sensing coil. This reduces undesirable effects due to long connecting leads, causes both coils to see the same thermal environment, and in some cases provides adjustments to the reference coil at the point of measurement.

Noncontacting designs are used more commonly than coupled designs. Relative proximity of the sensing coil to the measured object causes changes in the coil's inductance. Although measurements are feasible when the object is made of a highly conductive diamagnetic or paramagnetic material, they are more successful when the object is made of a ferromagnetic material and has a high permeability. The full-scale measuring ranges of such transducers are typically below 1.5 cm.

A design variation of a noncontacting inductive displacement transducer is illustrated in Figure 5-5. Displacements of the measured object,

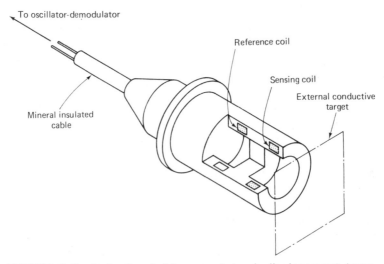

**FIGURE 5-5.** Inductive (eddy-current type) displacement transducer. (Courtesy of Kaman Instrumentation Corp.)

which must be conductive but need not be ferromagnetic, cause changes in *eddy currents*, which, in turn, cause changes in the impedance of the sensing coil. A reference coil is integrally packaged into the transducer. Transducers of this type have been designed to operate continuously at temperatures as high as 600°C. An associated electronics unit provides bridge excitation at RF frequencies as well as bridge balance adjustments and ac-to-dc conversion.

Somewhat related to the inductive position transducer is the *electromagnetic proximity sensor*, which contains a coil wound on a fixed core. This type of sensor cannot be used for steady-state displacement measurements, because it responds to a time rate of change in flux due to a rate of change in proximity to a metallic object (target). Such sensors are used as position switches (see Section 5.4.10; they can easily detect, for example, a door being closed), but are more frequently used as sensing coils in toothed-rotor tachometers (see Section 6.4.4) and turbine flowmeters.

### 5.4.3  Reluctive Displacement Transducers

The category of reluctive displacement transducers includes all transducers that convert displacement into an ac voltage change by a variation in the reluctance path between two or more coils (windings) in the presence of ac excitation to the coils. Several of the types illustrated in Figure 5–6 are very widely used, sometimes as transduction elements for transducers measuring quantities other than displacement. Most popular is the *differential transformer*, which is used for linear as well as angular displacement measurements (*LVDT* denotes a *linear variable differential transformer*, and *RVDT* a *rotary variable differential transformer*). The same holds true for the less widely used *inductance bridge*. The *induction potentiometer, synchro* ("selsyn"), *resolver, microsyn,* and *shorted-turn signal generator* are used to sense angular displacements only.

The operation of an LVDT is illustrated in Figure 5–7. Of the three coaxial windings, the center winding is the primary and the windings to each side of it are the secondaries, which, in their most elementary connection scheme, are connected together at one of their two terminals (as shown subsequently in Figure 5–9). When ac excitation is applied to the primary winding, and the ferromagnetic core (*armature*) moves within the coil assembly, the coupling between the primary and each of the two secondaries changes. As a result, the output voltage magnitude and phase at the secondary (output) terminals change from the *null*, which occurs when the core is centered between the two secondaries. Note that output changes tend to become nonlinear near the ends of core travel. The operation of an RVDT, such as the one illustrated schematically in Figure 5–6(b), is similar (see Figure 5–8). In addition to the basic winding configuration shown, there are a considerable number of other configurations, including split or center-

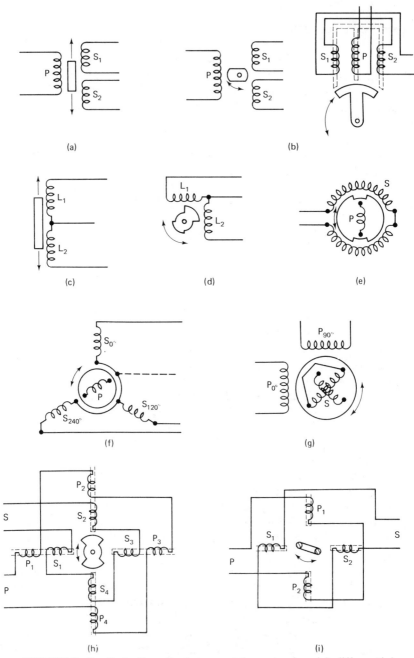

**FIGURE 5-6.** Reluctive displacement transducers: (a) differential transformer (linear); (b) differential transformers (angular); (c) inductance bridge (linear); (d) inductance bridge (angular), (e) induction potentiometer; (f) synchro; (g) resolver; (h) microsyn; (i) shorted-turn signal generator.

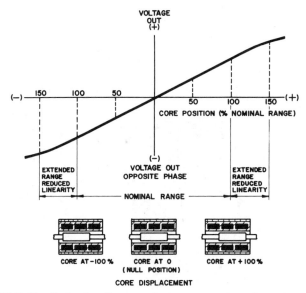

**FIGURE 5-7.** Output voltage and phase as a function of core position for a linear variable differential transformer. (Courtesy of Schaevitz Engineering.)

tapped primaries and multiple secondaries with different types of interconnections (e.g., those intended to reduce phase angle changes).

The construction of a typical LVDT displacement transducer is illustrated in Figure 5-9. The primary winding and the two secondary windings are wound over a hollow coil form made of a nonmagnetic and insulating

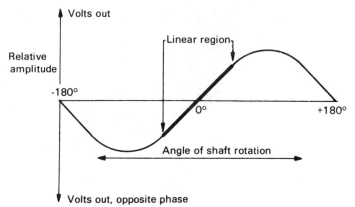

**FIGURE 5-8.** Output voltage as a function of shaft rotation for a rotary variable differential transformer. (Courtesy of Schaevitz Engineering.)

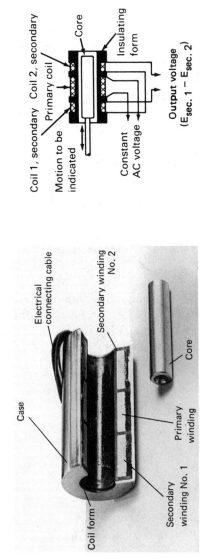

**FIGURE 5–9.** Reluctive (LVDT) linear displacement transducer. (Courtesy of Schaevitz Engineering.)

Coil 1, secondary   Coil 2, secondary

Core

Primary coil

Insulating form

Motion to be indicated

Output voltage
$(E_{sec.\ 1} - E_{sec.\ 2})$

Constant AC voltage

Electrical connecting cable

Secondary winding No. 2

Case

Core

Primary winding

Coil form

Secondary winding No. 1

material. The ferromagnetic core is threaded to accept one of a variety of sensing shafts. The winding assembly is potted within a cylindrical case, and the connecting leads are brought out of the potted assembly. The case is usually made of a ferromagnetic metal so that it acts as a magnetic shield. A transducer of similar design but incorporating integrally packaged dc–dc conversion circuitry is shown in Figure 5–10. This design uses a separate magnetic shield as well as a bore liner within which the armature (core) moves. The conversion circuitry, which also provides for internal and external connections, is mounted in the end of the case. The design illustrated operates from a ±15-V dc power supply and provides an output voltage of ±10 V dc. A variety of other excitation voltages can be accommodated by similar transducers, such as 12 V dc (automotive) and 28 V dc (aircraft and aerospace). Other output voltages can also be provided, such as the popular 0 to 5 V dc. As shown in the block diagram of Figure 5–10, an oscillator is used to convert the dc into the ac excitation required by the LVDT, whose output is then demodulated into dc and amplified. Circuitry to protect the electronics from the application of reverse polarity is included in this design. The output can be applied directly to a readout device, or fed to a telemetry set or computer.

Displacement transducers with spring-loaded shafts, such as the LVDT type shown in Figure 5–11, are sometimes referred to as "gage heads," because they are commonly used in machine tool inspection and gaging equipment. The sensing shaft is guided in a sleeve bearing whose type and fit govern transducer repeatability errors to a large extent. Various sensing-shaft tip configurations are available, as shown in Figure 5–12.

Angular-displacement transducers of the differential-transformer type are limited in range to about ±40°, since, for larger angles, the output becomes increasingly nonlinear (see Figure 5–13). In a typical design (see Figure 5–8) the primary and secondary windings are wound on a coil form (stator) and a cardioid-shaped ferromagnetic rotor changes the coupling from the (split) primary to each of the secondary windings. Appropriate shaping

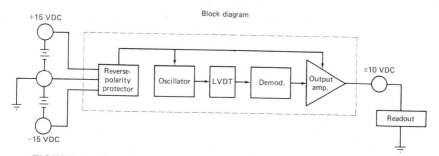

**FIGURE 5–10.** Reluctive (LVDT) linear displacement transducer with dc excitation and dc output ("DC-DC LVDT"). (Courtesy of Schaevitz Engineering.)

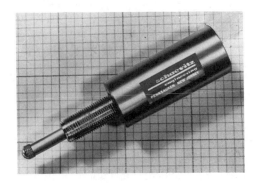

**FIGURE 5–11.** Reluctive linear displacement transducer with spring-loaded shaft ("LVDT gage head"). (Courtesy of Schaevitz Engineering.)

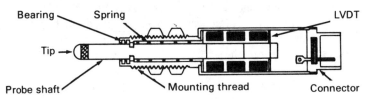

of the rotor aids output linearization. Integrally packaged dc–dc conversion circuitry is available for this type of transducer.

Inductance-bridge-type transducers are used for linear (Figure 5–6(c)) as well as angular (Figure 5–6(d)) displacement measurements. Two coils and the core are so arranged that the inductance of one coil increases while the inductance of the other coil decreases with motion of the core. The matched set of two coils forms two arms of a four-arm ac bridge. The other two arms are usually resistive. Dc–dc conversion circuitry can also be integrally packaged with such transducer designs.

The *induction potentiometer* (see Figure 5–6(e)) has been used in control systems. The coupling between the primary winding, on the rotor which sees the angular displacement, and the single secondary winding (or two windings in series), on the stator, changes with rotor rotation, providing an output which is reasonably linear up to a range of about ±35°.

In the *synchro* (see Figure 5–6(f)) the single primary winding on the rotor, which moves with the angular displacement to be sensed, interacts inductively with a three-phase stator whose windings are physically spaced 120° apart. In most applications the synchro used for sensing (*synchro transmitter*) provides an electrical output which is used to control the mechanical

**FIGURE 5–12.** Typical tips for gage heads. (Courtesy of Schaevitz Engineering.)

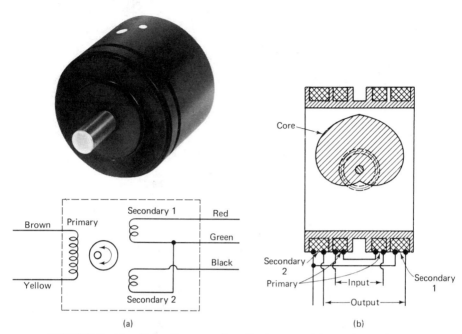

**FIGURE 5–13.** Reluctive angular displacement transducer (rotary variable differential transformer, RVDT): (a) schematic diagram; (b) cross section (simplified). (Courtesy of Schaevitz Engineering.)

position of the rotor in a second synchro (*synchro receiver*), which can then be used as a display device (see Figure 5–14) or to do other mechanical work. When used to drive a synchro receiver, the synchro transmitter operates as a *torque synchro*. When its rotor moves to a given angular position, the stator of the synchro receiver causes a torque to act on its rotor. This torque is reduced to zero only when the receiver's rotor assumes the same angular position. When the electrical output of a synchro is used for purposes other than driving a synchro receiver, the synchro transmitter can be referred to as a *control synchro*.

The *resolver* (Figure 5–6(g)) differs from the synchro mainly in the number and spacing of its windings. Both the two-phase rotor and the two-phase stator have windings spaced 90° apart. When used as a measuring transducer, the resolver has one of its two rotor windings shorted.

The *microsyn* (Figure 5–6(h)) and the related *shorted-turn signal generator* (Figure 5–6(i)) are more suitable as measuring transducers than are the synchro and resolver, although all four of these rotary transformer devices are mostly used in control systems. This is because the microsyn and shorted-turn signal generator operate off single-phase ac excitation and provide a single-phase output proportional to rotor displacement. Both devices have a four-pole stator, usually made of laminated iron plates. Each of the

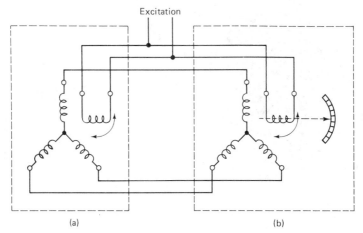

**FIGURE 5–14.**   Synchro transmitter-receiver system: (a) transducer synchro (transmitter); (b) readout synchro (receiver).

microsyn poles is wound with two coils, a primary and a secondary winding. The primaries are all connected in series. The secondaries are also in series, but are so connected that the voltage induced in coils $S_1$ and $S_3$ opposes the voltage induced in coils $S_2$ and $S_4$. The ferromagnetic rotor has a unique "butterfly" configuration. Angular displacement of the rotor, from its null position, causes a change in the reluctance path to coils $S_1$ and $S_3$ which is opposite to that to coils $S_2$ and $S_4$. The resulting unbalance creates a net output voltage in the secondary.

The four poles of the shorted-turn signal generator are each wound with only one coil. Opposing pairs constitute the primary and secondary winding set, respectively, at right angles to each other. The ring-shaped rotor is a single-turn shorted coil. Flux is induced in the rotor by the ac-excited primaries, and an output is produced in the secondaries when the rotor moves from its null position and produces flux linkages to them. The shorted turn can be machined integrally with the sensing shaft.

### 5.4.4  Potentiometric Displacement Transducers

Potentiometric displacement transducers are relatively simple devices in which a sliding contact (*wiper*) attached to or otherwise mechanically linked with the sensing shaft moves over a resistance element. The wiper is electrically insulated from the sensing shaft. Potentiometric transducers exist for linear- as well as angular-displacement measurements (see Figure 5–15). In linear-displacement transducers a second wiper and wiper bus (Figure 5–15(b)) are often used to prevent those problems that could be caused by a flexing, relatively long wiper lead. The output is a displacement-proportional fraction of the excitation voltage (applied across the " + " and " − "

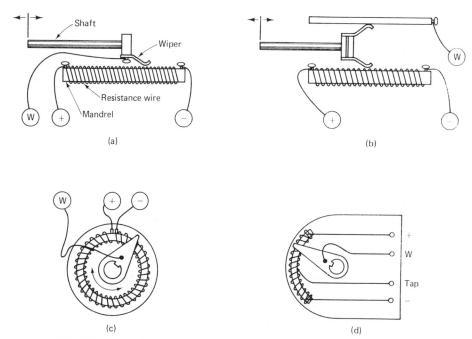

**FIGURE 5–15.** Basic types of potentiometric displacement transducers: (a) linear (basic); (b) linear (with wiper bus); (c) angular (single turn); (d) angular (sector, tapped).

terminals). The resistance element (*potentiometric element*) can be provided with one or more additional electrical connections (*taps*) at specified positions.

When the potentiometric element is wire wound (typically with a platinum or nickel alloy wire) the resolution steps are given by the number of wire turns per unit length. To obtain small steps, the total element resistance can be made large (5000 to 10 000 Ω) and thin wire (approximately 0.01 mm in diameter) would then be used. This, however, results in a variable, relatively high output impedence that may cause loading errors. Continuous-resolution ("infinite-resolution") potentiometric elements have been made of conductive plastic, a carbon film, a metal film, or a ceramic-metal mix ("cermet"). The mandrel, around which the winding is wound, is made of an insulating material or of an insulation-coated metal. To allow for over-travel beyond a specified displacement range, the mandrel can be extended and provided with conductive strips (or wire turns soldered together) that are electrically connected to (or part of) the respective excitation terminal.

Wipers exist in a number of different designs (see Figure 5–16) and are typically made from precious-metal alloys or spring-tempered copper alloys. Leaf-spring and dual wipers are used in transducers expected to see severe

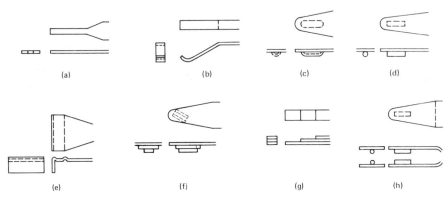

**FIGURE 5–16.** Typical wiper configurations: (a) plain; (b) plain, hook type; (c) dimpled; (d) rod type; (e) knife edge; (f) knife edge, offset; (g) leaf spring; (h) dual.

shock or vibration environments. The wiper contact force is typically adjusted to between 4 and 15g. The contact as well as the resistance element are sensitive to externally introduced contamination; hence, a good sliding seal for the sensing shaft and a good seal for the electrical connections are important.

Displacements sensed by angular-displacement transducers of the potentiometric type range between less than 10° and over 3500°. For small ranges, some internal mechanical amplification is often employed. Helical resistance elements are used in multiturn potentiometers for ranges in excess of 360°. Single-rotation transducers can be designed with mechanical end stops or for continuous rotation; in either case, the presence of excitation terminals at the ends of the potentiometric element limits their active range to about 357°. Linear-displacement transducers with very large measuring ranges (up to about 8 m) use a spring-loaded cable-and-pulley arrangement, either with the pulley rotation geared down to drive a single-turn rotary potentiometer or with the pulley rotation used to drive a multiturn potentiometer.

### 5.4.5 Strain-Gage Displacement Sensors

The operating principle of strain-gage displacement sensors is shown in Figure 5–17. The gages are attached to a flexure or bending beam whose tip can act as a sensing shaft, or the beam can be deflected by a plunger-type sensing shaft. The gages are mounted so that one (in a half-bridge configuration) or one pair (in a full-bridge configuration) responds to tension in the beam, whereas the other responds to compression. Both conductor (metal foil) and semiconductor gages have been employed in such sensors, which are usable for measuring very small displacements. Some sensors of this

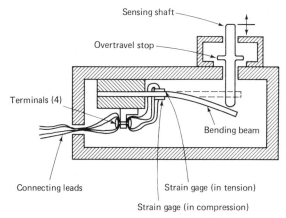

**FIGURE 5–17.** Basic design of strain-gage displacement transducer.

type have been built by laboratories that use them for their own work; commercially produced sensors are quite rare.

### 5.4.6 *Vibrating-Wire Displacement Sensors*

The vibrating-element principle (see Section 15.4.9) has been applied to sensing very small displacements (measuring ranges of 50 micrometers or less). Some sensors of this type have been custom built for certain applications, and some have been built in small quantities. No such sensors have been reported to be commercially produced and marketed.

### 5.4.7 *Electro-Optical Displacement Sensors*

With the exception of electro-optical encoders (see Section 5.4.8), electro-optical displacement sensing devices are generally of the noncontacting type and are used primarily for dimensional gaging.

When a surface of an object whose displacement (or whose dimension relative to its opposite surface) is to be measured is sufficiently reflective, a light source–sensor combination can be used to measure small displacements (or dimensional deviations). This principle is illustrated in Figure 5–18. The output of the sensor decreases exponentially with increasing distance to the measured object. Related designs use fiber optics to transmit and receive the light reflected by the measured surface; they can be used to measure very small displacements with fine resolution. When the measured surface is not inherently reflective, a reflector can be attached to it. Light beams can be of constant intensity (dc) or modulated (ac) at low or high frequencies (depending on the response time of the light sensor and the

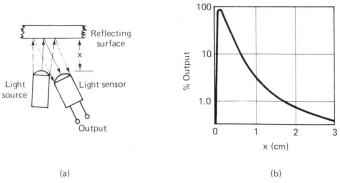

(a)                                                                    (b)

**FIGURE 5–18.** Displacement sensing by reflection method (basic):
(a) optical arrangement; (b) typical response.

application); they can also be chopped or pulsed. Infrared light-emitting
diodes (LEDs) are often used as a light source.

One of the methods employed in electro-optical dimensional gaging is
illustrated in Figure 5–19. In this edge view of a typical basic setup, the
height and/or width of an object can be measured in terms of the amount of
occultation of the beam between source and sensor. When the light sensor
is a multiline array of light-sensitive elements, or another type of imager
such as a vidicon camera, the shape of the measured object, and its variations
in height and/or CCD width, can be determined very accurately by processing the sensor output data and either displaying the resulting image or
comparing it with a model of the "standard" image. Such systems are very
useful for quality control of machined or otherwise fabricated parts.

The well-collimated, coherent, monochromatic light beams from lasers
have made these devices an important element in electro-optical dimensional
measuring systems. One of the methods employing a laser for the measurement of product thickness or, generally, distance between the sensing head
(laser light source and two light sensors) and a surface, is the *triangulation*
method. The spot of laser light on the surface is viewed by two light sensors
at the same angle, but from opposite directions. Surface position relative to
a reference position can be determined from the two light sensor outputs.
A variation of this method employs two laser beams at equal but opposite

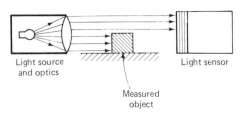

**FIGURE 5–19.** Dimensional gaging by occultation method.

angles (from the normal to the surface) and a light sensor viewing the intersection of the beams with a viewing direction normal to the surface. The imaging-type sensor detects the distance between the two laser spots on the surface, from which the distance to the surface (and, e.g., the thickness of the material whose bottom surface is at a known position) can then be determined. Laser scanning, which can be obtained by deflecting the beam with an oscillating mirror, is also used for dimensional gaging as well as flaw detection in the product of rolling mills.

Applying interferometric techniques to reflected-laser-light displacement and dimensional measuring systems resulted in development of the *laser interferometer* (Figure 5–20). The optical principle of this instrument is based on that of the Michelson interferometer, a principle used also in many other applications, such as infrared spectroscopy. The laser beam is split into two orthogonal beams by a beam splitter. One beam is directed at a fixed flat mirror which reflects it back to the beam splitter, and the other beam is either directed at a mirror which displaces along the direction of the beam, and whose displacement is to be measured, or, as shown in the illustration, at a reflector, from which it reflects toward another fixed mirror. In the latter case, it is the displacement of the reflector (or the object to which it is attached) which is being measured. The second mirror returns the beam along the same path, back to the beam splitter. Here the beam recombines and (optically) interferes with the reference beam. The interference is constructive when the difference between the two beam paths is an integral number of wavelengths; it is destructive when the path difference is an odd integral of half-wavelengths. As the reflector moves and changes

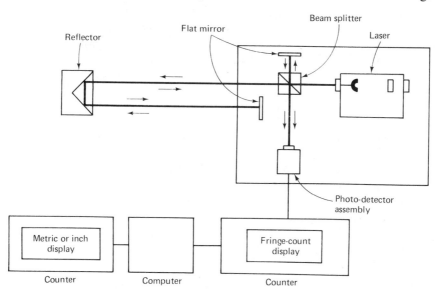

**FIGURE 5–20.**   Laser interferometer system.

the path length of the "measuring" beam, one complete interference fringe is detected by the photodetector (light sensor) for every wavelength change in the path length. The displacement is then determined from a fringe count, providing a digital output characterized by extremely high precision.

Advances in the development of the laser interferometer, such as the two-frequency laser and the holographic interferometer, have extended its measurement capabilities, which now also include measurements of angles (with resolution of a fraction of an arc-second), alignment, straightness, and flatness.

### 5.4.8  Angular and Linear Encoders

Linear as well as angular displacement can be sensed and converted into a digital output more simply than most other measurands and without using an analog-to-digital converter. The devices providing such a digital output in response to displacement changes are known as *linear encoders* and *angular encoders*, respectively.

**5.4.8.1 Encoder Types.**  Three different transduction methods are employed in encoders (see Figure 5–21). In the *brush-type* encoder the moving disk or strip contains conducting and insulating segments on its surface. The conducting segments are all connected to a common terminal. When the pick-off brush is in contact with a conducting segment, a contact closure occurs. When the brush is in contact with an insulating segment, an open contact results. When a source of excitation voltage is connected between the common contact and the associated signal processing circuitry, an output voltage (a "1") is seen during contact closure, and no output (a "0") is seen when the brush contact is open. In the *photoelectric* or *optical encoder* a transparent disk or strip is provided with a pattern of opaque segments on its upper surface. These interrupt a light beam and prevent it from illuminating a light sensor. Hence, an output "1" is produced when the light sensor is below a transparent segment, and a "0" results from the light sensor being below an opaque segment. In the *magnetic encoder* the disk or strip is provided with a pattern consisting of magnetized and nonmagnetized segments on its surface. A ferromagnetic core, provided with an input winding and an output winding, is placed above this surface. An input ("interrogate") signal is applied to the input winding. When the core is above a nonmagnetic segment, the core remains unsaturated and an output signal (a "1") is produced. When the core is above a magnetic segment, the flux from the segment causes the core to be saturated and no output signal (a "0") is produced.

Brush-type encoders have been essentially rendered obsolete by magnetic and, especially, optical encoders, both of which provide contactless

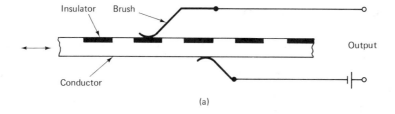

(a)

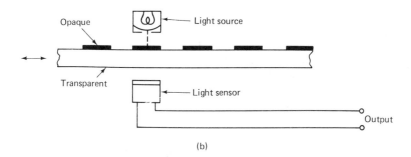

(b)

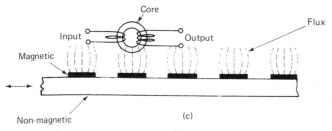

(c)

**FIGURE 5–21.** Encoder transduction methods: (a) brush type; (b) photoelectric type; (c) magnetic type.

sensing and, hence, a very long operating life. Optical encoders originally used small, ruggedized light bulbs as a light source; light-emitting diodes (LEDs) are now used for this purpose in virtually all modern designs. The opaque/transparent pattern is applied by sophisticated methods such as vacuum bombardment through a precisely cut mask on a glass substrate. Generally, techniques used to manufacture miniature printed circuits are often applied to both of these encoder types. Magnetic encoders tend to use a ferrite disk on which the magnetic pattern is inscribed very precisely. Techniques employed in the design and manufacture of high-capacity, modern digital tape recorders have been used to improve magnetic encoders. Sinusoidal ac as well as pulse signals are used in their input and output sides.

The major categories of encoders are the following:

1. The *incremental encoder* (see Figure 5–22) produces equally spaced pulses from one or more tracks. The pulses are fed to and accumulated in an up/down counter, and the count is indicative of displacement. The starting point can be arbitrarily selected, and the associated readout equipment can be zeroed for the new setting. Alternatively, an indexing pulse can be provided by an additional "1"-producing segment in the code pattern.

2. The *absolute encoder* (see Figure 5–23) produces a digitally encoded number, indicative of position, by an array of reading heads and a multitrack pattern on the code disk or strip. A variety of codes are used in addition to the binary code represented by the pattern shown in the illustration. Linear codes include the "V-scan," "Gray," and BCD (binary-coded decimal) codes; nonlinear codes such as sine, cosine, and tangent are also used. A brief introduction to these codes is presented next.

**5.4.8.2   Digital Codes and Coding Patterns.**   The most frequently used digital codes in absolute encoders are illustrated in Figure 5–24. The natural binary code, or just *binary code*, is the simplest code for use with arithmetic or comparison-computing circuitry. It is easily understood by personnel

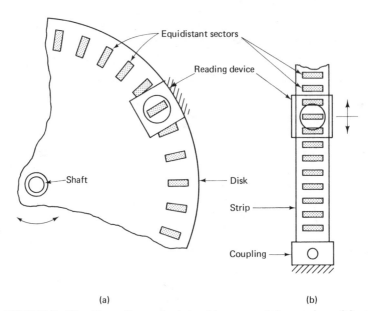

(a)                                                             (b)

**FIGURE 5–22.**   Operating principle of incremental encoders: (a) angular; (b) linear.

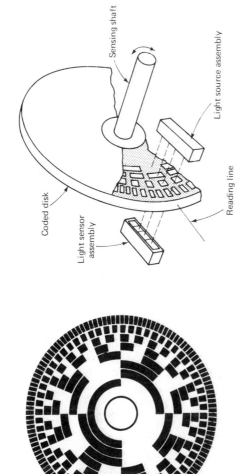

(a)

(b)

**FIGURE 5–23.** Basic absolute (photoelectric, angular) encoder: (a) typical code disk; (b) encoder elements.

| Arabic number | (Natural) Binary | | Gray (Binary) | | Binary Coded Decimal (BCD) | | | | | |
|---|---|---|---|---|---|---|---|---|---|---|
| | Digital number | Code pattern | Digital number | Code pattern | Digital number | | Code pattern | | | |
| | | | | | Tens | Units | Tens | Units | | |
| | 8 4 2 1 | $2^3\ 2^2\ 2^1\ 2^0$ | | $G_3\ G_2\ G_1\ G_0$ | 8 4 2 1 | 8 4 2 1 | $2^0$ | $2^3\ 2^2\ 2^1\ 2^0$ | | |
| 0 | 0 0 0 0 | | 0 0 0 0 | | 0 0 0 0 | 0 0 0 0 | | | | |
| 1 | 0 0 0 1 | | 0 0 0 1 | | | 0 0 0 1 | | | | |
| 2 | 0 0 1 0 | | 0 0 1 1 | | | 0 0 1 0 | | | | |
| 3 | 0 0 1 1 | | 0 0 1 0 | | | 0 0 1 1 | | | | |
| 4 | 0 1 0 0 | | 0 1 1 0 | | | 0 1 0 0 | | | | |
| 5 | 0 1 0 1 | | 0 1 1 1 | | | 0 1 0 1 | | | | |
| 6 | 0 1 1 0 | | 0 1 0 1 | | | 0 1 1 0 | | | | |
| 7 | 0 1 1 1 | | 0 1 0 0 | | | 0 1 1 1 | | | | |
| 8 | 1 0 0 0 | | 1 1 0 0 | | | 1 0 0 0 | | | | |
| 9 | 1 0 0 1 | | 1 1 0 1 | | 0 0 0 0 | 1 0 0 1 | | | | |
| 10 | 1 0 1 0 | | 1 1 1 1 | | 0 0 0 1 | 0 0 0 0 | | | | |

**FIGURE 5–24.** Digital code structures for absolute encoders.

working with digital electronics systems. It is based on the relative position of bits having a value of "0" (or "no") and those having a value of "1" (or "yes") in a data word of a given length. The data words in the illustration have a length of 4 bits. The shaded portions of the code patterns indicate a "1." In practice, these shaded portions would correspond to a contact in the case of brush-type encoders, a nonmagnetized segment in the case of magnetic encoders, and a transparent segment in the case of optical encoders. The binary code is based on powers of 2; hence, a "1" in the last bit—the *least significant bit*, or *LSB* (the $2^0$ bit)—indicates the arabic numeral "1" ($2^0 = 1$, and a "1" or "yes" indication in that bit position means "one"). Other arabic numbers are indicated by a "1" or "0" in the other bit positions. For example, the digital number "1001," in the binary code, is read as follows: bit position $2^3$ (which is 8): yes; bit positions $2^2$ and $2^1$ (which are 4 and 2, respectively): no; bit position $2^0$ (which is 1): yes. The resulting number, then, is 8 + 1, or arabic 9. The number of discrete increments obtainable from an encoder is given by $2^n - 1$, the " $- 1$ " because zero always exists as one of the numbers. If, for example, we want to know how many bits (and how many tracks on an encoder) are needed to indicate angular displacements of a full circle to better than 1 second of arc, we first convert seconds to numbers of increments for a full circle ($60 \times 60 \times 360 = 1\ 296\ 000$), and then look up a table of powers of 2 and find that the nearest power of 2 giving a resolution better than 1 in 1 296 000 is 21; that is, $2^{20} - 1$ is only 1 048 575, but $2^{21} - 1 = 2\ 097\ 151$. In angular encoders the LSB is normally the outermost track. Table 1–1 shows the number of increments versus the number of bits.

For error detection by associated computing circuitry, an additional track for a *parity bit* can be added to binary-coded disks and strips. This track is so coded that, for every position of the encoder, either an odd number of "1's" is always obtained (*odd parity*) or an even number of "1's" is always obtained (*even parity*).

The *binary-coded-decimal* (*BCD*) code is a combination of the binary system and the Arabic decimal system. Four-bit binary data words are so arranged that the last word indicates units in binary code, the next-to-last data word indicates tens, and so on. The difference between the BCD and natural binary codes can be seen if we examine the digital representations (and the code pattern) of the Arabic numeral "10" in Figure 5–24. A number such as 835 would be represented in binary by "1000 0011 0101."

A disadvantage of binary code disks is that two or more bits can change simultaneously during a single position change within the resolution of the encoder, as, for example, when going from 7 (0111) to 8 (1000). Certain other codes were developed to get around any problems that may be caused by this. One of these is the *Gray code* (also shown in Figure 5–24), named after Frank Gray of Bell Laboratories. In this code pattern the state of the LSB is changed only every two counts, rather than every count as in the binary code. This reduces possible reading ambiguities and permits tracks to be closer together, hence reducing the overall size of an encoder. When fed into computer circuitry, this code must first be converted into binary form.

The *V-scan* method has been used in absolute encoders to minimize reading ambiguities. It uses either a pair of brushes, one behind the other, on all except the LSB track, and external logic; or has each of the tracks (except the LSB track) of the code pattern split in half, with each half slightly displaced from the other so that one half is "leading" while the other is "lagging" (as is the case for the two brushes per track). The V-disk arrangement also requires external logic circuitry to prevent ambiguous readings.

Various methods have been used to improve the reading capability of incremental encoders, which, in their basic version, produce a simple train of equidistant pulses, with the pulse indicating a "1" and the baseline a "0." One method is to add a track and form a code pattern equivalent to a series of two-bit data words. Other methods involve the use of *interference patterns* (see Figure 5–25), usually obtained by placing a stationary reticle or other mask over the moving pattern. When the "$N + 1$" pattern is used, the rotating disk has $n$ segments and a stationary disk placed between the rotating disk and the reading devices has $n + 1$ segments. The reading devices (light sensors) are placed 180° apart. As the disk rotates, the output of each sensor is modulated quasi-sinusoidally, with the total number of periods equal to $n$. The outputs of the two light sensors can be used for multiplication as well as for direction sensing. The *moiré pattern* is formed by a stationary mask with a light bar pattern similar to that of the moving element but tilted at a small angle with respect to it. As the element (typically a strip in a linear

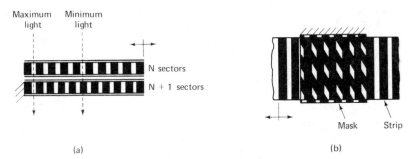

**FIGURE 5–25.** Interfering patterns used in encoders: (a) *N* + 1 pattern; (b) moiré pattern.

encoder) moves, the light bands, which have the appearance of a series of parallelograms, move up or down, depending on the direction of motion. A light sensor (with collimating optics) is placed in front of each of the light bands (arranged vertically). This system provides substantial improvements in resolution. Multiplication has also been obtained mechanically in angular encoders by using two disks and a gear train between them so that the "fast disk" rotates many times for each rotation of the "slow disk" in such a *multispeed encoder*. In addition to incremental and absolute encoders there are also "hybrid" encoders, which employ two different techniques, such as absolute encoding or moiré interference patterns for the less significant digits and incremental encoding for the more significant digits.

**5.4.8.3 Hardware Designs.** Angular encoders, incremental or absolute, are provided either with an axial sensing shaft or, when their size is relatively large, with a through hole, an annular rotor, and an annular code disk. The latter type is illustrated in Figures 5–26 and 5–27. This design has a separate electronics unit; some have integrally packaged electronics. The absolute encoder of the design uses V-scan as well as multiple reading heads on the LSB track, with averaging circuitry, to provide encoding up to $2^{22}$ bits. Both the multiple reading head and averaging techniques are also used in incremental encoders.

A shaft-type incremental optical encoder is illustrated by the three designs shown in Figure 5–28. This model uses LED illumination and dual reading heads; a labyrinth-type shaft seal provides environmental protection of the encoder. Integral electronics can furnish "up/down" (indicative of direction of rotation) pulses up to 216 000 pulses per revolution, equivalent to a resolution of 6 arc-seconds; additional external electronics can reduce the resolution to 3 arc-seconds (432 000 pulses per revolution).

An incremental linear optical encoder is shown in Figure 5–29. This design has an extruded aluminum enclosure and employs a glass scale with vacuum-deposited chrome lines and a reading head with an LED light source.

**FIGURE 5–26.** Angular encoder (through-hole type). (Courtesy of Itek Corp., Measurement Systems Div.)

Integral electronics generate quadrature outputs of either buffered sinusoids or TTL square waves. The scale density is 25 lines per mm, and optional electronics can provide a resolution as low as 1 micrometer. Working lengths between 10 cm and 122 cm are available.

### 5.4.9 Radar, Lidar, and Sonar Distance Sensors

The category of radar, lidar, and sonar distance sensors comprises devices which provide a measurement of distance by remote sensing. The measured distance is the distance between the sensor and a point or small area called the *target*. Most of these sensors emit pulses of electromagnetic energy directed at the target and then determine the distance to the target by measuring the time it takes a corresponding reflected signal (*echo*) to be received by the sensor. The number of pulses emitted per second is called the *repetition rate*. At least one type of laser sensor, however, emits continuous-wave (*CW*) energy and determines (relatively short) distances as a function of phase difference between the transmitted and received signals.

Electromagnetic radiation at ultrasonic frequencies (see Section 16.6.4) is used by *sonar* equipment, primarily to measure distances underwater. A single "transducer," usually piezoelectric, is typically used in the transmitting as well as the receiving mode. Related devices are used in a mapping mode (ultrasonic scan) in many applications, including biomedical ones; a

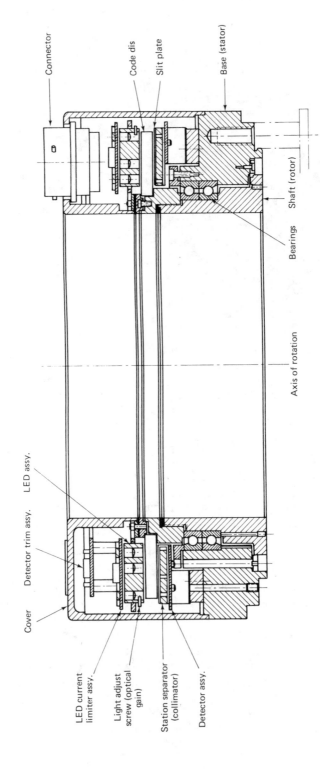

**FIGURE 5–27.** Sectional view of through-hole type of angular encoder. (Courtesy of Itek Corp., Measurement Systems Div.)

**FIGURE 5–28.** Angular encoders (shaft type). (Courtesy of Teledyne Gurley.)

well-known example of the latter is ultrasonic imaging of a fetus within the mother-to-be's uterus.

Very short wavelengths, extending into the microwave region, characterize sensors such as radar trackers and radar altimeters. Still shorter wavelengths, in the optical (typically the infrared) region (see Chapter 22), are used in lidar (laser) altimeters and related distance-measuring equipment. All these devices normally use separate transmitters and receivers. Critical parameters driving equipment design include the effective radiated power (the power level fed to the antenna or optical aperture, and the size of the antenna or optical aperture), reflectivity of the expected target, and antenna or optical aperture size as well as sensitivity of the receiver.

Distance can also be determined from observations of target velocity (see Chapter 6), by using the Doppler effect (see Chapter 12), or by distance determinations employing microwave interferometry (for space applications) together with appropriate calculations.

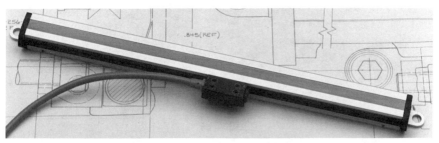

**FIGURE 5–29.** Linear incremental encoder. (Courtesy of Teledyne Gurley.)

### 5.4.10   *Position-Sensing Switches*

Directly actuated as well as noncontacting position-sensing switches are very widely used devices. Many of these employ operating principles (inductive, eddy current, electromagnetic, electro-optical) covered earlier in this chapter. The switch which has enjoyed the most popularity, and is only gradually expected to be replaced by nonmechanical devices, is the *electromechanical switch*, an example of which is shown in Figure 5–30. It is essentially a momentary-contact, single-pole double-throw (*SPDT*) push-button switch. When the plunger is depressed, it acts upon a spring linkage and causes the normally closed (*NC*) contact to open and the normally open (*NO*) contact to close. The switch is so installed that the measured object will actuate it when it attains the position that is to be measured.

Switches of this type can be equipped with various kinds of actuators to facilitate their operation. These include a roller actuator, a leaf-spring actuator, a lever-type actuator, and plunger extensions of various heights, some allowing for large amounts of overtravel, some in a waterproof boot, some with a roller at their tip. These are either an integral part of a switch or can be added to it. Hermetically sealed versions and completely waterproof switches are among the available design variations. Most such electromechanical position-sensing switches are designed for rugged use and for large numbers of actuations. The spring is typically made of beryllium copper and the contacts of silver or gold. Terminals can be of the screw, quickconnect, or solder type.

There are also a number of switch designs that do not rely on electromechanical means to open or close a contact. One of these is the *Hall-effect*

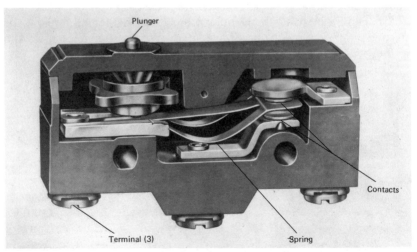

**FIGURE 5–30.**   Cutaway of electromechanical basic switch. (Courtesy of MICRO SWITCH, A Division of Honeywell.)

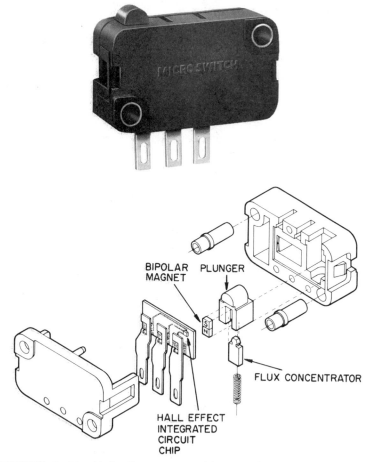

**FIGURE 5–31.** Hall-effect-type position-sensing switch (plunger operated). (Courtesy of MICRO SWITCH, A Division of Honeywell.)

switch (see Figure 5–31). This effect was discovered in 1879 by Edward H. Hall at the Johns Hopkins University. He found that a difference in potential would appear at opposite edges of a thin rectangle of gold through which a current was flowing when a magnet was placed so that its field was perpendicular to one face of the gold foil. Much later, when this effect was applied to semiconductors rather than gold, and much higher Hall voltages were obtained, a number of useful devices using the effect were developed. Since the voltage is proportional to the current flowing through the semiconductor, the effect lends itself to current measurement. Because the voltage is also proportional to the magnetic flux density, the device can be used either where the flux density itself is to be measured or where changes in it are to cause voltage changes for purposes other than flux measurement. The Hall-effect switch can also be used in devices where the relative proximity of the magnet to the semiconductor is to give an indication by a voltage change.

In the switch illustrated, actuation of the plunger causes the vertical displacement of a magnet, changing is proximity to the Hall generator. A flux concentrator is used to concentrate more of the available flux into the sensor. An integrated-circuit chip associated with the Hall generator provides signal conditioning, including a step-function generator with hysteresis to establish a "dead band," prevent erroneous actuations due to noise or vibration, and output logic suitable for interfacing. Besides the plunger-actuated switch, there are noncontacting versions of this design for which the measured object must be magnetic (or equipped with a magnet).

For the Hall-effect switch as well as the eddy-current type to be described momentarily, the typical motions of the measured object are either "head-on" (motion to and from the sensor along its longitudinal axis), "slide-by" (lateral or vertical motion past the sensing surface of the switch), or "rotary" (motion of magnetic "teeth" or poles along the rim of a rotating member past the sensing surface). Neither of these types, however, is sensitive to the rate of change of flux coupling, which is a drawback of the *electromagnetic* proximity sensor. The *eddy-current* proximity switch (see Figure 5–32) is similar to the eddy-current inductive displacement transducer. It is the (integrally packaged) circuitry that makes it into a switch rather than an analog-output sensor. The measured object must be metallic, but need not be ferrous. The measured object (the "target") tends to absorb

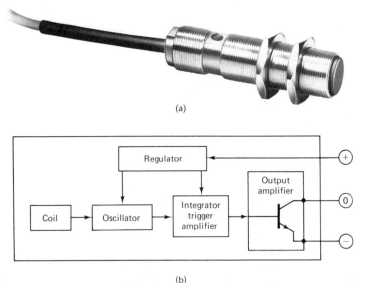

(a)

(b)

**FIGURE 5–32.** Eddy-current-type proximity switch: (a) sensor assembly (with integrally packaged electronics); (b) block diagram. (Courtesy of MICRO SWITCH, A Division of Honeywell.)

the electromagnetic field generated in the sensing coil due to the generation of eddy currents, which present a reflected load to the oscillator, reducing its signal level. This change in level is amplified by the integrator circuitry, which drives a Schmitt trigger coupled to an output transistor, thus providing the switching action. An (optional) LED, located in the wall of the rear portion of the housing, furnishes a visible output indication. The unit illustrated is fully shielded; unshielded versions are also available.

*Photoelectric* position-sensing switches employ a light source and a light sensor to detect the presence or absence of objects that block (or unblock) the light path or cause reflection of the light beam or light-scattering products to be incident on the light sensor. The two primary methods used are direct scan and retroreflective scan. In *direct scan* (Figure 5–33(a)) a light source and a light sensor ("photoreceiver") are positioned opposite each other, and the object to be detected passes between them. The object must be opaque enough as well as large enough to block the light beam sufficiently to cause switching operation. Collimating can be used for source and sensor to aid in the detection of relatively small objects. In *reflective scan* the light source and sensor are placed on the same side of the object to be detected, and the light beam is reflected toward the light sensor from an inherently reflective object or from a reflective target installed for scanning purposes. There are three types of reflective scan: retroreflective, specular, and diffuse. With *retroreflective scan* (Figure 5–33(b)) the light source and sensor are usually mounted in a common housing. The light beam is directed at a retroreflective target, which returns the light along the same path. Acrylic disks or tape, or even chalk, are used as retroreflectors; the bicycle-type reflector is a good example. The larger the reflector, the longer the light path can be. Alignment is not very critical, and retroreflective position-sensing systems are useful in the presence of vibration. They are also generally useful when the measured objects are relatively translucent, since the light beam is attenuated during both portions of its round trip.

In the *specular scan* technique the measured object must be highly reflective (e.g., polished metal, shiny plastic, or mirrors or mirrorlike surfaces; the Latin word *speculum* means "mirror"). As illustrated in Figure 5–33(c), the angle of incidence equals the angle of reflectance; hence, positioning of the light source and sensor, as well as their distance from the measured object, must all be accurately controlled. In *diffuse scan* the measured objects are matte rather than shiny, and the light sensor detects scattering products in a manner similar to nephelometry.

Light-emitting diodes (LEDs) are commonly used as light sources in photoelectric position-sensing switch systems. The use of infrared light, especially modulated infrared light, with light sensor circuitry tuned to the modulating frequency, has been found to be very effective in minimizing any undesirable effects due to ambient light. The light beam modulation is usually in a pulse mode: high-intensity infrared pulses emanating from an

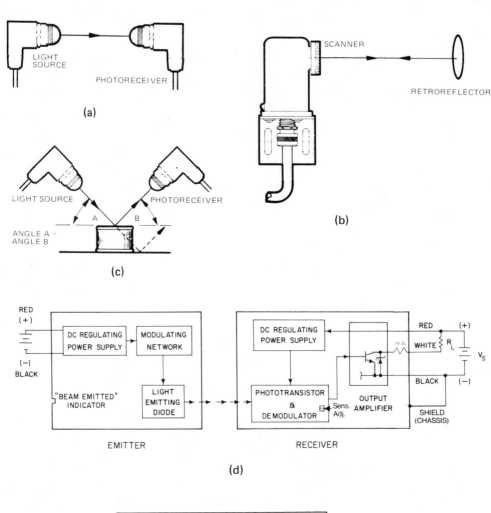

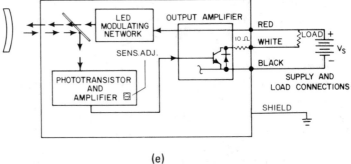

**FIGURE 5–33.** Photoelectric switch systems: (a) direct scan; (b) retroreflective scan; (c) specular scan; (d) functional block diagram of separate emitter/receiver system; (e) functional block diagram of self-contained system. (Courtesy of MICRO SWITCH, A Division of Honeywell.)

LED improve system efficiency. Figures 5–33(d) and (e) show block diagrams for transmittive and reflective position-sensing systems using modulated light.

## 5.5 TRANSDUCER CHARACTERISTICS

Among the typical mechanical design characteristics, the mountings of displacement transducers and many position sensors are particularly important, since they usually define the point relative to which displacement is sensed. Accordingly, they should be specified accurately. Some types of angular displacement sensors have standardized case outlines and, particularly, standardized mounting dimensions. This applies specifically to the "servo-mount" or "synchro-mount" configurations, each of which is usually a cylindrical case that is mounted by clamping it around a circumferential groove near the sensing shaft. The servo-mount "size" gives the outside case diameter (multiplied by 10 and rounded off to the nearest one-tenth of an inch, for models manufactured in the United States) as well as essential mounting dimensions. The case of a "size 11 synchro mount," for example, has a diameter of 1.062 in and a mounting-groove diameter of 1.000 in. Mounting torque or force needs to be specified for some applications.

*Overtravel* of the sensing shaft beyond the measuring range is often a requirement for linear and angular transducers. It is specified as the difference between each of the range limits and the points to which overtravel (which does not produce any further output changes) is required, and in many cases mechanical stops are set [e.g., "Overtravel: $-1.5$, $+3.0$ cm" (where " $-$ " means shaft retraction and " $+$ " means shaft extension)]. Similar overtravel requirements can be specified for angular-displacement transducers, expressed in degrees or radians. The maximum force or torque (*overload*) that may be applied to the sensing shaft when it reaches a mechanical overtravel stop without damaging or degrading the transducer may also have to be specified.

Sensing shaft *concentricity, alignment* (with respect to the case and its mountings), and *radial* as well as *axial play* should be considered for inclusion in a specification, and *backlash* should be specified (or known) for transducers containing gears. The force (or torque, for angular transducers) required to initiate shaft motion, as indicated by a measurable change in output (even after prolonged storage of the transducer), called the *starting force* or *starting torque*, must be considered, as well as the subsequently applicable *running force* or *running torque* (which is always lower than the starting force or torque). For spring-loaded shafts, the *holding force* (at any point after initial motion) should also be specified or known. Moreover, requirements for *shaft sealing* should be considered. For angular-displacement transducers, particularly angular encoders, the maximum allowable radial and axial, static as well as operating, *shaft loading* is often specified, as is the *moment of inertia* of the entire rotating mass.

Electrical design characteristics, in addition to those generally applicable, include such encoder-peculiar output characteristics as waveform, phase relationships, pulse shape (including rise and fall times), voltage levels for "0" and for "1," and excitation and load characteristics stated in terms appropriate for digital electronic interfaces.

Among performance characteristics, the definition of *range* is somewhat more complex for displacement transducers than for most other types of transducers, because it really needs the definition of some sort of reference point it can be based on. For displacement transducers, range has been variously stated as "electrical travel," "electrical stroke," "full-scale deflection," "linear range," "useful range," "useful stroke," and "total stroke." Since the specification of either a unidirectional or a bidirectional range is essential to all other performance specifications, it is important to specify a reference point for the range of essentially all linear and angular displacement transducers except for such devices as incremental angular encoders. This reference position can be stated either (1) by a precise dimension between a point on the sensing shaft of a contacting transducer, or a point in space near a noncontacting transducer, and a point on the case of the transducer; or (2) by a specific transducer output (frequently the zero-measurand output, or "null") at the reference point. Range polarity must also be defined; usually, it is increasingly positive with displacements away from the transducer or, for angular displacements, in a clockwise direction (clockwise motion of the sensing shaft, looking at the shaft end).

Output is frequently stated in terms of sensitivity rather than full-scale output or end points, leaving it to the user to set the end points. This makes other specifications, such as those for static performance characteristics, more difficult unless the "sensitivity" specification is appropriately supplemented. For transducers whose output is also a function of excitation, the nominal (or reference) excitation values (e.g., voltage and frequency) must be stated or added to "sensitivity" (e.g., "____mV/cm/V at 400 Hz), with tolerances assigned to sensitivity. For some types of transducers, output is stated in terms of phase angle (as such, or in addition to voltage amplitude); for some others it is stated in terms of the output or indication of associated signal-conditioning or display equipment in conjunction with which the transducer has been calibrated and must be used.

Encoder output is stated in total counts per turn (revolution) for angular incremental encoders, and in total counts per total length or per unit length for linear incremental encoders. When multiplication of counts is obtained by multiple reading heads and logic circuitry, such multiplication is usually stated. For absolute encoders, output is stated primarily in terms of resolution (number of bits in the data word generated) and type of code used (binary, BCD, Gray, etc.), any antiambiguity logic used (or required externally), and output polarity and notation (linear counts; degrees, minutes, and seconds; altitude, etc.). When two or more outputs are provided by a transducer, the interrelationship of these outputs (*output correlation*) may have to be stated.

Among static characteristics, hysteresis is typically very low for most types of displacement transducers. Resolution is specified primarily for wire-wound-element potentiometric transducers (as is friction error, for all potentiometric transducers) and, of course, for encoders. Repeatability and linearity should always be considered. The near-zero output at the reference position for bidirectional-range transducers is specified as tolerances for null voltage or "zero balance." The specification of phase shift is important for most types of reluctive transducers, in combination with statements of applicable load and excitation characteristics. Mounting error and, for a few types of transducers, attitude error should be accounted for.

The only essential dynamic characteristic is *maximum shaft speed* (above which damage or degradation of performance can occur). This is shown for sustained speed (*operating speed*) as well as speed changes, during run-up or during operation (*slewing* or *slew speed*), for both angular- and linear-displacement transducers (in revolutions per minute or units of length per second, respectively).

Criteria for selection of a displacement transducer for a given application are, primarily, the characteristics of the measuring system (low- or high-level signals, dc, ac, frequency modulated, digital) and measuring range; other essential criteria involve the physical requirements for appropriate installation, the material or surface characteristics of measured objects for noncontacting transducers, resolution requirements (particularly for encoders), stability and life expectancy, and such general characteristics as mass, size, and power requirements. The environment in which the transducer must operate, is, of course, always an important consideration; generally applicable environmental characteristics (temperature, vibration, etc., effects) should always be stated for displacement transducers.

## 5.6 APPLICATION NOTES

Displacement transducers and position sensors are used in a wide variety of applications, wherever linear or angular motion or position must be measured, where dimensions must be established or verified, and where the presence or absence of an object must be determined. The measured ranges of dimensions and displacements extend from micrometers (e.g., for laser interferometers) to millimeters (e.g., for some noncontacting types of transducers) to centimeters (for most types of displacement transducers) to somewhat over one meter (for a few types of displacement transducers) to anywhere between a few meters and hundreds of milions of kilometers (for pulsed or CW electromagnetic-energy remote sensors) and down to fractions of an arc-second for angular displacement transducers.

Many applications exist in materials and mechanical engineering, especially in machine design, construction and testing, and manufacturing and inspection. Examples are the motions and positions of linear and angular actuators, an area of increasing importance in the relatively new field of robotics. Other examples are in dimensional gaging, where sensors (usually inductive or electro-optical noncontacting types, for very small displacements) measure flatness, thickness, and size (deviations from a nominal size); shaft runout, concentricity, and eccentricity; and the axial motion of rotating elements. Although most of the transducers employed are used under room conditions, designs exist for operation under water, in nuclear reactor systems, in the vacuum of space, and at temperatures in excess of 500°C. Pulsed (and CW) electromagnetic-energy devices have many applications in long-dimension measurements. For example, sonar equipment is used to locate (measure the distance to) schools of fish, ocean depth (distance to the ocean floor), friendly and enemy submarines, and sunken ships. Similarly, microwave devices determine the location of distant objects on and above the surface of the Earth, including satellites and planetary spacecraft. Laser devices are finding new applications in space; however, they are also used in surveying, in construction, for precise alignment of structures and mechanisms, and for monitoring displacements of large structures.

There are many applications of linear and angular displacement transducers besides those in manufacturing and inspection. For example, on ships and aircraft they are used to monitor the position of control surfaces by sensing either the extension of a linear actuator or direct angular measurement. Angular measurements are essential for measurements of pointing— of telescopes, large antenna dishes, scientific remote sensors, and weapons. Angular encoders are typically used for pointing measurements where fine resolution is required. Since most modern control systems, such as computer-assisted manufacturing and robotics, are inherently digital, angular and linear encoders are increasingly preferred in these applications. The alternative is an analog-to-digital converter for each transducer.

Most of the measurements just cited as examples can be used not only for monitoring, but also for open-loop or closed-loop control. This holds true for position-sensing switches as well; for example, the switch that senses that an electromechanical operation has been completed (e.g., "door closed") also causes the driving motor to be turned off. Similarly, the interruption of a light beam can be used to count objects (or people) passing through it, or to detect objects taller than a nominal height moving on a conveyor belt; it can also be used to cause a door to open, or to cause the too-tall object to be ejected from the belt. There are many industrial applications for electro-optical position-sensing switches as well. For example, a break in a (textile or paper) web can be detected using the reflected-light principle (and the machine can then be turned off, or a cutter actuated to remove the defective material, and an alarm buzzer can be turned on). Similar events can be monitored (and a control action initiated) by using the reflectance or transmittance principle.

# 6

# Tachometers and
# Velocity Transducers

## 6.1  BASIC DEFINITIONS

*Velocity* is the time rate of change of displacement with respect to a reference system; it is a vector quantity. *Speed* is the magnitude of the time rate of change of displacement; it is a scalar quantity.

*Average velocity* is the total displacement divided by the total time taken by this displacement. *Average speed* is the magnitude of the average velocity vector. The *translational velocity* $v$ of a rotating member is the product of the radius $r$ of the member and its angular velocity $\omega$, or $v = \omega r$.

The interrelationships between linear and angular displacement, velocity, and acceleration are explained in Section 7.1.

## 6.2  UNITS OF MEASUREMENT

The SI unit for linear speed and linear velocity is the *meter per second* (*m/s*); commonly used multiples and submultiples are *km/s* and *cm/s*; linear speed is also expressed in *kilometers per hour* (*km/h*).

The SI unit for angular speed and angular velocity is the *radian per second* (*rad/s*); commonly used submultiples are the *milliradian per second* (*mrad/s*) and the *microradian per second* (*μrad/s*). A non-SI but SI-compatible unit is the *degree per second* (°/s).

Angular speed, particularly when applied to shaft speed or rate of rotation, is very commonly expressed in the (non-SI, English-language) unit *revolutions per minute* (*r/min*) instead of rad/s. One revolution $= 2\pi$ rad $=$ 360°. The abbreviation "rpm" is obsolete.

Refer to Section 5.2.4 for conversion factors applicable to units of length and angle.

## 6.3  SENSING METHODS

Methods in which the transducer or a portion of it is mounted directly to the measured object, as well as remote sensing methods, are employed to measure linear and angular velocity and speed.

*Linear velocity*, when measured while the transducer is mounted to the measured object, is most commonly sensed by electromagnetic devices in which a change of electromagnetic flux induces an electromotive force (*emf*) in a wire-wound coil. The induced emf, measured as an ac voltage, is proportional to the number of turns in the winding and to the change of magnetic flux per unit time, or

$$e = -N \frac{d\phi}{dt}$$

where $e$ = induced emf
$\quad N$ = number of turns in the winding
$\quad \dfrac{d\phi}{dt}$ = change in magnetic flux $\phi$ per unit time $t$

The negative sign in the equation indicates that the direction of the induced emf opposes the change of flux that produced it.

The flux change results from relative motion between the coil and a permanent magnet. The coil can be fixed, with the magnet moving axially within it, or the coil can be the moving element and the magnet the stationary element. This sensing method is usually applied to the measurement of bidirectionally fluctuating, or oscillatory, motion.

*Linear speed* is measured by sensing speed of rotation and then calculating the translational linear speed, based on a knowledge of the radius of the rotating sensing device. Linear speed $c$ is the product of speed in rotation $n$ and the radius $r$ of the rotating member, or $c = nr$, where $n$ is typically expressed in revolutions per second (*r/s*). This method is used, for example, in "measuring wheels," which can be attached to a spring-loaded arm on a vehicle such as a car so that they are pressed against the road and rotate with vehicle motion, or they can be pressed against a continuously moving material in a production facility, such as wire, paper, textiles, or steel sheet in a rolling mill, to provide an output indicative of the speed at which the material moves.

Since acceleration is the time rate of change of velocity, velocity can also be determined by *integrating the output of an accelerometer*. In many accelerometer applications, integrating circuitry is added to the signal-conditioning circuitry so that outputs are provided in terms of both acceleration and velocity.

*Remote sensing of linear velocity* is usually performed by devices utilizing the *Doppler effect*, the effect upon the apparent frequency of a wave train produced by relative motion between the source of the wave train and an observer. This effect is the cause for the apparent change in pitch of the sound made by a vehicle moving first toward, and then away from, an observer. Doppler shifts, as shifts in the spectral lines in the light emitted from a star ("red shift," "blue shift") are used to determine the relative velocity between the star and earth (where the light is observed). Shifts in the frequency of radio waves transmitted (or transponded coherently with earth-sent radio waves) from a spacecraft are used to determine the velocity of the spacecraft (and its position, by integration). In most Doppler-based remote velocity sensing, however, it is the shift in the frequency transmitted toward and reflected by a moving object that is used, with the observer at the location of the transmitting and receiving apparatus.

In the reflected-wave method, the frequency of the wave returned from an object traveling at a velocity $v$ is equal to $f \pm f_D$, where $f$ is the frequency of the transmitted wave (and of the reflected wave if the object were standing still) and $f_D$ is the frequency shift due to the Doppler effect, with the polarity given by the direction of motion of the object. The Doppler shift $f_D$ is given by $f_D = 2v'/\lambda$, where $v'$ is the velocity component of the object in the direction of propagation of the electromagnetic wave and $\lambda$ is the wavelength of this radiation. When the reflected wave is compared with the transmitted wave, the velocity of the object can be calculated. When the reflected wave is superimposed upon a portion of the transmitted wave, the Doppler shift can be recorded directly. Interferometric methods have been employed for such direct Doppler-shift determinations, using wavelengths in various portions of the electromagnetic spectrum. Examples are radio frequencies used by continuous wave (*CW*) or pulsed *Doppler radars* (such radar systems are used to police speeding vehicles) and optical frequencies used by CW or pulsed lasers ("laser Doppler velocimeters"). Frequencies in the sonic or ultrasonic range are used not only for determinations of the velocity of underwater vehicles, but also for measuring the velocity of sound through liquids.

*Vertical speed* sensing, on aircraft, is performed by differentiating the output of a pressure–altitude transducer (or radar altimeter) so as to indicate vertical speed as the time rate of change of altitude.

*Angular speed* sensing (*tachometry*) methods can be of the contacting type (e.g., a device mechanically linked or attached to a rotating shaft), or noncontacting (e.g., a rotating element interacting magnetically or optically with a stationary element), or remote (as exemplified by the stroboscope). The various types of transducers employing these methods are described in Section 6.4. The sensing objective is the measurement of the angular speed (rate of rotation) of either a shaft or another portion of a rotating system.

## 6.4 DESIGN AND OPERATION

### 6.4.1 Electromagnetic Linear-Velocity Transducers

Electromagnetic linear-velocity transducers are typically used to measure oscillatory velocity. The basic principle is illustrated in Figure 6–1. The device consists of a permanent magnet that can move back and forth within a coil winding. As the shaft moves, an electromotive force (*emf*) is induced in the winding and appears as a voltage across the winding terminals. The faster the shaft moves, the greater this voltage will be. The shaft can have a threaded end so that it can be attached to a moving object such as an oscillating piston. A two-coil version of such a transducer is shown in Figure 6–2. The windings are connected out of phase in a series-opposing arrangement. The opposing voltages are summed so that the total induced voltage is proportional to the core velocity.

A slightly different principle is used in the fixed-coil design shown in Figure 6–3; a transducer with this design is used to measure the vibratory velocity of the object it is mounted to. A permanent magnet is supported between two springs, and gold–palladium alloy bearing rings are pressed onto the ends of the cylindrical magnet to minimize friction as the magnet moves within a chrome-plated stainless-steel sleeve. Threaded retainers seal the ends of the mechanical assembly. The entire transducer is intended to be mounted to an object whose oscillatory velocity is to be measured. When

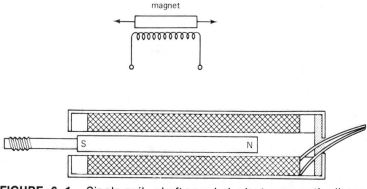

**FIGURE 6–1.** Single-coil, shaft-coupled electromagnetic linear-velocity transducer.

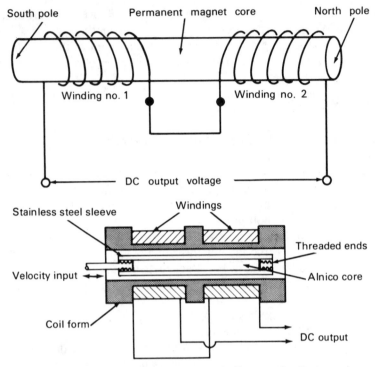

**FIGURE 6–2.** Dual-coil electromagnetic linear-velocity transducer. (Courtesy of Schaevitz Engineering.)

the frequency of this motion exceeds the natural frequency of the suspended system (about 15 Hz), the magnet remains in an essentially undisturbed position. The coil, fixed within the transducer case, moves relative to the now stationary magnet with the motion of the measured object. The resulting flux rate of change produces an output from the coil that is proportional to velocity. Other designs have the coil as part of a moving structure pivoted at bearings within the transducer assembly. The coil moves within a magnetic field established by the pole pieces of a fixed permanent magnet, and the resulting flux changes produce an output proportional to velocity in the moving coil.

### 6.4.2   Other Linear-Velocity Sensing Devices

Contacting-type sensors that measure linear velocity continuously in one direction are essentially limited to the *measuring wheel*, a wheel or disk of known circumference that is coupled to a tachometer (see Section 6.4.3). The wheel is usually carried by an arm attached to the moving object, and it is pressed against the surface over which the object moves. A typical

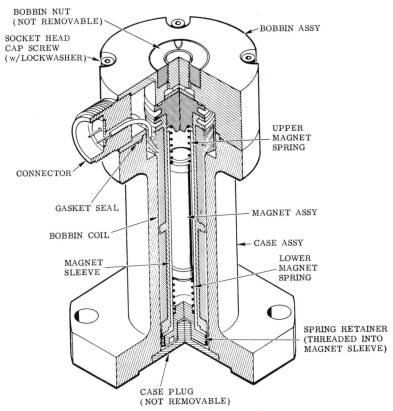

**FIGURE 6–3.** Electromagnetic linear-velocity transducer. (Courtesy of CEC Division/Transamerica Delaval.)

application is found in automotive road tests, where the wheel is pressed against the road. The wheel converts linear (translational) velocity into angular velocity, which can easily be measured by any of a large variety of tachometers.

Remote sensing of linear velocity is often performed by Doppler-effect devices, whose operating principle was described in Section 6.3; among such devices are Doppler sonars, Doppler radars (such as those used by police cars) and laser Doppler velocimeters.

### 6.4.3 Electromagnetic Tachometer Generators

Three types of electric generators are used as tachometers. Each provides an output voltage and, in the case of the permanent-magnet ac tachometer, an output frequency, proportional to angular speed.

*DC tachometer generators* use either a permanent magnet (*dc mag-*

*neto*) or a separately excited winding as their stator and a conventional generator winding on the commutator-equipped rotor. The brushes associated with the commutator require maintenance after a certain period of use. However, an advantage of these devices is that output polarity is indicative of direction of shaft rotation. The output of a dc magneto is typically 3 to 7 V, that of the stator-winding type about 10 to 20 V, per 1000 r/min.

*AC induction tachometers* operate as a variable-coupling transformers, with the coupling coefficient proportional to rotary speed. When the primary (input) winding, on the stator, is provided with ac excitation, an ac voltage at the excitation frequency appears across the output terminals of the secondary winding, which is also located on the stator. The rotor is usually either of the squirrel-cage type or cup-shaped and made of a high-conductance metal such as copper, copper alloy, or aluminum (*drag-cup tachometer*). Shaft rotation produces a shift of flux distribution which changes the coupling between primary and secondary so that the output voltage amplitude is proportional to angular speed.

*AC permanent-magnet tachometers* use the flux changes between a permanent-magnet rotor and a stator winding to provide an ac output which varies with rotary speed both in amplitude and frequency. This type of device has also been called an *ac magneto*. Signal-conditioning circuitry can be used to convert the output either into a non-amplitude-dependent frequency (or pulse-rate) output signal, or into a non-frequency-dependent dc amplitude-varying signal.

### 6.4.4   Toothed-Rotor Electromagnetic Tachometers

Toothed-rotor electromagnetic tachometers are very commonly used devices employing a rotor that resembles a gear in that it has tooth-like protrusions around its circumference. The protrusions are made of ferromagnetic material, as the entire rotor often is. The rotor is supported within a housing or frame in which a sensing-transduction coil (proximity sensor) is mounted in such a way that the gap between the "teeth" and the coil assembly is quite small. Transduction-coil assemblies can be of the Hall-effect type, or they can be of the inductive, eddy-current type, in which case the rotor material need not be ferromagnetic but only metallic (and the sensor output amplitude is not significantly affected by rotary speed). Most sensing-transduction coil assemblies, however, are of the electromagnetic type illustrated in Figure 6–4. The coil, which is made of thin magnet wire, is wound around an insulating bobbin form which is slipped over a pole piece attached to the permanent magnet. The assembly is potted within a shell, usually hermetically sealed. The front portion of the shell is often threaded for ease of sensor installation and the adjustment of the gap between pole piece and rotor teeth. The rotor or, at least, the rotor teeth are made of a ferromagnetic material, typically a magnetic steel. The pole piece conducts

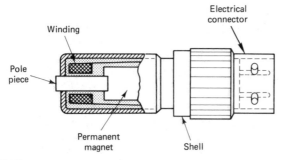

**FIGURE 6–4.**  Electromagnetic transduction coil assembly.

a magnetic-flux path from the magnet, creating a magnetic field in front of the sensor. When a ferromagnetic tooth passes through this field, the resulting flux change induces an emf in the coil. When the rotating object whose angular speed is to be measured is equipped with a single tooth, an output pulse is created once per revolution (see Figure 6–5)(a). With increasing angular speed, the number of pulses per unit time increases proportionally. The pulse rate can then be displayed [e.g., directly on an EPUT (events-per-unit-time) meter]. The increasing rate of change of magnetic flux also produces increasing pulse amplitudes, and the amplitude of the output signal is also indicative of angular speed, at least over a portion of the speed range, with varying amounts of linearity.

A greater number of teeth on the rotor produces more pulses per unit time (more pulses per revolution), as illustrated in Figure 6–5(b). Continuous

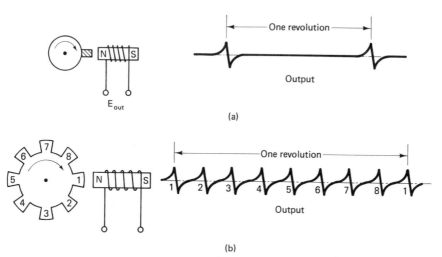

**FIGURE 6–5.**  Operating principle of electromagnetic frequency-output tachometer: (a) single tooth; (b) continuous teeth.

teeth are used most frequently. Other rotor shapes include the "Maltese cross" and specially configured cam. Typical configurations, and the waveshapes produced by them, are illustrated in Figure 6–6. It can be seen that gear teeth can be so machined ("fine tooth gear") that the output signal is very close to sinusoidal. This is desirable when a pure frequency output is required by the data system.

The electromagnetic coil assembly must be so installed that there is no magnetic barrier between the pole piece and the rotor. In fact, it is best to install the assembly so that there is no barrier of any sort between the pole piece and the rotor. If the rotor assembly must be sealed from the ambient environment, the coil assembly itself and its mounting provisions can usually provide this seal. If a barrier is still required, the magnetic path from the pole piece can be extended by mounting a soft-steel pin through the barrier (which must still be nonmagnetic) and pressing the pole piece against the outside end of this pin. The sensor must also be so positioned that the pole piece is as close as possible to the rotor teeth, without, however, causing potential physical contact, including binding due to thermal expansion or unevenness of tooth surfaces. The output pulse amplitude drops sharply with increasing gap (see Figure 6–7), since output is inversely proportional to the square of the gap distance.

The standard pole-piece configuration is cylindrical. However, depending on the application and on the tooth configuration, other shapes may be more suitable. Other available shapes are the conical and the chisel configuration (see Figure 6–8). The chisel point aids in providing a higher output

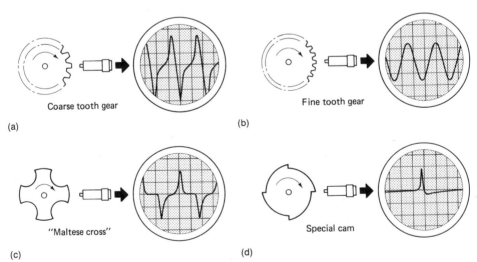

(a)         Coarse tooth gear         (b)         Fine tooth gear

(c)         "Maltese cross"         (d)         Special cam

**FIGURE 6–6.** Waveshapes produced by various rotor configurations: (a) coarse-tooth gear; (b) fine-tooth gear; (c) "Maltese cross"; (d) special cam. (Courtesy of Electro Corp.)

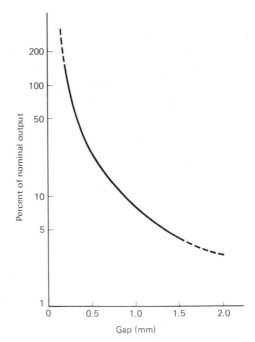

**FIGURE 6–7.** Variation of output amplitude with gap between pole piece and rotor teeth.

and better resolution in conjunction with fine gear teeth. The conical tip does the same, but to a lesser extent; however, it poses no requirements for orientation that apply to the chisel tip. Orientation also makes the gap adjustment more difficult.

Some angular-speed measuring systems require two sensing-coil assemblies, either to provide signals to two different portions of the system or (when the two coils are mounted 90° apart) to obtain a phase difference that can be converted into an indication of direction of rotation by means of digital logic circuitry. Coil assemblies have also been designed with integral signal-conditioning circuitry which converts the coil-output pulses of various shapes into square-shaped pulses suitable as inputs to digital data systems. Coil assemblies so equipped typically have three interface connections; an excitation " + " terminal (somewhere between 5 and 15 V dc), a common terminal for excitation and output, and an output terminal which

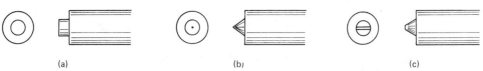

(a)  (b)  (c)

**FIGURE 6–8.** Pole-piece configurations: (a) cylindrical pole piece; (b) conical pole piece; (c) chisel pole piece. (Courtesy of Electro Corp.)

provides the square-wave pulses (with between 0 and 0.2 V indicating a "0" and between 3 and 14 V indicating a "1") referred to the common terminal.

When a dc signal is needed by the data system (which can be a simple system with direct analog display on a dc meter), a *frequency-to-dc converter* can be inserted between the sensing coil and the display device or data system. Such a converter converts the coil-output pulses into constant-energy pulses at the same rate. The pulse train is then electronically integrated to provide a dc voltage proportional to angular speed. The resulting signal may also be fed to a comparator to determine whether the angular speed is above or below a certain set point or narrow range of speeds.

### 6.4.5  Electro-Optical Tachometers

Angular velocity is often measured by photoelectric sensing devices, using either the transmittance or the reflectance method (each of which is described in Section 5.4.7). The transmittance method is typically used in tachometers that resemble an *incremental angular encoder* (see Section 5.4.8) with a continuous (360°) coding pattern. Solid-shaft as well as hollow-shaft designs are available. Depending on the geometry of the coding pattern, the output can be either a square wave (which is digital in nature and TTL-compatible) or sinusoidal. When quadrature output is needed to also indicate the direction of rotation, two readings heads are used.

The *reflectance method* is used in a variety of angular-speed sensing systems. The sensing head, which can be hand held or mounted, contains a light source that emits a collimated beam toward a reflective portion or spot on the rotating object and a light sensor that detects a light pulse whenever the beam is reflected back. Some rotating members have inherently reflective, symmetrically spaced members (e.g., spokes of a wheel). However, most rotating objects will require a piece of reflecting tape to be affixed to them (or several symmetrically spaced pieces of tape if count multiplication is required). The output of the light sensor is then a count of revolutions of the measured object which can easily be converted into r/s or r/min either by integration to produce an analog signal or by using digital logic for comparison with clock-generated pulses to produce a digital output signal.

Figure 6–9 shows a photoelectric tachometer with direct analog display on the portable, battery-operated indicator. The sensing head, which is connected to the signal-conditioning and display unit by a cable, is shown held by a clip at the top of the unit. Other designs, which frequently are also portable, provide a digital display, usually in r/min, mostly without the need for manual range selection. Fiber optics are also used to carry emitted and reflected light to and from the point of measurement, an advantage when this point is not readily accessible.

**FIGURE 6–9.** Portable photoelectric tachometry system. (Courtesy of AMETEK, Mansfield & Green Div.)

A different, but widely used, method of measuring angular speed is offered by the use of a stroboscope. This method includes the observer in the measuring loop. A stroboscope employs a cold-cathode gas-filled lamp as a source of high-intensity light flashes. The flashes are used to illuminate the rotating object. The number of flashes per minute is subject to continuous manual adjustment. The flash rate is adjusted until the rotating member appears stationary. This occurs when one flash is emitted for each cycle of motion, that is, for each full revolution of a rotating object. (The stroboscope can also be used to detect nonrotational motion.) The number of flashes per minute (corresponding to revolutions per minute) either is read out on a dial attached to the flash-rate adjusting knob, or it can be displayed in digital form, as on the stroboscope shown in Figure 6–10, which also shows the typical use of this sensing device.

### 6.4.6  *Other Types of Angular-Speed Transducers*

A number of other types of tachometers have been designed and used at various times, but are not used very commonly.

*Capacitive* tachometers use rotor plates attached to or machined in-

(a)

(b)

**FIGURE 6–10.**   Stroboscope tachometer: (a) stroboscope (with digital readout); (b) "freeze-motion" method of measuring angular velocity. (Courtesy of AMETEK, Mansfield & Green Div.)

tegrally with the shaft. One, two, or four stator plates are shaped to comply with the rotor electrode configuration. Capacitance changes occur periodically with shaft rotation. When the electrodes are connected into an ac bridge, the bridge output voltage will undergo corresponding changes.

*Strain-gage* tachometers employ a cantilever beam to which strain gages are bonded and an eccentric disk or cam attached to the rotating shaft. The beam is in continuous contact with the perimeter of the disk or cam and deflects once per revolution. By appropriate design of the disk or cam, a sinusoidal output can be obtained from the strain-gage bridge.

*Switch* tachometers involve the making and breaking of contacts. Magnetic reed switches can be used, actuated by a magnet attached to the rotating shaft or some other member. One design uses a pair of rotating contacts between which a capacitor is connected so that it is alternatingly charged with voltages of opposite polarity; the stationary contacts are connected across a dc source in series with a resistor, across which current pulses are then produced. Cam-actuated and yoke-actuated reed switches have been used in such circuits, or in other circuits where a temporary switch contact closure produces an output pulse.

## 6.5 TRANSDUCER CHARACTERISTICS

Design and performance characteristics for linear and angular velocity transducers include most of those generally shown and specified for transducers, with certain characteristics obviously not applicable to noncontacting and remote-sensing devices. Some characteristics deserve particular attention. For electromagnetic linear-velocity transducers, these include the frequency response (which always has a finite lower limit), the maximum displacement of the internal moving member (core or coil), the maximum acceleration, linearity (which can be specified as applicable at a reference frequency and over a range of linear velocities), and the type of damping employed. These transducers are self-generating, and their output voltage is proportional to instantaneous (vibratory) velocity. Sensitivity is usually specified as peak voltage per peak velocity (e.g., "50 mV (peak)/cm/s (peak)" at a selected reference frequency (e.g., 100 Hz), with a tolerance expressed either in mV or percent, and with the minimum value of the load impedance specified. The upper limit of the (velocity) range is always shown; it may be higher than the upper limit of the range for which linearity is specified, in which case it is useful to know the degradation of linearity up to the maximum velocity value. The mounting orientation must be specified (or the amount of attitude error for the affected axes should be shown). Besides the output impedance of the coil(s), the (ohmic, dc) output resistance should be given to facilitate continuity checks. The latter applies also to electromagnetic sensing-coil assemblies (proximity sensors) used in conjunction with toothed-rotor tachometers.

For remote sensing systems for the measurement of linear velocity, a number of characteristics peculiar to such systems (and different from those generally specified for transducers) are stated. One essential performance characteristic for such a system is the threshold (minimum detectable velocity) at a given distance, as well as the resolution at that distance (and preferably also the variation of these two characteristics with distance).

Essential characteristics for angular-speed transducers include the speed range, in r/min (the abbreviation "rpm" is to be phased out), the starting and running torque, linearity and the speed range over which it is applicable, and sometimes the moment of inertia. Sensitivity is shown in mV/r/min or $V/1000$ r/min for dc tachometers. The amount of noise in the output should also be shown. Sensitivity, phase characteristics, and harmonic content are called out for ac tachometers. Toothed-rotor tachometers with electromagnetic sensing-coil assemblies need specifications for variations in output voltage amplitude over a given speed range as well as for one revolution (because of possible nonuniformity in characteristics of teeth), and also output frequency in cycles or pulses per revolution, pulse characteristics for nonsinusoidal output, and harmonic content for sinusoidal output. Essential characteristics for encoder-type tachometers are similar to those cited for angular encoders (see Section 5.4.3.1). Frequency response should be shown for all types of tachometers, and maximum sensing distance should be shown, together with target size, for reflectance-type optical tachometers. Maximum total error ("accuracy") should be stated for tachometer systems, including a digital or analog indicator and signal conditioning.

## 6.6  APPLICATION NOTES

Linear-velocity transducers are most commonly used to measure vibration (in terms of vibratory velocity). Their output can be differentiated if acceleration data are desired, or it can be integrated when displacement is to be derived from it.

Remote sensors for linear velocity are sometimes employed in industry (typically using electro-optical techniques). Those employing the Doppler principle are exemplified by police-car radars as well as by satellite and spacecraft tracking stations.

Measuring wheels, which convert linear (translational) velocity to angular velocity, are used not only on moving vehicles but also for industrial monitoring, such as monitoring the speed of conveyor belts or of webs of paper, plastic products (sheet, filament), and textiles.

Tachometers are widely used in the measurement of the rotational speed of wheels, engines, pumps, turbines, generators, and other types of machinery and rotating mechanisms. In most applications it is possible to mount the tachometer other than as a link between the driving device and the driven device; if such an installation becomes necessary, special precautions (including safety measures) must be observed. However, even if the tachometer is driven separately by a mechanical linkage (e.g., a gear with or without a flexible shaft), care must be taken to prevent a failure of the tachometer or linkage to impinge upon the measured rotating mechanism.

# 7

# *Accelerometers*

## 7.1 BASIC DEFINITIONS

Accelerometers are used to measure acceleration, vibration, and (mechanical) shock. *Acceleration* is the time rate of change of velocity with respect to a reference system and is a vector quantity. The instantaneous values of velocity and acceleration are the first derivative and the second derivative, respectively, of displacement. Displacement, velocity, and acceleration can be linear (translational, rectilinear) or angular (rotational). Their interrelationship are given by the equations

$$v = \frac{dx}{dt} \qquad a = \frac{d^2x}{dt^2} \qquad \text{(linear)}$$

$$\omega = \frac{d\theta}{dt} \qquad \alpha = \frac{d^2\theta}{dt^2} \qquad \text{(angular)}$$

where $x$ = linear displacement
$\theta$ = angular displacement
$v$ = linear velocity
$\omega$ = angular velocity
$a$ = linear acceleration
$\alpha$ = angular acceleration
$t$ = time

*Oscillation* is the variation, usually with time, of the magnitude of a

141

quantity with respect to a reference system when this variation is characterized by a number of reversals of direction. *Vibration* (mechanical vibration) is an oscillation wherein the quantity is mechanical in nature (e.g., force, stress, displacement, velocity, acceleration). *Harmonic motion* is a vibration whose instantaneous amplitude varies sinusoidally with time. *Mechanical impedance* is the complex ratio of force to velocity during simple harmonic motion; it is a quantitative measure of the ability of a structure to resist a vibratory force.

*Periodic vibration* is vibration having a waveform that repeats itself in all its particulars at certain equal time increments. A *period* is the smallest increment of time for which the waveform of a periodic vibration repeats itself in all those particulars. A *cycle* is the complete sequence of magnitudes of a periodic vibration that occur during one period. The *frequency* of a periodic vibration is the reciprocal of its period. The *phase* of a periodic vibration is the fractional part of a period through which the vibration has advanced, as measured from an arbitrary reference point.

*Random vibration* is nonperiodic vibration whose magnitude at any given time can be described only in statistical terms. It is usually taken to mean Gaussian random vibration, whose instantaneous amplitude distribution follows a Gaussian ("normal error curve") distribution.

The *frequency spectrum* of a vibratory quantity is a description of the instantaneous content of its components, each of different frequency and usually of different amplitude and phase. The mean and rms magnitudes of a vibratory quantity are respectively defined by

$$\overline{A} = \frac{1}{t} \int_0^t A(t)\, dt \quad \text{and} \quad A_{\text{rms}} = \sqrt{\frac{1}{t} \int_0^t [A(t)]^2\, dt}$$

where $\overline{A}$ = mean magnitude
  $A(t)$ = vibratory quantity
  $t$ = time over which the averaging is done
  $A_{\text{rms}}$ = rms magnitude

The *power density* of random vibration is the mean square magnitude per unit bandwidth of the output of an ideal filter having unit gain and responding to this vibration. It is defined as

$$W(f)^2 = \frac{A(f)_{\text{rms}}^2}{\Delta f}$$

where $W(f)$ = power density
  $A(f)_{\text{rms}}^2$ = mean-square magnitude
  $\Delta f$ = bandwidth (usually chosen as 1 Hz)

A *power-density spectrum* is a graphical representation of values of power density displayed as a function of frequency so as to represent the distribution of vibration energy with frequency.

*Shock* (mechanical shock) is a sudden nonperiodic or transient excitation of a mechanical system. *Jerk* is the time rate of change of acceleration with respect to a reference system.

## 7.2 UNITS OF MEASUREMENT

Linear acceleration is expressed in *meters per second squared* $(m/s^2)$; angular acceleration is expressed in *radians per second squared* $(rad/s^2)$.

Linear acceleration (as well as shock and vibration amplitude) is often still expressed in $g$'s. The $g$ can be converted to units of the SI as well as the American Customary system. Note that the symbol, except for being in italics, is the same as that of the gram (g). The quantity $g$ is the acceleration produced by the force of the earth's gravity. Gravity varies with the latitude and elevation of the point of observation. The value of $g$ has been standardized by international agreement as follows:

$$1 \text{ standard } g = 9.806\ 65 \text{ m/s}^2 = 980.665 \text{ cm/s}^2$$

$$= 386.087 \text{ in/s}^2 = 32.1739 \text{ ft/s}^2$$

These can be simplified, where allowable, to

$$1\ g = 9.81 \text{ m/s}^2 = 981 \text{ cm/s}^2 = 386 \text{ in/s}^2 = 32.2 \text{ ft/s}^2$$

## 7.3 SENSING METHOD

All acceleration transducers (*accelerometers*) use a sensing method in which the acceleration acts upon a *seismic mass* (proof mass) that is restrained by a spring and whose motion is usually damped in a spring-mass system (Figure 7–1). When acceleration is applied to the accelerometer case, the mass moves relative to the case. When the acceleration stops, the spring returns the mass to its original position (Figure 7–2). The small black-and-white

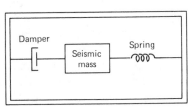

**FIGURE 7–1.** Basic spring-mass system of an acceleration transducer.

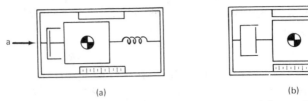

**FIGURE 7–2.** Displacement of seismic mass: (a) acceleration applied; (b) acceleration removed.

circular symbol is commonly used to denote the location of the gravitational center of seismic mass. If acceleration were applied to the transducer case in the opposite direction, the spring would be compressed rather than extended.

Under steady-state acceleration conditions the displacement $y$ (in cm) of the seismic mass is given by the acceleration $a$ (in cm/s$^2$) multiplied by the ratio of the mass $M$ (in grams) to the spring constant $k$ (in dynes/cm, where 1 dyne $= 10^{-5}$ N), or $y = aM/k$. Under dynamic (varying) acceleration conditions the damping constant becomes a factor in a modified version of this relationship. The spring extension (and mass displacement) is proportional to acceleration only below the natural frequency of the spring-mass system. The response characteristics of such systems are adequately described in available textbooks dealing with mechanical vibrations.

The linear displacement of the seismic mass or the spring, or the force exerted by the seismic mass on an elastic member, is the quantity that acts upon the transduction element of a linear accelerometer. The seismic mass must be restrained from motion along all but the sensing axis. Similarly, the seismic mass of an angular accelerometer, exemplified by a disc pivoted at its center and restrained by a spiral spring, responds to angular acceleration with an angular displacement. To summarize, before it is transduced, acceleration is first converted into either force or displacement.

## 7.4   DESIGN AND OPERATION

### 7.4.1   *Capacitive Accelerometers*

The capacitive transduction principle is employed in some accelerometer designs, typically those which use a diaphragm-supported seismic mass or a flexure-supported disk-shaped seismic mass as the moving electrode and either one or two fixed electrodes (stator plates). As acceleration is applied, the proximity between the moving and fixed electrodes changes. This results in a change of capacitance, a simple capacitance variation if one stator plate is used and a differential change in two capacitances if the moving electrode is located between two stator plates. Various signal-conditioning schemes have been applied to such transduction elements. The element can be connected into an ac bridge, and a change in transducer ac output voltage (or dc output if the signal is rectified) occurs. Or the element can be connected as the capacitor in the *LC* or *RC* portion of an oscillator circuit so that a change in transducer output frequency results in response to changes in applied acceleration. The capacitance changes in a two-stator transduction element have also been used in a switching circuit so that a pulse train is generated in which pulse width and distance between pulses are the result of the changes in the two capacitances. Single-stator elements have been used in some servo accelerometers. An experimental triaxial accelerometer design uses a freely moving spherical metallic mass and three stator elements arranged in three mutually orthogonal axes, in a spherical evacuated cavity containing the mass, to provide outputs in the form of capacitance changes for acceleration along any axis.

### 7.4.2 Piezoelectric Accelerometers

Piezoelectric transduction is used in a large variety of accelerometers that primarily measure vibration and, to some extent, shock. A typical design is shown in Figure 7–3. Acceleration is sensed along the longitudinal axis of the unit. The acceleration acts on the seismic mass, which exerts a force on the piezoelectric crystals (quartz, in this unit) which then produces an electrical charge. The quartz plates are preloaded so that either an increase or a decrease in the force acting on the crystals (due to acceleration in either direction) causes changes in the charge produced by them. Two or more crystals are connected for output multiplication when quartz, which has relatively low sensitivity, is used as the piezoelectric element. Ceramic crystals are also used in many piezoelectric accelerometers. Typical ceramic materials used for this purpose are barium titanate, a lead zirconate–lead titanate mixture, lead metaniobate, and more often, mixtures whose composition is considered proprietary by their manufacturers and which are designated by registered trade names such as Piezite (Endevco) or Glennite (Gulton).

Piezoelectric crystal materials differ in their essential characteristics, sensitivity, frequency response, bulk resistivity, and thermal response. Some ceramic elements (but not quartz) have been reported to exhibit a zero shift when exposed to an environment that includes high stress and high noise. The upper limit of the operating temperature range is given by the *Curie point,* or Curie temperature of the material. (The piezoelectric effect was discovered by Pierre and Jacques Curie in 1880.) At this temperature, ceramic elements, which are polarized by exposure to an orienting electric field during cooling after firing, lose their polarization. Curie points vary from 120°C for barium titanate to about 570°C for lead metaniobate; however, some proprietary ceramics with Curie points above 950°C have also been developed.

Of the various possible internal mechanical designs (seismic mass and

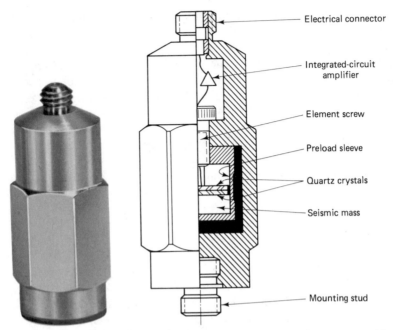

Electrical connector

Integrated-circuit amplifier

Element screw

Preload sleeve

Quartz crystals

Seismic mass

Mounting stud

**FIGURE 7–3.** Piezoelectric accelerometer (general-purpose, with built-in integrated-circuit amplifier). (Courtesy of PCB Piezotronics, Inc.)

crystal and their support), the center-mounted and inverted center-mounted compression-type designs and the shear design (in which the piezoelectric crystal is stressed in a shear mode instead of a compression mode) provide minimized *case sensitivity,* that is, sensitivity to acoustic noise, temperature transients, and strains in the mounting surface, all acting on the case of the accelerometer. Another important factor in mechanical design is minimization of sensitivity to transverse acceleration (acceleration along axes other than the sensing axis).

Figure 7–4 shows a variety of piezoelectric accelerometer designs. The units in the foreground (bottom of picture) are miniature accelerometers characterized by small size, low mass (1 to 2.3 g), and a flat frequency range ($\pm 5\%$), generally from 1 to 10 000 Hz. Most of the units in the center of the picture are general-purpose accelerometers with frequency ranges from 1 to between (generally) 3000 and 5000 Hz and including some low-range high-sensitivity models. The large unit at the right-hand side of the picture is a seismic accelerometer which has the low frequency range (0.025 to 800 Hz) required for seismic applications. The units at the extreme upper right and left of the illustration are triaxial accelerometers, assemblies of three accelerometers mounted in or on a single block and precisely aligned so that

**FIGURE 7–4.** General-purpose and special-purpose piezoelectric accelerometers. (Courtesy of PCB Piezotronics, Inc.)

their "$X$-, $Y$-, and $Z$-axes" are mutually orthogonal. One of the units near the bottom center of the picture is from a series of shock accelerometers that are characterized by low mass (4 g), a wide frequency range (0.25 to 8000 Hz), and a high acceleration range (to 100 000 $g$). All the models shown have a built-in integrated circuit amplifier and an output impedance of less than 100 ohms (see next).

All basic piezoelectric transducers are characterized by a relatively low output signal and a very high output impedance. This poses a requirement for amplifiers which also act as impedance converters. Before integrated-circuit amplifiers/impedance converters (which were small and light-weight enough to fit into the case of such accelerometers) became commercially available and were proven reliable, separate amplifiers had to be used. These were connected to the transducer by a coaxial cable to which stringent requirements were applied. These cables had to be kept as short as possible, thin, flexible, coaxial, shielded, insulated, sealed against moisture, of very low capacitance, and free from *triboelectric noise* (noise caused by friction between the conductor and the insulator, due to flexing of the cable). Even the electrical connectors at both ends of the cable had to be sealed against moisture during installation of the transducer. Amplifiers were of two types: *voltage amplifiers,* which consist of several cascaded amplification stages, and *charge amplifiers,* operational amplifiers with capacitive feedback. Such accelerometer systems are still in use, especially in very-high-temperature applications in which a built-in amplifier can malfunction; however, most modern piezoelectric accelerometers have a built-in amplifier which also acts as impedance converter and can operate from a dc power source. A typical system circuit diagram is shown in Figure 7–5.

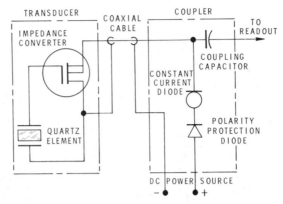

**FIGURE 7-5.** System diagram for piezoelectric transducer with built-in amplifier/impedance converter. (Courtesy of Sundstrand Data Control, Inc.)

### 7.4.3  Potentiometric Accelerometers

Potentiometric accelerometers are characterized by a high output signal and a relatively narrow frequency range, extending typically from zero to 20 Hz for lower ranges of acceleration to around 60 Hz for higher ranges. (The higher the range, the stiffer the spring and the higher the natural frequency.) They can also be produced at relatively low cost and are used in applications where slowly varying acceleration is to be measured with moderate accuracy and where cost is a significant factor.

The displacement of the spring-mass system in these transducer designs is mechanically linked to a wiper arm which moves along a potentiometric resistance element. The linkage can provide amplification of displacement if needed. The design shown in Figure 7-6 employs a conductive-plastic resistance element which is sufficiently short to allow direct attachment of the wiper arm to the mass. The unit shown is gas damped. Other designs use viscous damping or magnetic damping. Damping is needed to minimize any noise in the output that could be introduced by whipping of the wiper arm and large changes in instantaneous contact resistance between the wiper and the resistance element. Gas damping is often preferred because it does not require the additional mass of magnetic-damping provisions and because it is not as much subject to temperature effects as viscous damping (e.g., damping attained by using silicone oil).

It is fairly easy to design potentiometric accelerometers so that the effects of transverse acceleration are minimized. In some designs the mass is constrained from motion in any but the specified axis by having it move along a coaxial shaft. The design illustrated employs flexural ("E"-shaped) springs as well as having the mass slide within a precision-molded plastic block that is part of the transducer case. Mechanical stops are usually pro-

**FIGURE 7–6.** Potentiometric accelerometer. (Courtesy of Bourns Instruments, Inc.)

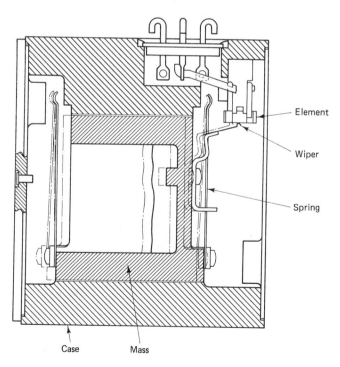

vided to limit mass displacement when overrange accelerations are seen by the unit. Placing the wiper arm at the center of the resistance element when no acceleration is applied allows bidirectional measurements.

### 7.4.4 Reluctive Accelerometers

Reluctive accelerometers comprise accelerometers of the LVDT (linear variable differential transformer) as well as the inductance bridge type. The principle of the LVDT accelerometer is shown in Figure 7–7. The differential transformer winding is the primary winding between the two secondary windings. The spring-mass system is constituted by the transformer core, which is the seismic mass, suspended inside the windings by parallel cantilever springs. The displacement of the ferromagnetic core, due to applied acceleration, results in output-voltage changes in the two secondary windings when an ac voltage is applied to the primary winding.

In many inductance-bridge accelerometer designs the displacement of the seismic mass causes the inductances of two suitably mounted coils to vary in the opposite direction; that is, the inductance of one coil increases while the inductance of the other coil decreases due to their relative proximity to a moving ferromagnetic armature. The two coils are connected as two arms of an inductance bridge, the other two arms of which are usually resistive. An ac output voltage variation is produced at the bridge output terminals when ac excitation is applied to the bridge input terminals, with variations in applied acceleration.

The ac output of reluctive accelerometers varies in phase as well as amplitude. It can be converted to dc by means of a phase-sensitive demodulator. An oscillator can be employed to provide the ac excitation when

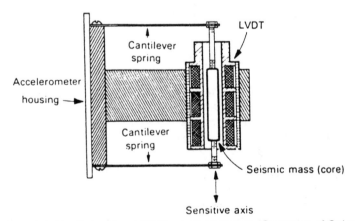

**FIGURE 7–7.** Reluctive (LVDT) accelerometer. (Courtesy of Schaevitz Engineering.)

operation of the accelerometer from a dc supply is required. The frequency of the excitation voltage must be substantially higher than the upper limit of the accelerometer's frequency range. The frequency range of inductance-bridge designs is usually higher than that of LVDT types because of the relatively smaller displacement of their seismic mass.

### 7.4.5 Servo Accelerometers

Servo accelerometers use closed-loop servo systems of the force-balance, torque-balance, or null-balance type to provide very close accuracy as well as a high-level output signal, but at a relatively higher cost than most other designs. Figure 7–8 illustrates the operating principle of one type of servo accelerometer. Acceleration causes the pendulous seismic mass to move. As soon as motion is detected by the position-sensing device, a signal is produced which acts as the error signal in the servo loop. The position sensor in this design is of the RF-excited eddy-current type. After demodulation and amplification, the signal passes through a passive damping network and is then applied to a torquing coil located at the axis of rotation of the mass. The torquer shown for this design is similar to the D'Arsonval movement used in electric indicating meters; the jewel bearings are also typical for such meter movements. The torque developed is proportional to the current applied to the coil, and it balances, in magnitude as well as direction, the torque acting on the seismic mass due to acceleration, preventing further motion of the mass. The current through the torquing coils is, therefore, proportional to acceleration. This current then passes through the stable load resistor. The voltage $E_o$ across this resistor is the output voltage of the transducer.

In angular accelerometers of the same design the seismic mass is balanced; in linear accelerometers (as illustrated) it is unbalanced. However, pendulous-mass linear accelerometers tend to be affected by angular acceleration. Such pendulosity errors are eliminated in the design shown in Figure 7–9, which is of the force-balance type. A rectilinear displacement of the axial seismic mass is sensed by a capacitive sensing element, and a forcing coil is used to attain servo balance. The forcer coil current, and hence the voltage across a resistor through which this current passes, is proportional to acceleration. The range and frequency response of servo accelerometers can be established within reasonable limits by the characteristics of the amplifier and damping network, rather than being dependent on primarily mechanical characteristics of the transducer. In addition to reluctive and capacitive displacement sensors, strain-gage and photoelectric sensors have been used in some servo-accelerometer designs. A frequency-modulated or digital output can also be provided, and some designs are available with a discrete output (acceleration switch) at one or more set points.

(a)

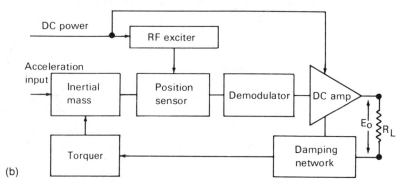

(b)

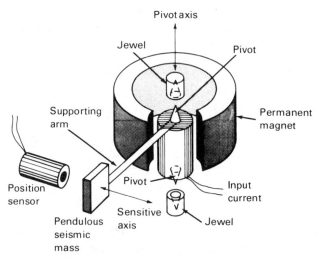

(c)

**FIGURE 7–8.** Servo accelerometer with pendulous mass and reluctive displacement sensor: (a) accelerometer; (b) block diagram ($R_L$, load resistor); (c) simplified mechanism. (Courtesy of Schaevitz Engineering.)

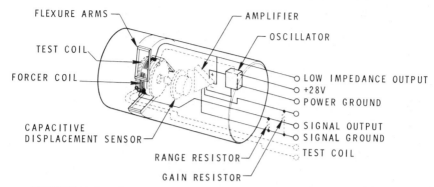

FLEXURE ARMS

AMPLIFIER

OSCILLATOR

TEST COIL

FORCER COIL

LOW IMPEDANCE OUTPUT

+28V

POWER GROUND

SIGNAL OUTPUT

SIGNAL GROUND

TEST COIL

CAPACITIVE
DISPLACEMENT SENSOR

RANGE RESISTOR

GAIN RESISTOR

**FIGURE 7–9.** Servo accelerometer with nonpendulous mass and capacitive displacement sensor. (Courtesy of Sundstrand Data Control, Inc.)

### *7.4.6  Strain-Gage Accelerometers*

In strain-gage accelerometers the displacement of the spring-mass system is converted into a change in resistance, due to strain, in four arms of a Wheatstone bridge (strain-gage bridge). The strain gages can be of the unbonded metal wire, bonded metal wire or foil, or bonded semiconductor types. The term "piezoresistive" has been applied to transducers incorporating semiconductor strain gages. The gages are mounted either to the spring, to a separate stress member additional to the spring, or between the seismic mass and a stationary frame.

The unit shown in Figure 7–10 contains a cylindrical seismic mass and has two strain windings wound between four insulated posts attached to each of two cross-shaped springs, one above and one below the mass. The windings contribute to the spring action. As the mass is displaced (vertically, in the illustration), the tension in one pair of strain gages increases and the tension in the opposite pair of windings is reduced. The windings are electrically connected as a four-active-arm bridge, whose output, when dc excitation is applied to the bridge, is proportional to acceleration and whose polarity indicates the direction of acceleration. This design employs viscous damping, using silicone oil, to provide a damping ratio of about 0.7 at room temperature.

A similar principle (two gages acting in tension, two in compression) is used in the accelerometer shown in Figure 7–11. This unit, however, has two semiconductor gages bonded to both sides of a cantilever spring to which the seismic mass is attached. Since semiconductor strain gages are significantly more sensitive than metal foil or metal wire gages, the spring-mass system of such devices can be made stiffer and the mass displacement can be kept smaller, so that the frequency response as well as the output signal amplitude can be considerably higher. Temperature-compensation resistors

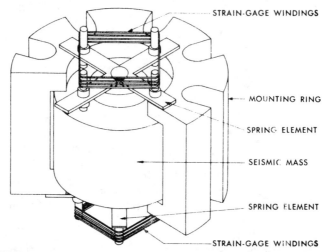

**FIGURE 7–10.** Unbonded-strain-gage-accelerometer. (Courtesy of Transamerica Delaval, CEC Div.)

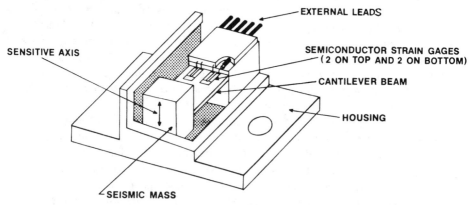

**FIGURE 7–11.** Bonded-semiconductor-strain-gage accelerometer. (Courtesy of Entran Devices, Inc.)

are provided within the transducer case. Such accelerometers can be made very small in size and mass.

### 7.4.7   *Vibrating-Element Accelerometers*

At least one accelerometer design has been produced in which a very small displacement of the seismic mass varies the tension of a tungsten wire, within a permanent magnetic field, whose other end is fixed. A current is passed through the wire, and the presence of the field causes the wire to vibrate at its resonant frequency (much like a string in a piano). The wire is connected into a circuit that produces an output signal in the form of frequency modulation (deviation from a center frequency). With proper design, construction, and burn-in, such transducers can provide very close repeatability; however, they were found to be quite temperature sensitive and difficult to produce, and hence, quite costly.

### 7.4.8 Acceleration Switches

Acceleration switches provide a discrete output at one or more set points. Normally open or normally closed switching is available. Various methods are used, including using magnetic switches actuated by seismic-mass displacement, replacing a potentiometric element by a contact strip, and providing an electronic gate for an otherwise analog output. In some simple acceleration switches with a single set point, a pair of contacts gets closed with a predetermined displacement of the seismic mass.

## 7.5 ACCELEROMETER CHARACTERISTICS

The two most critical characteristics of accelerometers are range and frequency response. Next in order of importance is sensitivity (or full-scale output). Piezoelectric accelerometers offer the highest ranges (to 100 000 $g$ for some shock accelerometers) as well as the highest frequency response (to over 20 kHz flat response, $\pm 5\%$). However, their sensitivity is relatively low. In many newer designs this problem is overcome by a built-in integrated-circuit amplifier, but other models still need a separate amplifier. Their frequency response also drops off rapidly below 1 to 2 Hz (except for some seismographic transducers, whose response extends to below 0.1 Hz), and piezoelectric accelerometers are generally not usable for measuring steady-state or even slowly varying acceleration.

Semiconductor-strain-gage accelerometers have a somewhat lower frequency response limit (usually to several kHz), but they do respond to steady-state accelerations. Their sensitivity is considerably greater than that of (basic) piezoelectric units, but an amplifier (now often built in) is still needed for many of their applications. Potentiometric accelerometers have the lowest frequency response (from dc to generally between 15 and 30 Hz, depending mostly on range); however, they provide a high-level output signal. Other types are somewhere between semiconductor-strain-gage and potentiometric transducers as far as frequency response is concerned. Servo accelerometers can be used for very low measuring ranges (well below 1 $g$ for full-scale output) and have very favorable accuracy characteristics.

Accelerometers other than piezoelectric or semiconductor-strain-gage accelerometers often employ viscous or magnetic damping to suppress resonance peaks that might result in erroneous output; they are then typically designed with a damping ratio of 0.7 (see Figure 2–26). When viscous damping is used, the damping ratio can be affected by temperature variations. High-frequency-response accelerometers can be equipped with (built-in or separate) electronic filters when the data transmission system requires such limitations (typically, limitations on channel bandwidth).

Besides frequency response, range, and sensitivity, there are a number of other characteristics whose relative importance is mostly application dependent. These include transverse sensitivity, mounting error, temperature errors, temperature-gradient errors, sensitivity to high ambient sound-pressure levels, and overrange capability. Indeed, the mass of an accelerometer becomes an important characteristic when it becomes a significant fraction of the mass of a structural member the accelerometer is mounted on.


## 7.6   APPLICATION NOTES

Low- to medium-frequency-response accelerometers are used primarily in moving vehicles of all types, from automobiles and trucks to ships, aircraft, and spacecraft, to determine acceleration and deceleration along the axis the vehicle is propelled or along the actual axis of motion. (There can be differences between these two.)

High-frequency-response accelerometers are used mostly for vibration and shock testing. The shock or vibration can be actual (as in vehicle road tests or in operating rotating machinery) or simulated by testing machines. Accelerometers are mounted at selected points on the test object, and output signals are either displayed directly on a fast-response oscillograph or recorded for later playback, display, and analysis. The test object can range in size from a small mechanical part to a large, complex structure. Figure 7–12 shows an auto body model instrumented for vibration analysis. The vibration is applied to the model by a vibration exciter located adjacent to the right edge of the platform. (A vibration exciter can be compared to a very powerful loudspeaker without its cone.) Vibration analysis is based on observations of vibration amplitude at various frequencies. It is a very useful tool in many fields and is applied to stationary as well as moving objects. Among its many uses is rotating-machine analysis, during which such critical parameters as shaft misalignments, rotor imbalance, and bearing wear can be determined.

Included in the many areas in which shock testing is employed are crash testing of automobiles and aircraft, and determining the effects of explosions or earthquakes on buildings and other large structures. Quite different forms of shock are sensed by seismic (seismographic) accelerometers, which have a frequency response extending downward to below 0.1 Hz together with a relatively high sensitivity. These are used not only in geological studies (including earthquake monitoring), but also for subsurface resource exploration such as petroleum prospecting. In prospecting oper-

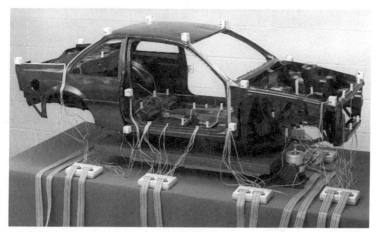

**FIGURE 7–12.**  Auto body model instrumented for structural vibration analysis. (Courtesy of PCB Piezotronics, Inc.)

ations, a group of such accelerometers is distributed over several locations on the surface near a point where either a surface impact shock is applied or, more commonly, below which a detonation is set off. The resulting shock waves travel through the subsurface to the accelerometers, which then provide information about subsurface discontinuities and other properties.

# 8

# *Gyros and Other Attitude Sensors*

## 8.1 BASIC DEFINITIONS

The concepts of attitude, bearing, and elevation are related to angular position; the concept of attitude rate is related to angular velocity.

*Attitude* is the relative orientation of a vehicle or an object represented by its angles of inclination to three orthogonal reference axes (axes at right angles to each other). *Attitude rate* is the time rate of change of attitude; it is often referred to as just "rate" and sometimes as "rate of rotation."

A *bearing* is a direction, at a reference point, direction given by the angle (in the *horizontal* plane) between a reference line and the line between the reference point and the point whose bearing is specified. As an angle, bearing is usually measured clockwise from the reference line. In most cases where bearing is used, the reference line is drawn between either a vehicle or a ground station and magnetic North.

*Azimuth* is a more generic term for bearing that is often used in such areas as celestial navigation as well as artillery (horizontal pointing of a weapon), tracking (horizontal pointing of a directional antenna), and pointing of airborne and spacecraft instruments. The reference line involved needs to be defined, since it may be different in different applications.

*Elevation* is similarly defined as the direction, at a reference point, given by the angle (in the *vertical* plane) between a reference line and the line between the reference point and the point whose elevation is specified.

## 8.2 UNITS OF MEASUREMENT

The units used for attitude, azimuth, and elevation are the *radian* (*rad*) or its submultiples such as the *milliradian* (*mrad*); the SI-approved alternative units, the *degree* (°), the *minute* (′), and the *second* (″) are also in common use. The terms "arc-minute" and "arc-second" have been declared obsolete by the BIPM, the international standards agency that regulates the use of units. Bearing is commonly expressed in degrees, minutes, and seconds, while attitude rate is expressed in *radians per second* (*rad/s*) or its submultiples such as *milliradians per second* (*mrad/s*).

## 8.3 SENSING METHODS

Methods of sensing attitude can be categorized on the basis of the reference system with respect to which the orientation (or pointing) of a vehicle or body is determined. In the case of a vehicle, there is always at least one design drawing that defines its own reference system. This system is expressed in terms of its three mutually orthogonal axes, the *X-, Y-, and Z-axes*, which are alternatively identified as the *pitch, yaw, and roll axes*, as in the example shown in Figure 8–1.

### 8.3.1 Inertial-Reference Sensing

The motion of a rotating body obeys Newton's first law of motion in an inertial frame of reference in which no forces are exerted on the body and the body is not accelerated. Unless acted upon by some unbalanced torque, a rotating body will continue turning about a fixed axis. This law is applied in the *gyroscope* (*gyro*), basically a wheel rotating within a frame. If the frame is affixed to a fixed portion of a vehicle, and the vehicle changes its attitude, this attitude change can be sensed as the change in angle between the frame (the "moving" portion) and the axis of rotation of the wheel (which is "fixed," at least theoretically).

### 8.3.2 Gravity-Reference Sensing

The force of gravity acting on a mass can be used to establish a vertical reference axis; the common plumb bob uses this principle. If the mass is suspended within a frame, and the frame is affixed to a fixed portion of a vehicle or body, changes in attitude can be sensed as changes in the angle between the frame (the "moving" portion) and the axis of suspension of the mass (which is "fixed").

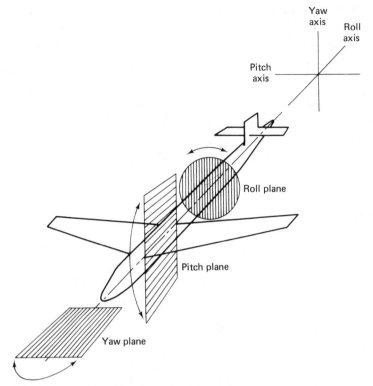

**FIGURE 8–1.**   Identification of vehicle planes and axes for attitude reference.

### 8.3.3   Magnetic-Reference Sensing

If a magnetic field remains fixed in position, its poles can establish a reference axis. A bar magnet aligns itself to the poles of a magnetic field, such as that of the earth, as exemplified by the action of a compass needle. When the bar magnet can freely rotate within a frame, changes in attitude can be determined as the angle between the pole-to-pole axis of the magnet (which is "fixed") and the frame (which "moves around" the bar magnet). If the frame is attached to a vehicle, the vehicle "moves around" the bar magnet.

### 8.3.4   Optical-Reference Sensing

A reference axis for the orientation of a vehicle can be established by aiming an electro-optical sensing device mounted to the vehicle at a light-radiating celestial body whose position is known at the time of measurement. Alternatively, the sensor can be aimed at a point where a step change in light emission, reflection, or refraction occurs, as seen from the vehicle, to establish such a reference axis. As the vehicle attitude changes, the sensor provides an output indicative of the change in the angle of beam incidence (or change in the angle to a light discontinuity) from the previously established reference axis. Some electro-optical sensing devices are specially designed to operate in this manner.

### 8.3.5  Radio-Reference Sensing

The interaction between a beam of electromagnetic radiation (at radio frequencies) emanating from a point of known position at a known direction and with known characteristics on the one hand, and a receiving device on the vehicle, capable of detecting changes in the angle of incidence of the beam, on the other hand, can be used to sense vehicle attitude.

### 8.3.6  Flow-Stream-Reference Sensing

The direction along which a fluid flows past a vehicle moving in the fluid can be used as reference for the attitude of the vehicle, provided that the vehicle itself does not alter the direction of the flow stream. This condition can be satisfied by placing a sensing element so designed as to align itself with the flow stream well ahead of the forward end of the vehicle. The attitude sensed in this manner is usually referred to as the *angle of attack;* its measurement is limited to use on high-speed aircraft and rockets and to rapidly descending atmospheric-entry vehicles.

## 8.4  DESIGN AND OPERATION

### 8.4.1  Attitude Gyros

In a simple basic gyro (see Figure 8–2) a rapidly spinning rotor, on an axle, is supported within a rotatable frame (*gimbal*) which is supported by bearings attached within a case. The axis about which the gimbal is free to rotate (*gimbal axis*) is perpendicular to the axis about which the rotor spins (*spin axis*). Since the spatial position of the spin axis remains fixed while the rotor is spinning, an attitude change of the case (angular displacement about the gimbal axis) results in an angular displacement between the gimbal shaft and the case. This displacement can be transduced by any angular-displacement transduction element (e.g., an annular potentiometric element with a wiper arm attached to the gimbal shaft).

The angular momentum of the gyro rotor is the essential stability characteristic of a gyro. The greater the angular momentum, the greater is the force required to deflect the spin axis from its position. The angular mo-

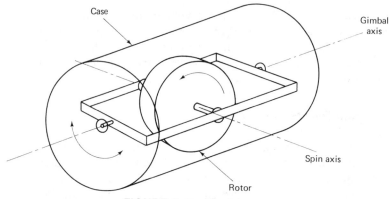

**FIGURE 8-2.**   Basic gyro.

mentum $H$ is the product of the rotor angular velocity $\omega$ and the moment of inertia $I_r$ of the rotor, or

$$H = \omega I_r$$

These quantities are vector quantities; that is, they have magnitude as well as direction. The relationship poses basic design criteria for gyros. Since a large angular momentum is desirable, the rotor spin rate should be high and the moment of inertia of the rotor should be large. These characteristics must then be traded off against gyro mass, size, and the spin-motor power requirements. The moment of inertia of a hollow cylinder is roughly double that of a solid cylinder of the same mass and diameter. Hence, most gyro designs employ a rotor in the shape of a wheel with a thick rim and thin web so that the shape of a hollow cylinder is approached.

When a torque is applied about any axis other than the spin axis, the spin axis will rotate about an axis which is perpendicular to the axis about which the torque is applied and which is also perpendicular to the spin axis. If, for example, a torque were applied in the plane of the gimbal frame of the gyro shown in Figure 8-2, the rotor would tend to balance this torque by deflecting angularly about the gimbal axis. The rotation of the spin axis produced by a torque is known as *precession*. The precession rate is constant when the torque is held constant. The mechanism will rotate about the axis of the applied torque only when it is not free to precess.

When precession of the spin axis from its intended position is caused by unwanted torques, it is called *drift*. Such torques can be caused by a number of sources inherent in the design and construction of the gyro, such as friction in gimbal bearings, a rotor radial mass unbalance, an unbalance within the gimbal assembly, forces exerted by electrical leads, and friction within the transduction element. They can also be caused by magnetic interaction and by external environmental conditions.

The mechanism illustrated in Figure 8–2 represents a *single-degree-of-freedom* gyro: the number of orthogonal axes about which the spin axis is free to rotate is one. In most practical applications a two-degree-of-freedom gyro (*free gyro*) is used; in free gyros the spin axis is free to attain any attitude. This is accomplished by adding a second gimbal which supports the bearings of the inner gimbal (which, in turn, supports the rotor). The inner and outer gimbal axes are orthogonal to each other. A transduction element is then located at both the inner and outer gimbal bearings. The outputs of the two transduction elements ("pick-offs") can represent attitude in any two of the three vehicle planes (pitch, yaw, roll). Potentiometric and reluctive transduction elements are used in many of the simpler gyro designs. Capacitive and photoelectric transduction are used in more advanced designs to detect spin-axis deflections.

Most gyros are equipped with a device that enables the gimbal(s) to be preset (locked) in a specified reference position. This process is known as *caging*, and the device is called a caging mechanism. After spin-up of the rotor in the desired axis, the gyro is *uncaged*, such as by means of an electrical solenoid. Caging/uncaging is not to be confused with *gimbal lock*, a condition that can occur when the inner gimbal aligns itself parallel to the outer gimbal, causing a loss of inertial reference. The reference can be re-established by caging the gyro, repositioning the spin axis in the desired reference axis, and then uncaging the gyro.

Many gyros are equipped with *torquers*, devices such as rotary solenoids which exert controlled amounts of torque on a gimbal. Torquers are used to correct for drift rates due to known sources. Closed-loop torquer control can be attained by feeding an error signal, due to drift rate, into a servo amplifier that powers the torquer.

DC as well as ac motors are used to drive the gyro rotor. The rotor of the drive motor can also act as the gyro rotor. In some designs characterized by a short operating life (e.g., in rockets) hot gas from a pyrotechnic charge has been used, in conjunction with a small turbine, to provide the spin function. AC induction or synchronous motors tend to produce lower drift in gyros.

A variety of other means have been devised to reduce drift. The spin assembly, within a sealed enclosure, floats in a viscous liquid to give the assembly neutral buoyancy (*floated gyro*). This technique eliminates purely mechanical suspension and reduces bearing friction. In the *gas-bearing* gyro pressurized gas is used as a flotation fluid. When the gyro rotor is freely suspended (bearingless), any retarding forces are minimized and the (non-powered) rundown time of the gyro is very long. In the *electrostatic gyro* this is accomplished by suspending the spherical rotor in an electrostatic suspension system. In the *cryogenic* gyro a spherical rotor, made of a material that becomes superconductive at the temperature of liquid helium, is maintained at that temperature. It is freely suspended within a magnetic-field coil system by interaction between this field and the magnetic field due to current flow in the superconductor. The *tuned-rotor gyro* consists of three concentric annular structures that are also concentric with the spin axis. The outer structure is the rotor, which is attached to the two inner structures, the gimbals, by flexures. The unit is so designed that the resonant frequency of the flexures is tuned to the rotational frequency of the rotor. Under these conditions the torques imposed by the flexures (which take the place of bearings) become negligible, and drift is minimized. This type of unit is relatively simple to construct. The (synchronous ac) motor can be external to the spin assembly, which is usually back-filled with argon to minimize friction in it.

Specialized versions of the two-degree-of-freedom gyro have been designed for specific vehicle applications. The *vertical gyro* is a pitch/roll attitude sensor in which a two-axis erection system maintains the spin axis in a vertical (gravity-referenced) position so that both gimbal axes are in a plane parallel to the ground surface. The *directional gyro* is a yaw-attitude transducer in which the rotor spin axis is maintained in the horizontal position (parallel to the ground surface). The gyro is installed on the vehicle so that the outer gimbal is parallel to the yaw axis. The direction of the spin axis (in the horizontal plane) is established upon uncaging and can also be referenced to the coordinates of the earth's magnetic field, such as by slaving the gyro to a compass synchro-transmitter. The directional gyro then operates as a *gyro compass*.

### 8.4.2  Attitude-Rate Gyros

The time rate of change of attitude is measured by the attitude-rate gyro (*rate gyro*, "rate sensor") directly in terms of angular velocity. A rate gyro (see Figure 8–3) is essentially a single-degree-of-freedom gyro whose gimbal is elastically restrained and whose motion is damped. When the rate gyro senses changes in attitude rate about its measurand axis (*input axis*), it produces an output signal due to gimbal deflection about the *output axis* by precession. Attitude rate is a vectorial quantity; hence, the output signal represents both the magnitude and direction of the attitude rate.

When a constant level of attitude rate is sensed, a relatively simple relationship defines the static response of a rate gyro, since restraining torques due to damping and to gimbal inertia can be neglected. This relationship is

$$\theta = \frac{H\Omega}{K}$$

where $\theta$ = gimbal deflection angle (output angle)
$H$ = gyro angular momentum
$\Omega$ = attitude rate
$K$ = restraining-device spring constant

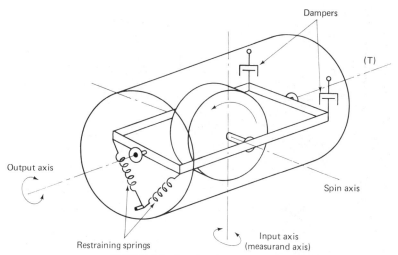

**FIGURE 8–3.** Basic rate gyro.

The gimbal deflection, or an equivalent displacement about the axis defined as the gimbal axis, can be transduced by various means, ranging from simple potentiometric or reluctive elements to a vibrating wire responding to Coriolis forces. The "restraining springs" can take the form of leaf springs, a torsion bar, torsional springs, flexures, magnetic springs, or other, non-mechanical, devices. Among damping devices are those of the dashpot, paddle-in-liquid, viscous-shear, hydraulic-bleed, and eddy-current varieties; the damping function can also be controlled electronically. Servo operation, using a torquer in the control loop, can be attained in torque-balance rate gyros in a manner similar to that described for servo accelerometers.

Electromechanical rate gyros are subject to the same problems (drift and a limited operating life due to the use of a mechanical rotor suspension and mechanical bearings) as are discussed for free gyros (attitude gyros) in Section 8.4.1. Two relatively new types of gyros overcome these problems without introducing complexities in the design of electromechanical gyros; they use circulating light beams generated by a laser instead of using a motor-driven rotor. The *ring-laser gyro* originally employed a ring-shaped cavity in which two oppositely directed traveling waves could circulate many times. A frequency difference between the two beams would occur that was a function of attitude rate applied perpendicularly to the plane of the traveling beams. This frequency difference was sensed by a photodetector in terms of interferometric patterns; the resulting fringe counts provided an inherently digital output. Later developments resulted in a triangular shape instead of a ring, with mirrors used to assure continuing beam travel (see Figure 8–4). A problem of frequency lock-in at very low attitude rates, causing the output to drop to zero in this region of the bidirectional range, was overcome by dithering the mirrors, i.e., causing them to oscillate electromechanically.

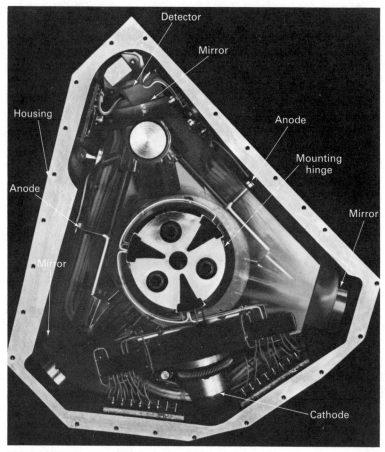

**FIGURE 8–4.** Ring laser gyro. (Courtesy of Kearfott Guidance and Navigation Div., The Singer Co.)

This very tiny mirror motion is the only mechanical motion in a ring-laser gyro.

The *fiber-optic gyro* (fiber-optic rotation sensor) uses a very long (between 1 and 5 km) coiled optical fiber to establish the path along which two beams travel in opposite directions. A beam splitter splits the laser beam into the counterpropagating light beams. The fiber terminates at a photodetector. At zero applied attitude rate, the two beams are 180° out of phase, so that the combined outputs of both beams cancel and the detector output is zero. As the applied attitude rate increases, the apparent path lengths of the two beams change (one path gets "shorter," the other "longer"), the phase difference changes, the combined outputs no longer cancel, and the photodetector output increases. When the attitude rate is large enough to cause the difference in apparent path length to be half the wavelength of the laser source, the detector output is maximum. The output of a fiber-optic gyro is, therefore, inherently analog, but can, of course, be digitized by an analog-to-digital converter.

### 8.4.3 Rate-Integrating Gyros

The (attitude-) rate-integrating gyro is a design derivative of the rate gyro. It is a single-degree-of-freedom gyro without the elastic restraint of the gyro, having, instead, viscous restraint of the spin axis about the output axis (Figure 8–5). Precession of the gimbal produces an output signal that is proportional to the integral of the attitude rate sensed about the input axis. The viscous shear in the damping liquid, which fills the case, including, particularly, the gap between the spin-assembly case and the external case, provides the integration effect. Various modes of operation can be obtained, depending on the type of command signal applied to the torquer on the output shaft. The term "hermetic integrating gyro" (HIG) has been applied to this design because the case must be hermetically sealed to contain the liquid used for floatation as well as damping.

The related *double-integrating gyro* has no elastic or viscous restraint of the gimbal about its output axis. The dynamic behavior is established primarily by the inertial properties of the gimbal. The output signal is proportional to the double integral of the attitude rate of the case about the input axis.

### 8.4.4 Gravity-Referenced Attitude Transducers

The force of gravity can be used to establish a very useful vertical reference, as anyone who has used a plumb bob or bubble level knows. Transducers used to measure angular displacement from the vertical are relatively simple designs employing a solid or liquid mass upon which gravity acts and which

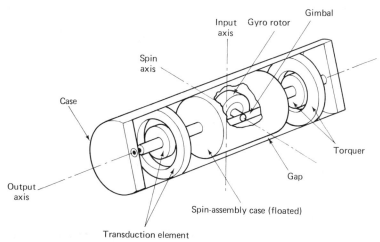

**FIGURE 8–5.** Rate-integrating gyro.

is either linked to a transduction element or is an active part of it. Since angular deviation from the vertical is often called "inclination," and such transducers have also been referred to as *inclinometers*.

The simplest of these designs is the *pendulum*-type transducer. It consists of a weight, in the shape of a flat segment of a circle, which is mounted so that it can rotate by means of a ball bearing to which it is attached at its tip. The ball bearing is at the center of a nonconductive disk. A curved potentiometric element is mounted to this disk, and a wiper arm is attached to the weight. A slip ring forms the connecting contact for the wiper arm. The weight remains vertical. As the case, to which the nonconductive disk is attached, is rotated, the wiper arm slides over the potentiometric resistance element.

The *electrolytic-potentiometer* type of transducer resembles the liquid-filled glass tube in a carpenter's level. The small, curved tube is partially filled with a liquid electrolyte, and the "common" or "wiper" contact extends completely through the tube. The "end terminals" protrude partially into the tube, one at each end. As the tube is angularly deflected from a true horizontal position (orthogonal to true vertical), the resistance to the common electrode increases from one electrode and decreases from the other. The device needs to be ac-excited to avoid polarization of the electrolyte. Designs of this type can have a full-scale range around ±1°.

Small variations from true horizontal have also been measured by a *capacitive* transducer. Two metallic half-disks are suspended closely above the surface of a pool of mercury contained in a cylindrical cavity. (The internal configuration needs to be circular.) As the device is tilted, the capacitance between the mercury, which acts as the common "rotor" electrode, and one of the "stator" electrodes increases while it decreases between the mercury and the other "stator" electrode. This device can be connected into an ac bridge circuit so that changes in capacitance are converted into ac output changes.

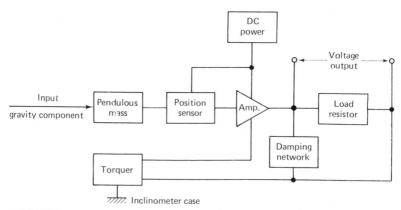

**FIGURE 8–6.**    Torque-balance inclinometer: block diagram. (Courtesy of Schaevitz Engineering.)

A *torque-balance, reluctive inclinometer* produces an output proportional to the sine of the angle of deviation from the vertical. It employs a pendulous mass and operates in a manner similar to that of a torque-balance servo accelerometer (which, essentially, the inclinometer is). As illustrated in the block diagram of Figure 8–6, motion of the mass is detected by a (reluctive) position sensor whose output is amplified and fed to a torquer, i.e., a coil in a magnetic field, which counteracts the torque applied to the pendulous mass due to deviations from the vertical. The current supplied to the coil is thus proportional to changes in the gravitational force. When the current is passed through a load resistor, a voltage output is produced. Typical inclinometer designs of this type are shown in Figure 8–7.

### 8.4.5 Compass-Type Sensors

Sensing devices related to the compass are used on most seaborne and airborne vehicles to indicate *bearing*, the vehicle's attitude in the terrestrial horizontal plane with reference to the earth's magnetic field. The reference is a line between the vehicle and magnetic north. "True north" can be established from this indication together with knowledge of the global varia-

**FIGURE 8–7.** Typical torque-balance inclinometers. (Courtesy of Schaevitz Engineering.)

tions in the magnetic field. A simple version of such a "north seeker" uses a bar-magnet assembly to move a wiper arm across a potentiometric element. Other designs use reluctive, notably synchro, transduction and may operate in a closed-loop servo mode. A pendulous suspension is commonly used to stabilize the sensing mechanism in a true horizontal plane. In the *gyrocompass* the combined use of friction and a gravitational restoring torque causes the equilibrium position of the rotor's spin axis to align itself with the north-south reference line. The *flux-gate compass* employs three cores, having an excitation as well as an output winding, which are so arranged that they are balanced in the absence of an external magnetic field. The magnitude and direction of the earth's magnetic field then causes unbalance voltages proportional in magnitude to the field's magnitude, and proportional in phase to the direction of the field. A phase-sensitive detection circuit or indicator is used to provide an output indicative of direction, or indicate direction, respectively. The *induction compass* uses a rotating coil as sensing element. The voltage induced in the coil is at its maximum when the rotational axis of the coil is perpendicular to the magnetic field.

### 8.4.6  Angle-of-Attack Transducers

Angle-of-attack transducers, which are used on high-speed airborne vehicles, sense attitude with reference to the direction of the fluid stream ambient to the vehicle. Some designs use a single, rotatable, wedge-shaped vane for this purpose. The unit is so installed that the sharp edge of the vane points forward and then, while in flight, aligns itself with the direction of the ambient air stream. The transduction element responds to vane rotation. The *ogive* angle-of-attack transducer uses a freely swiveled projectile-shaped body with four fins, one every 90° around the body. As the finned body aligns itself with the airstream, its relative motions about the pitch and yaw axes are detected by two transduction elements, one for each axis. The ogive transducer is usually installed at the tip of a boom extending well forward of the nose of the vehicle. Such an installation is also typical for *pressure-differential* angle-of-attack transducers, which use a pair of pressure ports, tubulated to a differential-pressure transducer, to sense the direction of airflow relative to a given vehicle axis. Note that only the angle about the pitch axis is properly referred to as the angle of attack. The angle about the yaw axis is more correctly termed the *angle of sideslip*. Some versions of angle-of-attack transducers have been designed to operate in water rather than in air.

### 8.4.7  Celestial-Reference Attitude Sensors

Electro-optical sensors are commonly used for attitude determinations on spacecraft, but sometimes also in research-balloon payloads and on special-applications aircraft. They sense attitude with reference to a line between the vehicle and a light-emitting or light-reflecting target whose position is known. Typical targets are the sun, certain stars, and the light-to-dark boundary (*limb*) of the earth or another planet. The choice of the target determines the name of the sensor—for example, *sun sensor, star sensor* (or "star tracker," or, on spin-stabilized spacecraft, "star pipper"), and *limb sensor* or *horizon sensor*. The sensors employ photodetectors of the segmented or array type. Charge-coupled devices (CCDs) are used in advanced star trackers because they allow imaging of the target-body region, and pattern recognition schemes can then be used for attitude information in one or even two axes. The most common use of such sensors is for closed-loop attitude-control systems.

### 8.4.8  Radio Navigation Sensors

The most widely used sensor in this category is the *radio compass*, essentially a radio receiver with a highly directional rotatable antenna ("loop antenna"). When the antenna is rotated so that the received signal is at a maximum, the position (and yaw attitude) of the vehicle relative to the line between the vehicle and the radio transmitter can be determined as a function of the angular position of the receiving antenna. When the expected deviations of the vehicle's attitude from a received radio beam are relatively small, the receiving antenna can be fixed and the variations in the output signal from the receiver can be used to indicate attitude deviations. This principle is used in an aircraft's *glide path* receiver, which receives a radio beam of known direction from a transmitter located at an airport and indicates deviations (in pitch) from this beam during the aircraft's final approach. The same principle is used on some spacecraft either to keep a movable (*articulable*) antenna, which is used for receiving as well as transmitting, pointed at the source of the beam or to keep the spacecraft in an attitude referenced to the direction of one or more radio beams. The latter is effected by using received-signal deviations as error signals in a closed loop ("beam rider") attitude-control system.

## 8.5  SENSOR CHARACTERISTICS

General performance characteristics to be considered for attitude and attitude-rate transducers comprise primarily range and full-scale output, zero-measurand output (or output at a reference level of measurand), linearity, hysteresis and repeatability, threshold, resolution, and stability. Drift rate is probably the most essential performance characteristic of gyros. Depending on design and construction, gyro drift rates can be between less than 0.1 and over 100 degrees per hour. Other characteristics to be considered for electromechanical gyros and rate gyros are run-up and run-down time of the rotor, overall warm-up time, caging and uncaging provisions and the time required for these, response time, and erection rate. For electro-mechanical rate gyros, damping ratio and natural frequency often become important. Typical specifications for ring-laser gyros include drift rate, but broken down into bias error and "random walk"; scale factor (transfer function) and its stability; linearity; and asymmetry. The alignment of most types of attitude sensors is usually quite critical: not only must the external alignment of the sensor on a vehicle or object be well established, but the alignment of critical internal elements, and the change of this alignment due to thermal effects, can also become important. Among environmental performance characteristics, the effects of acceleration, shock, and vibration are the most critical for gyros and other electromechanical attitude and attitude-rate transducers. Acceleration and vibration can cause errors in gyros due to *mass unbalance* (lack of coincidence of the center of supporting forces and the center of mass) and *anisoelasticity* (inequality of compliance of the rotor and gimbal assembly to forces in different directions). Operating temperatures are critical to electro-optical sensors.

## 8.6   APPLICATION NOTES

Gyros are used primarily for inertial navigation, guidance, and control on vehicles operating underwater, on the oceans, in the air, and in space. Compass-type and radio navigation sensors are similarly employed in ships, on some surface vehicles, and on aircraft. Angle-of-attack transducers are used primarily for flight testing of high-speed aircraft and atmospheric entry bodies. Celestial-reference attitude sensors are usually found on satellites and other spacecraft. Most spacecraft attitude-control systems rely on celestial as well as inertial sensors for their input data.

The applications of gravity-referenced attitude transducers, mainly those of the pendulum type as well as the inclinometer types, are an exception to the ones just mentioned. They are used wherever something like a high-precision plumb bob or bubble level would be used, but where a continuous electrical output is needed for often very small deviations from the true vertical. Thus, they are used on structures of many types; on heavy construction equipment such as cranes; on pipelines; for bore-hole mapping; for seismic studies, including those used for earthquake prediction; and, generally, on any sort of structure that must be kept level, i.e., normal to the true vertical, or whose deviation from the true vertical must be known.

# 9

## Strain Gages

### 9.1 BASIC DEFINITIONS

*Strain* is the deformation of a solid resulting from stress. It is measured as the ratio of dimensional change to the total value of the dimension in which the change occurs. *Poisson's ratio* is the ratio of transverse to longitudinal unit strain. *Stress* is the force acting on a unit area in a solid.

The *modulus of elasticity* is the ratio of stress to the corresponding strain (below the proportional limits). It is defined, by Hooke's law, as

$$E = \text{constant} = \frac{s}{\epsilon}$$

where $E$ = modulus of elasticity
$s$ = stress
$\epsilon$ = strain

The tensile and compressive moduli of elasticity are defined as

$$E_t = Y = \frac{s}{\epsilon} = \frac{F_t/a}{\Delta L/L} \quad \text{and} \quad E_c = \frac{s}{\epsilon} = \frac{F_c/a}{\Delta L/L}$$

where $F_t$ = force (tension)
$F_c$ = force (compression)
$a$ = cross-sectional area (normal to direction of application of force)

$\Delta L$ = elongation or contraction (of solid along the direction of application of force)

$L$ = original length of solid

$Y$ = Young's modulus ("stretch modulus of elasticity")

$E_t$ = tensile modulus of elasticity

$E_c$ = compressive modulus of elasticity

The *elastic limit* is the maximum unit stress not causing permanent deformation of a solid.

*Torsion* is the twisting of an object, such as a rod, bar, or tube, about its axis of symmetry. Torsional deflection and shear stress (due to torsion) are defined subsequently in mathematical form to show proportionality. The units used for the equations are U.S. customary units because of the familiarity of engineers in the U.S.A. with the equations in this form, including the constants (32, and 16, respectively). In the SI equivalents of the equations, the constants will be different but the proportionalities will be the same.

The *torsional deflection* of a solid cylindrical shaft is

$$\theta = \frac{32TL}{\pi d^4 E_s}$$

where $\theta$ = helical angle of deflection

$T$ = torque, lb-in

$d$ = shaft diameter, in

$L$ = length of shaft under torsion, in

$E_s$ = shear modulus of elasticity, $lb_f/in^2$

The *shear stress,* due to torsion, in a solid cylindrical shaft is

$$S_s = \frac{16T}{\pi d^3}$$

where $S_s$ = maximum shear stress, $lb_f/in^2$

$T$ = torque, lb-in

$d$ = shaft diameter, in

## 9.2   UNITS OF MEASUREMENT

Strain is expressed as the (dimensionless) ratio of multiples of the same length unit, for example, microinches per inch, or, generally, in percent (for large deformations), or, most commonly, in microstrain ($\mu\epsilon$), the ratio of $10^{-6}$ of a length unit to this length unit.

The SI unit for stress as well as for modulus of elasticity is the *newton per square meter (N/m²)*.

## 9.3.  SENSING METHODS

Although strain has been measured by reluctive, capacitive, and vibrating-wire sensors and is still often determined by photo-optical (*photostress*) methods, the most commonly used device for such measurements is the resistive strain sensor or *strain gage*. It consists essentially of a conductor or semiconductor of small cross section which is mounted to the measured surface so that it undergoes the small elongations or contractions due to tension or compression stresses, respectively, in that surface. As a result, the strain gage undergoes a corresponding change in resistance due to stress (*piezoresistive effect*).

The resistance change of a strain gage is usually converted into a voltage by connecting two or four matched gages as arms of a Wheatstone bridge (see Section 9.6.2); in such applications the Wheatstone bridge is sometimes called a *strain-gage bridge*.

Excitation is applied across the bridge, and the bridge output voltage is then a measure of the strain sensed by each strain-sensing gage (*active gage, active arm*). Gages are sometimes connected into the bridge, but are so mounted that they do not change their resistance with strain (*dummy gage*); in this mode they are used for bridge balance or compensation. Strain transduction, therefore, is performed by the active arms of a strain-gage bridge.

*Stress* is related to strain by *Hooke's law:* the modulus of elasticity is the ratio of stress to strain. Hence, if the modulus of elasticity of the surface (material) is known, and the strain is measured, the stress can be determined. This holds true, at least theoretically, when a single gage is used. When two or more gages are used, the stresses in the sensing directions of each of the gages can be determined by calculations; however, the equations are of varying degrees of complexity, depending on the combination and orientation of the gages.

The sensitivity of a strain gage is called the *gage factor*, the ratio of the unit change in resistance to the unit change in length, or

$$\mathrm{GF} = \frac{\Delta R/R}{\Delta L/L} = \frac{\Delta R/R}{\epsilon}$$

where GF = gage factor
$\Delta R/R$ = unit change of resistance
$\Delta L/L$ = unit change in length
$\epsilon$ = strain

## 9.4  DESIGN AND OPERATION

### 9.4.1  Metal-Wire Strain Gages

Metal-wire strain gages are included here mostly for historical reasons, since they were the earliest form of strain gage. They consist of thin (less than 0.025 mm in diameter) wire applied in a zigzag pattern to a dimensionally stable substrate (*carrier, backing, matrix*). The wire is cemented to the substrate. Some types, called "surface transferable" are only loosely applied to the carrier, and then have to be removed from the carrier and cemented to the measured surface, which takes extremely careful handling. Others are encapsulated in insulating material and then sandwiched between two thin metal plates; this makes them weldable to the measured surface. Since metal-foil gages are much easier to produce, in a variety of sizes and configurations, they have essentially obsoleted the wire gages.

### 9.4.2  Metal-Foil Strain Gages

Refinements in photoetching techniques, followed by improvements in producibility paralleling those of printed circuits, permitted the development and mass production of low-cost metal-foil gages. They can be made in a variety of sizes, down to about 0.2 mm in gage length. The photoetching process also lends itself to the design and fabrication of strain gage rosettes (see shortly) and other multielement gages.

The nomenclature for gage dimensions is illustrated in Figure 9–1. The strain direction is along the *sensing axis*, which is along the gage length. The axis along the gage width is the *transverse axis*. The strain gage terminates in the (relatively large, as illustrated) *tabs*, to which wires can be attached. Some strain gages have solder dots on their tabs, to facilitate soldering; others are available with preattached leads. The gages are usually made by photoetching thin, heat-treated metal-alloy foil a few micrometers thick. The process involves removal of unwanted metal by etching so that the desired shape is obtained. Foil gages exist in a variety of geometries, the most common of which, for general-purpose gages, has a grid pattern whose length is larger than its width. Typical single-element gages are illustrated in Figure 9–2.

The metal used for the foil is generally selected to have a high gage factor, a low temperature coefficient of resistance, high resistivity, high mechanical strength, and a minimum thermoelectric potential at the junction with the leads. The most commonly used material is a copper-nickel alloy such as Constantan. Selection of foil material by temperature coefficient allows two materials to be used for one gage, each material having a temperature coefficient of the same magnitude but opposite polarity. Such self-temperature-compensated gages are typically made by using two half-grids

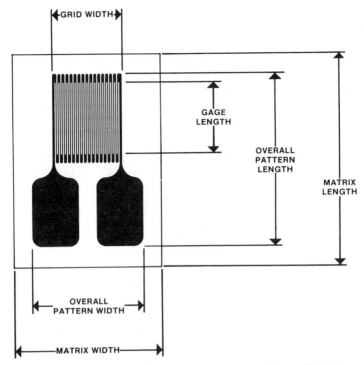

**FIGURE 9–1.** Strain-gage dimensional nomenclature. (Courtesy of Measurements Group, Inc.)

connected in series. Annealed Constantan is used for high-elongation or post-yield (see Section 9.4.5) gages. Nickel-chromium alloys are used for self-temperature-compensated gages and for strippable (surface-transferable) gages that must operate at very high temperatures. An iron-chromium-aluminum alloy is used in gages intended for service at very high or very low

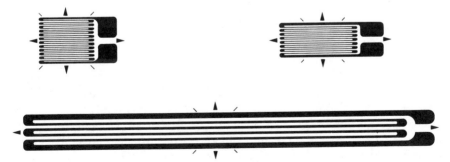

**FIGURE 9–2.** Metal-foil strain gages. For size comparison, the gage shown on the bottom is 5 cm long. (Courtesy of Measurements Group, Inc.)

temperatures. Leads are typically made of copper (often nickel clad or stainless-steel clad), silver (sometimes nickel clad), or nickel-chromium alloys.

Materials used for the carrier matrix (backing) of strain gages include polyimide, epoxy, or epoxy-phenolic resins with or without glass-fiber reinforcement, depending on the application, especially as regards the operating temperature. Fully encapsulated gages are also available. The backing materials of these are generally very thin (less than 0.1 mm thick). Fully encapsulated gages can be sandwiched between thin metal plates to make them weldable.

When strain must be measured simultaneously in more than one direction, multielement gages called *rosettes* can be used. Foil gages, and the processes used to form them, are particularly suitable for rosettes. Figure 9–3 shows some examples, all of which exist in a variety of sizes, down to 1 mm or a few mm for a complete rosette. The 90° rosettes measure axial and transverse strains simultaneously. A variation of this design is the "stress gage," in which the two elements have different resistances. The resistances are so selected that their ratio provides a sensor whose combined output is proportional to "stress," and the output of the axial element alone is proportional to "strain." Figure 9–3(b) shows a "shear gage," whose geometry enables the two elements to be aligned in the direction of shear strains, such as those encountered in torsion bars of torque transducers.

Three-element rosettes are particularly useful for determinations of the magnitude and direction of principal strains under complex loading conditions. The 45° rosettes provide greater angular resolution than the 60° rosettes; however, they are normally used where the approximate direction of the principal strains is known. Equations as well as computer programs have been developed for determinations of principal stresses and strains when such rosettes are used. The "diaphragm gage" (Figure 9–3(g)) is intended for use on such sensing elements as the diaphragms of pressure transducers. Multielement gages are also available for other sensing elements such as bending beams.

### 9.4.3 Deposited-Metal Strain Gages

*Thin-film* strain gages are applied directly to the measured surface, which is first coated with an insulating substrate, by evaporative or bombardment methods. They have been applied primarily to pressure-sensing diaphragms. The thin-film techniques provide gages, in four-active-arm configurations, which are not only very small in size, but are also very uniform in their characteristics. Several major manufacturers of pressure transducers have been producing thin-film strain-gage transducers with very satisfactory results, including performance and long-term stability (see Section 15.4.6).

*Flame-sprayed* strain gages have been applied to structures which are to be exposed to extremely hostile environments, such as the exterior sur-

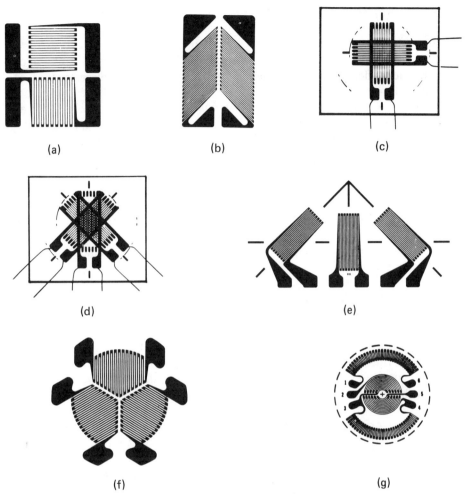

**FIGURE 9–3.** Examples of foil-gage rosettes: (a) two-element, 90° planar; (b) two-element, 90° shear planar; (c) two-element, 90° stacked; (d) three-element, 45° rectangular, stacked; (e) three-element, 45° planar; (f) three-element, 60° planar; (g) four-element, full-bridge diaphragm configuration. [(a), (b), (e), (f) courtesy of BLH Electronics, Inc.; (c), (d), (g) courtesy of Measurements Group, Inc.]

faces of rockets. A substrate is first flame-sprayed on the measured surface by propelling ceramic particles, at a high temperature and with high speed, onto it. A strain gage, typically a thin-metal-wire gage, is then applied to the substrate and bonded and encapsulated to it by a second application of flame-sprayed ceramic. Such gages have performed acceptably at operating temperatures up to about 1200°C.

### 9.4.4  Semiconductor Strain Gages

When experiments during the early 1950s confirmed that the piezoresistive effect is much larger in semiconductors than in conductors, a number of government and commercial laboratories undertook the development of semiconductor strain gages. Since then a variety of such devices, with satisfactory and controllable performance characteristics, have become commercially available. The gage factors of semiconductor strain gages are between 50 and 200 and are typically around 125, whereas the gage factors of metal strain gages are no greater than 5 and are typically around 2. However, semiconductor gages tend to be more difficult to apply to measured surfaces and, since they are basically made of thin silicon, tend to be fragile. Their strain measuring ranges are usually limited to about 5000 $\mu\epsilon$, whereas the ranges of metal gages extend to about 40 000 $\mu\epsilon$. The resistance change of semiconductor gages with applied strain is inherently nonlinear, maximum operating temperature is more limited, and temperature compensation is more laborious.

Because of their ability to provide relatively large output signals (when connected into a bridge circuit) in response to small strains, however, semiconductor gages have found quite a few applications, especially in transducers. In these (e.g., acceleration, force, and pressure transducers), they offer the very significant advantage of requiring much lower sensing-element (e.g., diaphragm, bending beam) deflections than metal gages. This permits the use of stiffer and generally more stable sensing elements, and stiffer sensing elements also provide a far higher frequency response. Also, transducer manufacturers can train personnel to install the gages properly, and their engineering staff can work out methods for nonlinearity and temperature compensation suitable for production operations. Compensation techniques, including those involving variations of doping and resulting carrier concentration in the semiconductor, are now quite well established.

Typical semiconductor strain gage configurations are shown in Figure 9–4. The gages can be bare (surface transferable) or encapsulated on a carrier. Dual-element and full-bridge rosettes can be made so that they are self-compensating for apparent strain. Dimensions are typically very small: for example, the overall lengths of bare gages are between about 1 and 5 mm. Leads are typically made of gold or silver wire or nickel ribbon. Compensation for thermoelectric potentials generated at their junction with the (silicon) gage is quite important, since silicon develops very high thermoelectric potentials.

By diffusing the dopant directly into selected portions of silicon wafers or diaphragms (see Figure 9–5), it is possible to manufacture complete sensing/transduction elements. Because of the low density and high stiffness of silicon, such elements are characterized by a very high natural frequency. Gage matching and compensation can be effected by appropriate control of

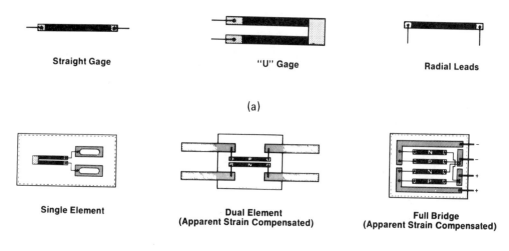

**Straight Gage**     **"U" Gage**     **Radial Leads**

(a)

**Single Element**     **Dual Element**
**(Apparent Strain Compensated)**     **Full Bridge**
**(Apparent Strain Compensated)**

(b)

**FIGURE 9–4.** Typical semiconductor-strain-gage configurations: (a) bare gages; (b) encapsulated gages. (Courtesy of Kulite Semiconductor Products, Inc.)

dopants and diffusion techniques, and a number of manufacturers have applied their expertise to these materials and methods. Several manufacturers have, in fact, mastered the techniques associated with "integrated pressure sensors" to the extent of mass-producing small pressure sensors that can be sold at a relatively very low price.

### 9.4.5 Special-Application Gages

*Post-yield gages,* typically made of special annealed Constantan with a high-elongation carrier, are used when large post-yield strains are to be measured, that is, those occurring after the test specimen has been loaded beyond the yield point. Such gages have strain measuring ranges between 10 and 20% (between 100 000 and 200 000$\mu\epsilon$). They will exhibit substantial zero shifts, however, when subjected to cyclic strains of large magnitudes.

*Fatigue-life gages* have the appearance of foil strain gages and are installed in the same manner. The material of the foil, however, is chosen and treated to give this gage its unique characteristic: cyclic loading will cause irreversible changes in its resistance. The ohmic value of the resistance change depends on the magnitude of the strain applied during each loading cycle and on the number of loading cycles. If the strain is held to a constant amplitude over a complete test in which the specimen is subjected to cyclic loading, the resistance changes will accumulate; the cumulative change of gage resistance is then a function of the number of loading cycles and, hence, the strain history of the specimen. The characteristics of the gage must be

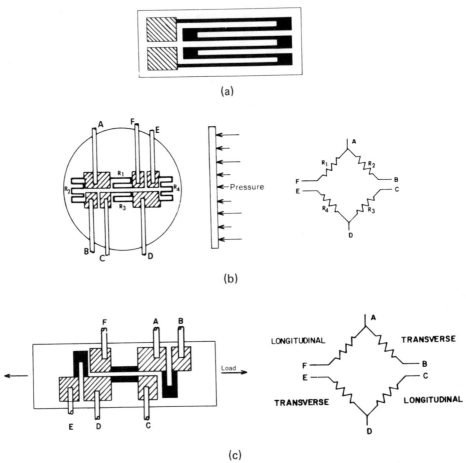

(a)

(b)

(c)

**FIGURE 9–5.** Diffused semiconductor-strain-gage sensors: (a) single-element gage; (b) integrated-bridge silicon pressure-sensing diaphragm; (c) integrated-bridge silicon bending beam. (Courtesy of Kulite Semiconductor Products, Inc.)

chosen so as to match the characteristics of the specimen material for the range of strains to be applied.

In many applications, such as in tests of large structures, it is necessary to have the fatigue-life gage experience a strain greater than that experienced by the measured surface. Mechanical multipliers (*strain multipliers*) can then be used. These multipliers contain the fatigue-life gage on a center section of special design and are then bonded to the measured surface at their ends. Strain multiplication between 2× and 20×, depending on choice of multiplier, is generally available. A regular strain gage may be included in the multiplier.

Data indicative of impending fatigue damage can be based on the results of a prior test in which the sensor is installed on an exact-replica specimen of the measured surface, which is then cycled in a testing machine until cracking occurs. The cumulative resistance change in the fatigue-life gage at the point where cracking starts is noted.

*Crack-propagation gages* have very thin foil grids consisting of a number of closely spaced parallel conductors connected to a single common bus tab at each end. The gage is installed on a portion of measured surface where a crack is expected to occur, with the conductors bridging the expected crack. As the crack propagates under the gage, each of the conductors is broken sequentially. Since the conductors are connected in parallel, the resistance measured across the bus tabs increases as each of the strands breaks until a completely open circuit is seen after the last strand has broken.

## 9.5 STRAIN-GAGE CHARACTERISTICS

Gage dimensions usually include not only the grid length (gage length) and grid width, but also the overall length and width, including the terminals, of the gage itself. For gages bonded to, or encapsulated within, a carrier (matrix), either the overall size of the carrier is shown, or, if the carrier is provided with trim marks, dimensions between those marks are given. Tolerances should be included.

The materials and nature of the gage itself, of the carrier, and of either integral leads or terminals should always be stated. For multielement gages (rosettes or active/dummy gage combinations), the nature and purpose of the gage pattern should be described. When gages are self-temperature-compensated for the thermal coefficient of expansion of a measured surface, the material of the measured surface the gage is matched to must be stated.

The gage resistance is shown, in ohms with a tolerance, for individual gages as well as for each gage in a rosette. The maximum excitation current allowable for a gage should be known for a given application. For weldable gages and some encapsulated gages, the insulation resistance is often stated.

The gage factor is always shown for a strain gage, together with its tolerance. The strain range should be shown (in percent, in microstrain, or in other units, such as $\mu in/in$), with the understanding that this is the range over which stated tolerances for linearity, hysteresis, and creep will apply. Additionally, an overrange capability ("strain limit") can be specified. Transverse sensitivity, where relevant, should also be stated. An operating temperature range is always specified for a gage or rosette, and the thermal effects on gage factor, and often on gage resistance, are also shown. Allowable maximum temperature limits, usually meant for short-term exposure, can be included.

The effects of thermal expansion of the strain gage and of differential thermal expansion between the gage and the measured surface, often referred to as "apparent strain," should be known, even for self-temperature-compensated (thermal expansion) types of gages. Actually, the apparent strain as well as the temperature gradient error can be stated with reasonable accuracy only for a gage installed by a specified method on a well-defined specimen. The following definitions can be applied to strain gages in strain measuring systems:

1. *Indicated strain* is the output of an installed strain gage as indicated on the data display device and as corrected for all errors introduced by the measuring system except the installed gage.

2. *Real strain* is the deformation of the measured surface due to applied loads as well as thermal changes.

3. *Apparent strain* is the difference between real and indicated strains.

4. *Thermal strain* is the deformation of an unrestrained specimen due to a change in temperature alone.

5. *Mechanical strain* is the difference between real and thermal strains.

6. *Thermal output* is the indicated strain resulting only from apparent and thermal strains.

These definitions are useful for an understanding of the various phenomena occurring during an actual strain measurement, but are not typical of definitions shown in gage specifications. In most of these, apparent strain is defined as the indicated strain produced by a gage mounted to an unrestrained specimen which is subjected to a change in temperature.

In addition to the foregoing characteristics, the cycling life of a gage (sometimes referred to as "fatigue life") is often shown.

## 9.6   APPLICATION NOTES

### 9.6.1   Gage Installation

Except for weldable and embeddable gages, the usual method of installing a strain gage is by means of an adhesive ("cement"). Recommendations for installing welded and embedded gages are usually stated by, or available from, the gage manufacturer. Gage manufacturers also provide instructions for the bonding of gages by adhesives. Several manufacturers can furnish not only the various adhesives needed, but also special installation tools. Some of them even offer training of installation technicians.

Strain gages differ from most other sensing devices in that their installation is intended to be permanent: they cannot be removed and reused. Proper installation is, therefore, essential. The following are generally the major steps in procedures for bonding strain gages by means of an adhesive:

1. *Surface preparation.* The surface to which the gage (or rosette) is to be bonded must be cleaned completely and should be roughened slightly; this is best done immediately before gage installation.

2. *Adhesive application.* The proper adhesive must first be selected; adhesives can be of the solvent-release type, contact-setting type, epoxy, phenolic, polyimide (or similar materials), or, for high-temperature applications, ceramic cements. Each of these requires different application methods and cure cycles. Cure temperatures vary with the type of adhesive used and can be as high as 320°C (for ceramic cements); cure time can be as short as 5 min and as long as 70 h, depending on adhesive, gage, and installation.

3. *Clamping and curing.* Most installations require clamping of the installed gage, using a metal plate with a strip of nonadhesive plastic between gage and plate and applying pressure to the plate. The pressure must be applied evenly over the entire gage installation, and for installations on curved surfaces the plate may have to be preformed. While clamped, the installation is cured.

4. *Moisture-proofing.* After curing and removal of clamping devices, a moisture-proofing compound should be applied over the installed gage.

The electrical connections between the gage leads and system cabling are frequently made by affixing small insulated terminals to the measured surface near the gage. Such terminals can be welded to the surface, or they can be attached by an adhesive, in which case their installation is part of the gage installation.

### 9.6.2   Strain-Measurement Circuits

The resistance changes in strain gages are converted into voltage changes by passive networks. The voltage is then amplified for signal transmission or display. Excitation is supplied to such networks from a well-regulated power supply. The output of networks used with semiconductor strain gages may be large enough, for some applications, to obviate the need for amplification.

A simple voltage-divider circuit (Figure 9–6) can be used when only the variable component of dynamic strains needs to be measured. The dc voltage drop across the strain gage (shown as a variable resistor with the symbol $\epsilon$) is removed from the output since the coupling capacitor only passes the ac variation in this voltage drop.

Wheatstone-bridge (*strain-gage bridge*) circuits are used for all other

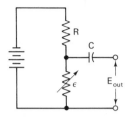

**FIGURE 9–6.** Simple voltage-divider circuit for dynamic-strain transduction.

types of strain measurements (i.e., static and dynamic, or only static, measurements). Excitation may be dc or ac. AC excitation is often preferred since the resulting ac bridge output tends to be simpler to amplify. The bridge network is normally balanced, externally or within the bridge, so that the bridge output is zero when the strain acting on the active gage(s) is zero.

The most commonly used Wheatstone bridge circuit used as a strain-gage bridge is the *four-active-arm* bridge, in which each of the four arms is "active," which means that each arm contains a gage that senses strain. A *one-active-arm bridge* has been used occasionally, often with a *dummy gage* in the adjacent arm of the bridge. The latter does not sense strain, but otherwise has the same characteristics as the strain gage (used in the active arm). Both gages are exposed to the same temperature, and, by virtue of their connection into the bridge, the dummy gage provides temperature compensation for the active gage.

A *two-active-arm* strain-gage bridge (Figure 9–7) is sometimes used in applications such as on bending beams, where one gage will sense tensile strain whereas the gage on the opposite surface senses compression strain when a force is applied to the beam. Tensile strain causes gage resistance to increase, and compression strain causes gage resistance to decrease, as indicated by the upward and downward arrows, respectively, in the schematic. The magnitude of these two strains is usually equal; hence, with the gages connected as shown, the bridge output will be double that obtained from a one-active-arm bridge. If the gages have identical characteristics and see the same temperature, they will also tend to compensate each other for thermal effects on gage resistance and for differential thermal expansion between gage and beam. The same compensation, but with bridge output four times that of a one-active-arm bridge, is provided by a *four-active-arm*

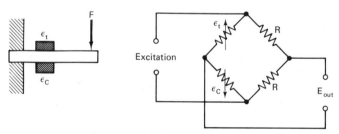

**FIGURE 9–7.** Two-active-arm strain-gage bridge.

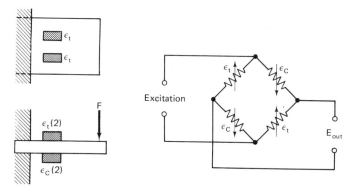

**FIGURE 9–8.** Four-active-arm strain-gage bridge.

bridge (see Figure 9–8), where two adjacent gages sense the tensile strain $\epsilon_t$ and two gages on the opposite surface sense the compression strain $\epsilon_c$ and are connected as shown in the schematic. The four-active-arm bridge is the most frequently used strain-gage bridge.

The use of bridge connection for linearizing the output of two or four matched semiconductor strain gages sensing strain in the same direction (all tensile or all compression) is illustrated in Figure 9–9. The increase and decrease in gage resistance in this case is due to the gage factor of one gage being positive ($P$) whereas the gage factor of the other gage is negative ($N$) as achieved by different doping techniques. The resistive legs ($R$) in the two-active-gage bridges are made up by matched resistors.

Strain-gage bridges used in transducers are almost invariably four-active-arm bridges. The passive networks used in conjunction with strain gages usually contain additional provisions to compensate for bridge unbalances

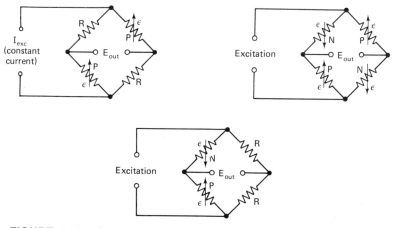

**FIGURE 9–9.** Compensation for nonlinearity of semiconductor gages.

and thermal effects, as illustrated in Figure 9–10. Temperature compensation is provided by temperature-sensitive resistive elements of which one ($R_{ZC}$) is intended to compensate for thermal zero shift, while the others ($R_{SC}$) are intended to compensate for thermal sensitivity shift. They are all in close thermal contact with the strain-gaged mechanical element. Resistor $R_{ZA}$ is selected to provide zero bridge output at zero measurand (*zero balance*). Resistors $R_{SA}$ are selected to adjust the full-scale output of the transducer to a nominal value. These resistors and temperature-sensitive resistive elements are selected and installed after the behavior of the strain-gage bridge in combination with the sensing element is first established, when adherence to close specified tolerances is required. More than one resistor may have to be used for each of the adjusting and compensating elements shown, or one or more of the elements may be omitted and replaced by a jumper when the precompensation tests indicate that there is no need for them.

External *shunt calibration* can be used in measuring systems, using strain-gage bridges either as such or in transducers, for system-calibration purposes. This is achieved by connecting a resistor of appropriately selected

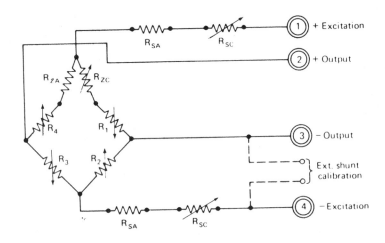

| Symbol | Function |
|--------|----------|
| $R_{ZA}$ | Zero balance adjustment |
| $R_{ZC}$ | Compensation for thermal zero shift |
| $R_{SA}$ | Sensitivity adjustment |
| $R_{SC}$ | Compensation for thermal sensitivity shift |
| $R_1$-$R_4$ | Strain-gage bridge (transduction element) |

**FIGURE 9–10.** Typical compensation and adjustment network used in strain-gage transducers.

value in parallel with one of the bridge arms (typically across the two terminals shown in Figure 9–10). Connecting this resistor in this manner causes a bridge unbalance which can be made to equal 50% (or a different specified percentage) of the nominal full-scale output of the bridge. The shunt calibration resistor is disconnected for the normal operation of the strain-gage bridge.

### 9.6.3 Typical Applications

Strain gages are used primarily for testing and secondarily for (long-term) monitoring. Examples of the latter are strain gages permanently installed or embedded in concrete in buildings, bridges, dams, and roadbeds. In test laboratories strain gages are used widely for materials testing, including tensile and compression testing of test specimens. In larger test facilities they are used to determine stresses in objects ranging from turbines and propellers to full-scale models of aircraft, spacecraft, automotive bodies and chassis, rockets, watercraft, and various types of large equipment and machinery, either while they are operating (often at load extremes) or while vibration is applied to them artificially. Entirely different examples are given by laboratory tests on teeth, bones, and prosthetic devices. Such tests and experiments are generally classified as *experimental stress analysis*. Further, as can be seen in other chapters of this book, strain gages are very popular transduction elements, especially in force, torque, pressure, and acceleration transducers.

The selection of a set of strain gages and the method of installing them depends primarily on the material and geometry of the measured surface, on the expected thermal environment, on the area available for the gage installation, and on the nature of the strain measurement intended (dynamic or static, unidirectional or bidirectional). Other important criteria include accuracy and stability requirements, expected maximum elongation, strain range and expected overrange, gage factor, gage resistance, test duration, and number of loading cycles.

Planning a strain-gage installation includes considerations of all wiring and connections, especially those portions exposed to the test environment, which may include temperature extremes, vibration and shock, and electromagnetic interference. Grounding and shielding require special attention, since signal levels are typically low. Finally, the availability of skilled technicians and proper installation tools should not be overlooked.

# 10

# Force Transducers
# (Load Cells)

## 10.1  BASIC DEFINITIONS

Force transducers are used for force measurements as such, as well as for mass determinations ("weighing"). *Force* is the vector quantity necessary to cause a change in momentum. When an unbalanced force acts on a body, the body accelerates in the direction of that force. The acceleration is directly proportional to the unbalanced force and inversely proportional to the mass of the body. Force is related to mass and acceleration by Newton's second law: $F = ma$. This law is expressed in the *absolute system* of units as

$$F = kma$$

where $F$ = force (acting on mass $m$)
   $m$ = mass
   $a$ = acceleration
   $k$ = proportionality constant depending on units used

*Mass* is the inertial property of a body; it is a measure of the quantity of matter in a body and of the resistance to change in motion of the body. *Weight* is the gravitational force of attraction; on earth, it is the force with which a body is attracted toward the earth (mass times the local acceleration due to gravity).

## 10.2 UNITS OF MEASUREMENT

### 10.2.1 SI Units

The unit of force is the *newton* ($N$); the interrelation of the newton with basic SI units is $N = kg \cdot m/s^2$.

The unit of mass is the *kilogram* ($kg$). This is an exception to the usual assignment of SI units to measurands: the kg is a multiple of the *gram* ($g$), i.e., 1 kg = 1000 g; however, the gram is not designated as an SI unit.

### 10.2.2 Conversion Factors

The following table gives conversion factors for force, mass, and weight:

| To Convert From | to | Multiply by |
|---|---|---|
| dyne | N | $1.000 \times 10^{-5}$ |
| kilogram-force, kilopond | N | 9.807 |
| ounce-force | N | 0.2780 |
| ounce-mass | kg | $2.835 \times 10^{-2}$ |
| poundal | N | 0.138255 |
| pound-force ($lb_f$) | N | 4.448 |
| pound-mass ($lb_m$) | kg | 0.4536 |
| slug | kg | 14.594 |
| ton (long, 2240 $lb_m$) | kg | $1.016 \times 10^3$ |
| ton (short, 2000 $lb_m$) | kg | $9.0718 \times 10^2$ |
| tonne (metric) (t) | kg | $1.000 \times 10^3$ |

## 10.3 SENSING METHODS

Force transducers generally employ sensing elements that convert the applied force into a deformation of an elastic element; this deformation is then converted into an output signal by a transduction element. Two characteristics of elastic deformation are used to sense force: local strain and gross deflection. A maximum level of each occurs at some point (but not necessarily the same point) in a sensing element. The transduction element is then either of the type that responds to strain or of the type that responds to deflection.

Force-sensing elements are made of homogeneity-controlled materials, usually some type of steel, and are manufactured to very close tolerances. Basic design parameters of force-sensing elements include size and shape, material density, modulus of elasticity, sensitivity in terms of local strain and gross deflection, dynamic response, and effects of loading by the trans-

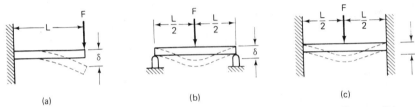

**FIGURE 10–1.** Beam force-sensing elements: (a) cantilever; (b) simply supported; (c) restrained.

ducer on the measured system. Deflection characteristics and measured-system loading are always more or less modified by an attached mass such as a coupling or a portion of the transduction element.

Sensing elements are almost invariably beams, proving rings, columns, or diaphragms.

Typical configurations of *bending beams* are shown in Figure 10–1. The maximum deflection δ of a beam occurs at the point of application of the force, except for the cantilever beam, where it occurs at its free end. The point of maximum strain in a cantilever beam (of constant section in both planes) is at the fixed end. Cantilever beams can also have a "constant-strength" configuration, that is, a triangularly or parabolically tapered shape, narrowest at the point of application of the force, in one plane, and a constant section in the other plane; the strain is then constant along the top or bottom of the beam. In a simply supported beam the point of maximum local strain is at the point of application of the force. In a restrained (fixed-end) beam, maximum strain occurs at both of the fixed ends as well as (in the opposite direction) at the point of application of the force.

In *proving rings* (sometimes called "proving frames" when of the flat configuration), as illustrated in Figure 10–2, maximum deflection occurs at the point of application of the force. Maximum strain also occurs at this

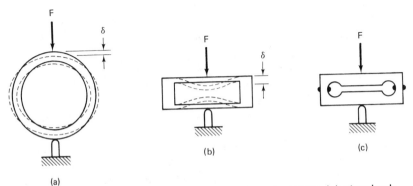

**FIGURE 10–2.** Proving-ring force-sensing elements: (a) standard; (b) flat (proving frame); (c) flat with stress concentration holes (dumbbell-cut proving frame).

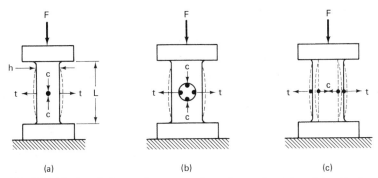

**FIGURE 10-3.** Column force-sensing elements: (a) solid cylinder;
(b) rectangular with stress-concentration hole; (c) hollow cylinder
(strain tube).

point, but strain of almost equal magnitude also occurs at points 90° of arc
in either direction from the point of application of the force, and strain sensing is more convenient at these points. Deflection is emphasized in the standard and flat proving rings. Local strain (indicated by black dots) is emphasized in the stress-concentration configuration. The standard proving ring
is usually of square or rectangular cross section.

   *Column* force-sensing elements (see Figure 10-3) normally have their
point of maximum deflection at their vertical center and their maximum
strain at their lateral center. Their characteristics are given primarily by
their height-to-width ratio $L/h$ and, for the hollow cylinder, by wall thickness.
Compression ($c$) and tension ($t$) forces are usually sensed as strain (hoop
strain in the hollow cylinder), and sometimes as changes in magnetic characteristics or the natural frequency.

   When used as force-sensing elements, *diaphragms* (described in Section 15.3.2) have good deflection characteristics and inherently good lateral
stability. Maximum deflection as well as maximum strain occur at the center
of the diaphragm, and force is always applied at that point.

## 10.4 DESIGN AND OPERATION

### 10.4.1 *Capacitive Force Transducers*

A few force transducer designs employ the capacitive transduction principle
and deflecting sensing elements. The electrodes are typically connected as
the frequency-controlling element in an oscillator so that the transducer provides a frequency-modulated output, which can be sinusoidal or square
wave. At least one design, however, uses the force-balance method and
demodulation to provide a dc output.

## 10.4.2  Reluctive Force Transducers

Reluctive transduction elements, usually LVDT types, but sometimes of the inductance-bridge type, are used in force transducers to convert the deflection of sensing elements into an electrical output. LVDT elements respond to the application of bidirectional axial force loads (tension and compression) with changes in output voltage amplitude as well as with a phase reversal when the force changes direction. The sensing and transduction elements are matched so that the linear range of the LVDT corresponds to the deflection of the sensing element for a specified range.

Figure 10–4 shows typical examples of LVDTs used for force trans-

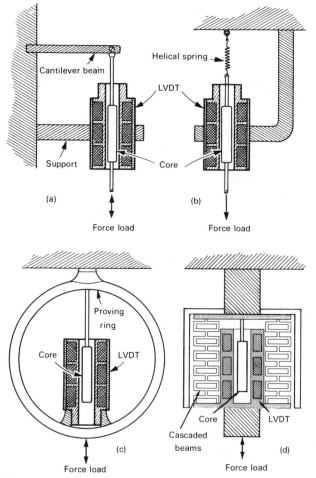

**FIGURE 10–4.** Reluctive force transduction: examples of LVDT combined with an elastic member. (Courtesy of Schaevitz Engineering.)

duction. Only the extension spring (Figure 10–4(b)) is limited to unidirectional (tension) force measurements. The proving ring (Figure 10–4(c)) and multiple-beam configurations, such as the one shown in Figure 10–4(d), are used in commercially available force transducers. Proving rings, however, cause the size of the transducer to be relatively large. They also exhibit poor lateral stability, and that imposes a requirement for accurate alignment of the applied force vector with their vertical centerline.

Reluctive force transducers have also been designed to use the magnetoelasticity in the stress-concentration section of a rectangular column sensing element to change the coupling between two windings, with variations of permeability within the column material due to variations of stress at that location. Essentially all reluctive force transducer designs can be equipped with optional dc-to-ac excitation conversion and ac-to-dc output conversion circuitry.

### 10.4.3 Strain-Gage Force Transducers

Strain-gage force transducers are the most commonly used type of force transducer, to the extent that, to many workers, the term "load cell" usually implies this type of transducer. Many designs exist for the measurement of compression or tension forces or of both ("universal"). Bonded metal-foil, and sometimes metal-wire, gages are most frequently used; semiconductor strain gages are also found in force transducers. Measuring ranges extend from 10 N (1 kg$_f$, 2.2 lb$_f$) to 5 MN (500 t, 1 000 000 lb$_f$). Various beam configurations, flat proving rings (proving frames), and columns are used as sensing elements, the latter typically for high ranges. Many models incorporate overload stops. Threaded male or female end fittings are used; for compression sensing, a load button that has a spherical surface can be threaded on or into an end fitting so that forces need not be applied only axially. Other optional end fittings (sometimes furnished integrally machined or permanently assembled) include rod end bearings and clevis fittings.

Typical designs are illustrated in Figure 10–5. Their primary application is in weighing; however, they are equally usable in other types of force measurement. The general-purpose design (Figure 6–5(a)) is a universal load cell (i.e., it measures tension as well as compression force); in weighing applications, available ranges are between 0 to 23 kg and 0 to 4.5 t. The low-profile design (Figure 6–5(b)) has weighing ranges between 0 to 450 kg and 0 to 4.5 t. The double-ended shear beam design (Figure 6–5(c)) is hermetically sealed and intended for use in truck scales; its weighing ranges are between 0 to 23 t and 0 to 45 t. All three designs are waterproof, and their electrical characteristics are typical for transducers using metal-foil strain-gage bridges: about 3 mV/V full-scale output, with a recommended excitation voltage of 10 V dc, and input as well as output impedances of 350 ohms.

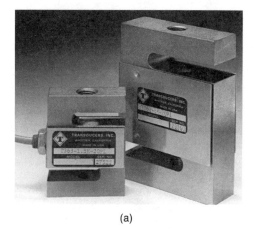

(a)

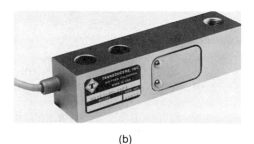

(b)

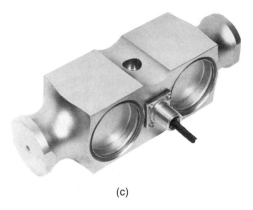

(c)

**FIGURE 10–5.** Typical strain-gage load cells: (a) general-purpose designs; (b) load cell used in platform scales for weighing between 4.5 kN (1000 lb$_f$) to 45 kN (10 000 lb$_f$); (c) load cell used in vehicle scales, weighing range from 225 kN to over 1.1 MN (50 000 lb$_f$ to over 250 000 lb$_f$). (Couresty of Transducers, Inc.)

A force transducer of somewhat different design is shown in Figure 10–6. The sensing element combines shear stress with a cantilever beam of special design. The maximum bending moment, normally largest at the fixed end of a cantilever beam, is reduced to one-half, without affecting the shear force, by having an extra beam rigidly attached to the free end of the cantilever beam. The strain gages are attached to a portion of the beam that has been milled out in an "I" shape. The force is intended to be applied at

**FIGURE 10–6.**   Strain-gage force transducer with shear-force sensing element (cut-section view, with one pair of gages exposed). (Courtesy of AB Bofors, Electronics Div.)

a point directly over the gages. The transducer is normally installed at its fixed end (the left end, not completely shown in the figure), which also contains the electrical connections (receptacle or cable).

Among other versions of strain-gage force transducers are *force washers*, which are annular transducers suitable for placing under a bolt head or a nut and typically employing a hollow-column sensing element, often just one ($\frac{1}{4}$ bridge) or two ($\frac{1}{2}$ bridge) strain gages. A *load beam* is a suitably machined (bending) beam to which strain gages (usually four) are attached and encapsulated, often surrounded by a housing. The load beam has provisions for mounting at both of its ends.

### 10.4.4   Piezoelectric Force Transducers

Although some measure of near-static response can be obtained from piezoelectric force transducers when they are used with charge amplifiers, their application is really in the measurement of rapidly fluctuating forces (dynamic force measurement), including impact forces and the measured objects' response to them. These transducers respond only to compression forces; however, they can be mechanically preloaded so that a compression force is continuously exerted on the crystal, constituting a "static level." They can then respond to bidirectionally varying forces.

Figure 10–7 illustrates a piezoelectric force transducer. The applied force is exerted through a preloading member (as indicated by threads) on a stack of quartz crystals separated by thin electrodes. This design includes circuitry which converts the charge, generated by the crystals across a high

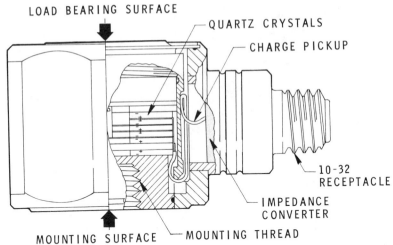

LOAD BEARING SURFACE

QUARTZ CRYSTALS

CHARGE PICKUP

10-32 RECEPTACLE

IMPEDANCE CONVERTER

MOUNTING SURFACE — MOUNTING THREAD

**FIGURE 10–7.** Piezoelectric force transducer with integral imped-ance-converter circuitry. (Courtesy of Kistler Instrument Corp.)

impedance, into a voltage at low output impedance. Preloading results in some models have a higher range in compression than in tension. This is also true for most designs of piezoelectric *force links* (see Figure 10–8), which sandwich a quartz force washer (load washer) between two end nuts. A beryllium–copper stud fastens the end nuts and preloads the force washer for tension measurements. The end nuts are threaded (female thread) to accept external fittings as required by their installation. Ranges extend from 4.5 kN (1000 lb$_f$) to 120 kN (25 000 lb$_f$) in tension, with compression ranges typically 1.5 times those values.

Piezoelectric *force washers* are force transducers of annular shape which contain one or more annular piezoelectric crystals sandwiched be-

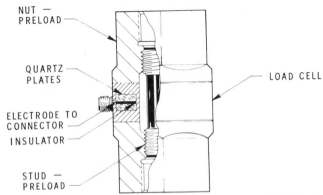

NUT — PRELOAD

QUARTZ PLATES

ELECTRODE TO CONNECTOR

INSULATOR

STUD — PRELOAD

LOAD CELL

**FIGURE 10–8.** Piezoelectric force link. (Courtesy of Kistler Instru-ment Corp.)

tween metallic annuli. In piezoelectric force transducers the combination of metal rings or disks and the crystals themselves form the elastic member which responds to applied forces. Most force washers (load washers, load rings) respond only to compression forces. However, they can be preloaded in their installation to also measure tension.

Quartz, in the crystalline form used for piezoelectric elements, is anisotropic (i.e., its properties are different along different directions). Thus, depending on the axis of the crystal along which the slices are cut, disks are obtained which are sensitive only to compression forces (longitudinal piezoelectric effect) or to shear forces in one specific direction (shear piezoelectric effect). These characteristics enable such crystals to be used in *multicomponent force transducers* (see Figure 10–9). With reference to the coordinate system shown in Figure 10–9(a), two rings containing disk-shaped crystals can be combined to measure compression ($F_z$) and torque ($M_z$) simultaneously (Figure 10–9(b) and (c)). One ring holds crystals so cut as to utilize the longitudinal effect (compression measurement), while the other ring (bottom) holds shear-effect disks which are so oriented that they respond to torque.

Compression-sensitive elements can also be combined with shear-sensitive elements in which the disks are so oriented that they respond to $F_x$ or to $F_y$ (two-axis force transducers). Instead of the disk-ring configurations, whole quartz rings can be used in multicomponent force transducers, usually in pairs to obtain greater sensitivity. For example, three pairs of whole quartz rings are used in a three-component (three-axis) force transducer.

Unless piezoelectric force transducers incorporate integral impedance-conversion (charge-to-voltage conversion) circuitry, they are usually connected to a charge amplifier, i.e., an amplifier whose input stage is a dc amplifier, with high input impedance and capacitive negative feedback, which converts the changes in charge (transducer output) into voltage changes. The input stage is then usually followed by an operational amplifier, often one having adjustable negative feedback that permits adjustment of the amplification to the transducer's sensitivity. Voltage amplifiers, with relatively low input impedance, can be used to further amplify the output of transducers with integral impedance converters. Such amplifiers can either be rack mounted or otherwise installed in the data-gathering area, or remotely mounted near the transducer, for unattended operation. In the latter case they must often be packaged to withstand the same environment as the transducer, including impact shocks.

### 10.4.5   Vibrating-Element Force Transducers

Some force transducers have been designed using a wire between two points of a deflecting sensing element. The wire is always in some degree of tension, is located in a permanent magnetic field, and is caused to vibrate at its

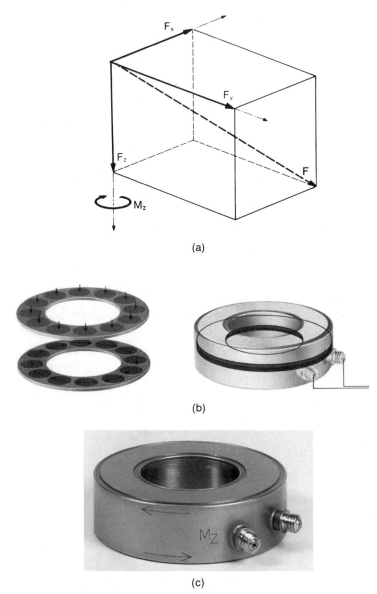

**FIGURE 10–9.** Multicomponent force transduction with example of two-component transducer: (a) positive Cartesian coordinate system as used for referencing piezoelectric multicomponent transducers; (b) compression-sensitive (top) and shear-sensitive (bottom) quartz elements in concentric rings arranged for simultaneous measurements of $F_z$ and $M_z$; (c) two-component ($F_z$ and $M_z$) force transducer. (Courtesy of Kistler Instrumente A.G.)

resonant frequency as the frequency-controlling element in a feedback oscillator. The transducer, therefore, has a frequency-modulated output. A second wire, not affected by sensing-element deflection, can be added. The second wire is adjusted to vibrate at that frequency at which the first wire vibrates when no force is applied. As force is applied, the frequency difference between the two wires (beat frequency) will increase with increasing force. A transducer using the beat-frequency design provides a frequency output rather than a frequency-modulated output. A frequency output is relatively easy to display on a frequency counter either as such, or converted, on the basis of the calibration curve of the transducer, into units of force or mass.

## 10.5 TRANSDUCER CHARACTERISTICS

The first characteristic to determine for force transducers is whether compression, or tension, or both are to be measured, together with the range (measuring range) and required overload capability (overrange). The range of force transducers (load cells) is sometimes referred to as "capacity."

Force connections and mountings should be defined, in detail, as to dimensions, threads, and, for many applications, material of the force connection. When compression buttons (load buttons) are used, their curvature and hardness should be known. The maximum allowable angular misalignment of axial force applied to the transducer and, preferably, also the effects on performance of varying amounts of off-axis force application should be specified.

Overload ratings for force transducers are specified for two conditions: the *safe overload rating* is the overrange force that can be applied to the transducer without causing subsequent performance degradation; the *ultimate overload rating* is the (maximum) force that can be applied without causing structural failure of the transducer (or portions of it). *Side loads* can be specified in the same manner.

Besides the usually specified electrical characteristics (which depend largely on the type of transduction used as well as on the use of any additional conditioning circuitry within the transducer) and performance characteristics, *creep* and *creep recovery* are sometimes required to be known. *Creep at load* is the change in output occurring with time under rated load with all other conditions remaining constant. *Creep recovery* is the change in no-load (zero-measurand) output occurring with time after removal of rated load at which creep at load was determined.

The normally specified dynamic characteristics are influenced by the amount of maximum sensing element deflection, at the upper range limit (*deflection factor*). The deflection factor is sometimes specified to facilitate determining relative dynamic characteristics. (A low deflection factor is typical for fast response.)

Among environmental characteristics, one that must be considered for force transducers in some applications (besides the usually specified environmental characteristics) is *ambient-pressure error.* Any case deformations due to changes in ambient pressure can cause sensing element deflections that are not indicative of force. Ambient-pressure error can become quite apparent, particularly in hermetically sealed transducers when they are exposed to pressures approaching a vacuum or when they are exposed to high ambient pressures (e.g., in underwater applications).

## 10.6   APPLICATION NOTES

Force transducers are used in two important areas: force measurement as such (measurement of quasi-static or dynamic forces) and weighing (mass determination). Force measurements are made for testing purposes as well as for operational measurement and control. Testing is performed for research, development, and verification, in general, and for quality control (product testing). Examples of general test applications are tensile-strength tests; determinations of coefficient of friction; weld strength testing; thrust measurement in aeronautical and rocket engine test stands; and measurements of the operating force required of vehicle controls, of control levers in various applications, of the gripping force of tools, and even of small forces such as depression of keyboard keys and push buttons and tone arms on phonograph records. Examples of product testing are tests for the peel strength of adhesives, bond strength of printed-circuit laminates, pullout force of wires in connectors (and of plugs inserted into receptacles), and crush resistance of containers.

Cable tension measurements use force transducers incorporated in rollers, pulleys, hooks, lifting eyes, threaded studs, and cable sockets. Among many examples of cable tension measurements are those made in cranes, towing (including underwater), buoy moorings, hoists and winches, and structural anchoring. Monitoring and control of industrial processes include such measurements as mill roll pressure and web, wire, tape, thread, and filament tension. In cable tension and industrial force and tension measurements, alarm signals can be produced when the transducer output signal reaches a preselected level. Also, in industrial processes the transducer is often connected as an element in a closed-loop control circuit.

Most of the preceding types of measurement involve slowly varying (quasi-static) forces. Dynamic (fast-response) force measurements are typically made in machine tools such as lathes, drills, and milling and grinding machines. Multicomponent dynamic force measurements (or combined force-torque measurements) are often made to measure cutting, drilling, thread-cutting, milling, countersinking, and reaming forces. Such measurements are also made in an entirely different field: biomechanical analyses using platforms where leg forces, motor response during running or walking, forces exerted during certain work-associated movements, and even posture control can be studied. One of the applications of such study results in the fitting of prosthetic devices. In many research and experimental force-testing applications, dynamic forces are measured in addition to (or sometimes instead of) quasi-static force measurements.

The field of weighing offers the largest single application to the manufacturers of force transducers (and in this field they are almost invariably called *load cells*). Until not too long ago, load cells were widely used in weighing heavy objects or materials: bulk material, tanks, trucks and trailers, and freight cars. The economic importance of accurate weighing of loaded or delivered bulk materials (in tanks or other containers, or fed into hoppers or bins) is quite obvious, as are highway-associated scales used as a basis for paying fees. The next development, of equal economic importance, was the mass production of similar weighing machines (scales) for much lighter materials, and we see these now in essentially all supermarkets and other retail outlets. Finally, more and more households (and most medical offices) now have a load-cell-based electronic scale (generally more accurate and reliable than the old mechanical scales). The household and supermarket weighing machines even provide a digital readout. Digital conversion of load cell outputs (more than one load cell is often used for determining mass—especially large masses) not only makes an easy-to-read display, but can also be fed into a computer which can combine and average the outputs, or can combine them with other information that aids in dealing with such essentials as inventory control and billing. Weighing need not be done by platforms (the large platforms used for trailers and freight cars are known as *weighbridge*), but can be derived instead from cable tension measurements in cranes or hoists. The *tare* weight (mass of the empty container or scale) must always be subtracted from the total weight. (Most digital weighing machines will do this automatically, but, when important, still provide a readout.)

Finally, we briefly mention *flexures*. In order to prevent measurement errors due to nonaxial loading of force transducers, mainly errors due to side loads and torsional loads, precisely machined flexures can be installed between the load cell and the load. Such flexural pivots are available for force-measuring ranges up to 23 MN ($5 \times 10^6$ lb$_f$).

# 11

# *Torque Transducers*

## 11.1 BASIC DEFINITIONS

*Torque* is the moment of force. That is, it is the product of the force and the perpendicular distance from the axis of rotation to the line of action of the force. (*Note:* This distance is referred to as *lever arm*.) Torque is defined as

$$T = Fl = I\alpha$$

where $T$ = torque
$\phantom{where }F$ = force
$\phantom{where }l$ = length of lever arm
$\phantom{where }I$ = moment of inertia
$\phantom{where }\alpha$ = angular acceleration

The *moment of inertia* of a body is the body's resistance to angular acceleration. It is defined as

$$I = \sum mr^2 = \frac{T}{\alpha}$$

where $I$ = moment of inertia
$\phantom{where }m$ = mass (of one particle of the body)
$\phantom{where }r$ = crank length (between the particle and the axis of rotation)

The moment of inertia of a solid cylindrical shaft of mass $M$ and radius $r$ about its own axis is $I = \frac{1}{2}Mr^2$.

## 11.2   UNITS OF MEASUREMENT

### 11.2.1   SI Units

The SI unit for torque is the *newton-meter* $(N \cdot m)$. The SI unit for mass moment of inertia is the *kilogram-meter squared* $(kg \cdot m^2)$.

### 11.2.2   Conversion Factors

See Section 10.2.2 for conversions to N; see Section 5.2.4 for conversions to m.

For mass moment of inertia:

to convert pound-feet squared to $kg \cdot m^2$, multiply by $4.214 \times 10^{-2}$;

to convert pound-inches squared to $kg \cdot m^2$, multiply by $2.9264 \times 10^{-5}$;

to convert slug-feet squared to $kg \cdot m^2$, multiply by $1.3558$.

## 11.3   SENSING METHODS

Torque sensing elements are generally *in-line* elements inserted between a driving shaft (power source) and a driven shaft (power sink). They are commonly referred to as *torsion bars* (or *torque bars*). When torque, the product of force $(F)$ and moment arm $(l)$, is applied to a cylindrical shaft (see Figure 11–1) at one end while the other end of the shaft is held fixed, the shaft will be subjected to torsional twisting. If a line were scribed on the surface of the shaft parallel to the axis of rotation, this line would become a (very small) portion of a helix. The deflection angle $\theta$ at the input side of the shaft is proportional to the torque and shaft length $L$ and inversely proportional

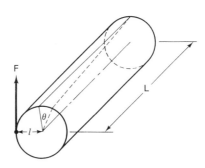

**FIGURE   11–1.** Shaft   twisting due to torque.

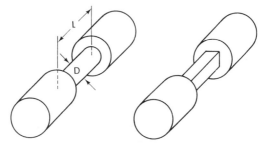

**FIGURE 11–2.** Round and square torsion bars.

to (theoretically) the fourth power of the diameter $D$, or $2l$, as well as to the modulus of shaft rigidity. The shaft is also subjected to shear stress. Hence, in-line torsion-sensing elements exhibit angular displacement as well as strain, both of which can be converted into output signals by appropriate transduction methods.

Torsion bars typically incorporate a necked-down section, either round or square in cross section (see Figure 11–2). The square configuration is favored in strain-gage torque transducers. It makes the gages easier to mount, enables solder joints between the gage and lead wires to be located at the corners of the shaft (where torsion strains are zero), provides a relatively higher resonant frequency, and is less affected by side loads.

*Reaction torque* is the torque required to prevent the stator of a rotary device from turning with the rotor (e.g., the housing of a generator from rotating with the driven rotor). It can be measured by mounting the housing on bearings so that it is free to rotate and then attaching a lever arm (*reaction arm*) to the housing and connecting a force transducer between the end of that arm and mechanical "ground," the surface to which the outer portion of the bearing is attached (see Figure 11–3). One or more force sensors can be used in such *reaction-torque dynamometers*. Reaction-torque sensors can also have the sensing and transduction elements incorporated within their case, so that they measure torque between a "driven" and a "fixed" mounting flange.

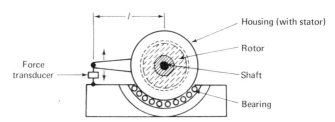

**FIGURE 11–3.** Reaction-torque sensing.

## 11.4 DESIGN AND OPERATION

Most torque transducers use strain-gage transduction, some use reluctive transduction, and some of relatively recently developed design use the difference in angular displacement between the two ends of a torsion bar to obtain either a phase-difference measurement (photoelectrically, electromagnetically, or inductively) or a variable-illumination measurement.

### 11.4.1 Reluctive Torque Transducers

Torque measurements can be made using the angular deflection of a torsion bar to actuate a linear variable differential transformer (LVDT) through a stiff arm attached radially to the torsion bar, or through a cantilever beam attached similarly. Another method uses a rotary variable differential transformer (RVDT) to sense the angular displacement of the torsion bar or of two facing disks mutually spring loaded in the plane of rotation. However, better results have been obtained with designs using changes in coupling introduced either in a torsion bar or in a member linked to it, mainly because this obviates the need for slip rings since the primary and secondary windings can then be located on the stationary rather than the rotating member of the device.

One such design is the *torsional variable differential transformer (TVDT)* illustrated in Figure 11–4. The torsion bar (or shaft) is made of nonmagnetic material, but has attached to it three tubular sections (A, B, C) of a ferromagnetic material, with gaps between the center section (B) and the two end sections (A and C) at 45° to the shaft axis. When torque is applied to the shaft ends, one gap will tend to widen and the other gap will

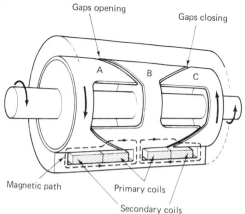

**FIGURE 11–4.** Torsional variable differential transformers. (Courtesy of Lebow Products, Eaton Corp.)

tend to close. The reluctance in the gaps will increase and decrease correspondingly and change the coupling between the primary and secondary windings, so that one secondary will show a voltage increase whereas the other secondary will show a voltage decrease. Associated signal-conditioning circuitry then uses these outputs to produce a signal whose magnitude is proportional to the torque and whose polarity is indicative of the direction of the torque.

Several torque transducer designs use the variations in permeability, due to stress, in a ferromagnetic shaft. The principle underlying this form of operation is illustrated in Figure 11–5. An applied torque creates internal stresses which alter the permeability of the material, as shown, and thus change the coupling between a primary winding $P$ and two secondary windings $S_1$ and $S_2$ in an X-shaped coil assembly or an assembly of three multipole-coil rings. The production of output signals from the secondaries, as well as the conversion of the signals into a signal indicative of torque magnitude as well as direction, is similar to that described previously for the TVDT.

### 11.4.2  Photoelectric Torque Transducers

In photoelectric torque transducers, light beams, code patterns, and light sensors are used to convert the differential angular displacement between the two ends of a torsion bar, due to applied torque, into an output signal.

A design using the phase-displacement method is illustrated in Figure 11–6. Each end of the torsion bar (torque bar) carries a code pattern made of light-reflecting strips. The patterns are illuminated by collimated light beams, and the reflected light is sensed by a light sensor (photocell). The output of each light sensor, then, is a pulse train. With differential angular deflection between the two coded ends of the torsion bar, due to applied torque, the two pulse trains show a difference of phase (synchronism) rela-

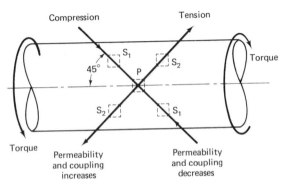

**FIGURE 11–5.** Operating principle of variable-permeability reluctive torque transducer.

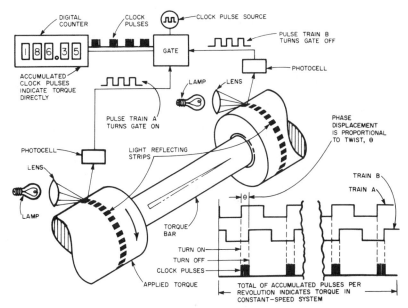

**FIGURE 11–6.** Photoelectric torque transducer using phase-displacement method. (Courtesy of Vibrac Corp.)

tive to each other. This phase difference is proportional to torque and can be converted into an analog output. However, the nature of the signals lends itself to providing a digital output, as illustrated. The leading edges of one train of pulses gate oscillator clock signals into a digital counter, while the leading edges of the other pulse train gate the clock pulses off. The counter's accumulated total in a given sampling interval, assuming constant shaft speed, is then proportional to phase (time) displacement between the two pulse trains and, hence, proportional to torque. More complex provisions are required when the shaft speed is not constant.

Photoelectric torque transducers using the variable-illumination method (see Figure 11–7) provide an analog output proportional to torque without the use of conditioning or digital circuitry. Two identical code disks, of the same type as used in incremental angular encoders, are attached to the two ends of a torsion bar. A light beam passes through both disks and thence to a light sensor. Relative angular displacement between the two disks, due to torque, changes the amount of "window area," the total area provided by transparent segments of both disks, and hence the amount of light incident upon the light sensor. Two light-source and sensor combinations are used to double the effect.

The output of the two parallel-connected light sensors is adequate for direct display on commercially available microammeters. Figure 11–8 shows a torque transducer using this principle, together with its major elements.

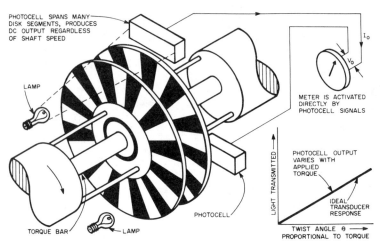

**FIGURE 11–7.** Photoelectric torque transducer using variable-illumination method. (Courtesy of Vibrac Corp.)

The two disks are initially positioned so that they provide a 50% overlap between transparent segments at zero applied torque. Clockwise torque will then increase the window area and resulting output current, whereas counterclockwise torque reduces the window area and output current. This enables the output signal to be also indicative of torque direction. The light

**FIGURE 11–8.** Variable-illumination torque transducer and its major internal elements. (Courtesy of Vibrac Corp.)

sensors span several disk segments; this minimizes flicker and assures a very low harmonic content in the output current regardless of shaft speed. A bias source with an adjustable resistor can be used to neutralize the zero-torque output current and then provide a bipolar output current that can be displayed (e.g., on a center-zero meter).

### 11.4.3 Electromagnetic and Inductive Phase-Displacement Torque Transducers

The operating principle of electromagnetic and inductive phase-displacement torque transducers is similar to that described for photoelectric torque transducers employing the phase-displacement method. However, the optical code pattern is replaced either by a magnetic code pattern or by gearlike teeth. A sensing coil assembly is placed in close proximity to each of the two torsion-bar shaft ends so equipped. If the sensing coil is of the electromagnetic type, the teeth must be made of ferromagnetic materials and the transducer requires some finite minimum shaft speed for proper operation. When the sensing coil assemblies are of the inductive, eddy-current type, the teeth can be of any metallic material. As for the photoelectric transducer described, each coil assembly produces a pulse train, and the phase difference between the two pulse trains is proportional to torque.

### 11.4.4 Strain-Gage Torque Transducers

Transducers using strain gages to respond to shear stresses in torsion bars, due to applied torque, are widely used to measure a wide range of torques in numerous applications. Round torsion bars and, more frequently, square torsion bars, or their design derivatives, have strain gages, typically metal-foil gages, bonded to them (see Figure 11–9). The solid round shaft is still used, at times, for measuring ranges above about 56 N·m (500 lb$_f$·in). Hollow circular shafts provide more bending strength. Flats, machined into the hollow shafts, facilitate the proper mounting of strain gages. The cruciform torsion bar tends to produce relatively high amounts of strain at low torques. The hollow cruciform configuration lends itself particularly well to measurements of small torques. Each of the four bars in such a *torsion frame* is subjected to torsion as well as bending in response to an applied torque, providing a relatively high torsional sensitivity together with good bending strength. The solid square torsion bar is used in high-range torque sensors; it offers ease of gage attachment and increased bending strength compared to round shafts.

Semiconductor strain gages are used in some torque transducer designs. Regardless of the type of gage used, the four strain gages are always connected as a four-active-arm bridge. The four terminals of this bridge circuit can be connected to slip rings. Brushes, typically made of silver

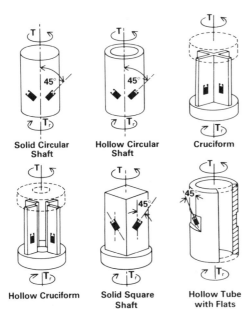

Solid Circular Shaft    Hollow Circular Shaft    Cruciform

Hollow Cruciform    Solid Square Shaft    Hollow Tube with Flats

**FIGURE 11–9.** Torsion bar configurations for strain-gage torque transducers. (Courtesy of Lebow Products, Eaton Corp.)

graphite, are in contact with the slip rings and provide the electrical connections between the rotating strain-gage bridge (which may include bridge-adjustment and temperature-compensation resistors) and the housing of the torque transducer. A design employing slip rings is illustrated in Figure 11–10. This particular design includes optional features such as a toothed-rotor electromagnetic tachometer and a device that permits lifting the brushes off the slip rings when no measurements are being made, to extend the operating life of both sets of contacts. The contact force of the brushes must be large enough to avoid noise due to brush bounce and variations in contact resistance, but small enough to avoid introducing measuring errors in the torque measurement. Slip rings and brushes require periodic cleaning as well as brush replacement.

The need for brushes and slip rings is obviated when a rotary transformer is used for noncontacting coupling of excitation and output to and from the rotating strain-gage bridge (see Figure 11–11). The carrier system electronics contains an oscillator which provides ac excitation, at 3 kHz or higher, to the excitation primary in the stationary component, from which it is coupled to the excitation winding of the rotary component (Figure 11–11(a)). The output of the strain-gage bridge is similarly coupled to an amplifier-demodulator in the electronics module. The transformers are a pair of concentrically wound coils, with one coil rotating within (or beside) the stationary coil (Figure 11–11(b)). High-permeability cores aid flux concentration and improve the coupling between coils. Figure 11–11(c) shows the layout of the major components in a rotary-transformer type of strain-gage torque sensor.

An alternative method of noncontacting coupling (Figure 11–12) uses a single pair of loops for inductively coupling excitation to the transducer and coupling an output-signal-modulated RF signal back into the stationary electronics. The rotating electronics module is in the form of a split collar which is clamped to the shaft and to which the strain-gage bridge is connected.

*Reaction-torque* transducers measure the torque required to restrain the housing (or stator) of a rotary device from turning with its rotary member. This torque is detected by a force sensor attached between mechanical "ground" (e.g., the top of the test bench) and a "moment arm" leading to that portion of the reaction-torque sensor mounted to the device being tested. The sensor can be provided with a coupling shaft, or it can have a hollow configuration (Figure 11–13). The housing of the latter is mounted to mechanical "ground," and the annular sensing flange is bolted to the housing

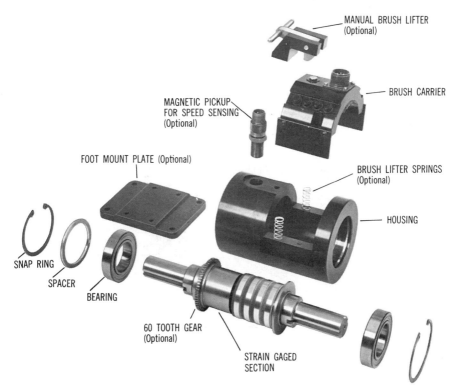

**FIGURE 11–10.** Strain-gage torque transducer with slip rings. (Courtesy of Lebow Products, Eaton Corp.)

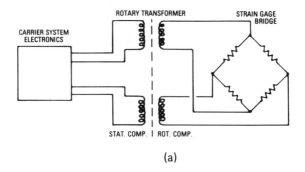

(a)

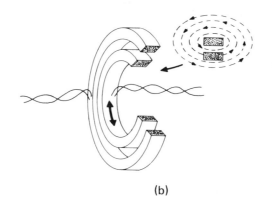

(b)

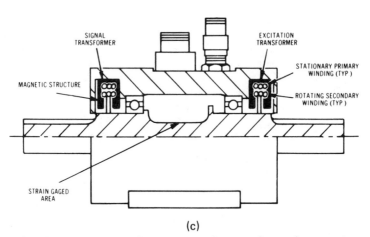

(c)

**FIGURE 11–11.** Use of rotary transformers in strain-gage torque transducers: (a) functional schematic diagram; (b) coupling between a stationary and a rotating coil; (c) internal construction of transducer. (Courtesy of Lebow Products, Eaton Corp.)

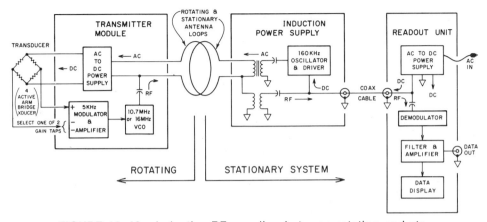

**FIGURE 11–12.** Inductive RF coupling between rotating and stationary portions of a torque measuring system; rotating loop is contained, together with rotating electronics, in a collar of insulating material, clamped to strain-gaged shaft (torsion bar). (Courtesy of ACUREX Autodata.)

of the device under test whose drive shaft extends through the torque sensor. The moment arm–force sensor combination is typically achieved by torque-sensitive flexures whose strain is transduced by strain gages. The angular displacement of the coupling shaft or sensing flange of a reaction-torque transducer is on the order of 0.2 to 1.0° for full-range reaction torque.

**FIGURE 11–13.** Reaction-torque sensor (with hollow configuration). (Courtesy of Lebow Products, Eaton Corp.)

## 11.5   TRANSDUCER CHARACTERISTICS

Among the mechanical design characteristics of particular importance to torque transducers is the type of coupling between the driving, transducer, and driven shafts. Rigid couplings include primarily machined keyways with a connecting key that is retained by a housing. Flexible couplings are used when a power transfer system can induce high bending loads. For transducers using bearings, bearing life, load, and lubrication require attention. When brushes and slip rings are used, brush life and slip-ring contamination have to be watched. Other mechanical design characteristics which are often required to be known (or specified) are the moment of inertia, torsional stiffness, shaft runout, and axial end float.

Torque transducer performance characteristics are generally the same as those specified for electromechanical types of transducers. However, there are certain dynamic characteristics that need attention, viz., the *critical speed* of the shaft of the sensor (at which the shaft becomes dynamically unstable and irreversible damage can result) and the sensor's response to *torsional vibration*, the oscillatory or fluctuating components of the torque being measured. The *maximum speed* of a torque transducer is always given as well below the critical speed. Torque measuring ranges extend from below 0.1 N·m (about 10 $oz_f$-in) to over 340 kN·m ($3 \times 10^6$ $lb_f$-in), full scale.

## 11.6   APPLICATION NOTES

Applications of torque transducers are primarily in monitoring, testing, and evaluating rotating components. In-line torque transducers are installed, for example, in the main drive shafts of ships for continuous torque monitoring, from which propulsive power can be determined. Examples of rotating devices that are tested and evaluated using torque transducers are motors, pumps, blowers, fans, clutches, brakes, bearings, gearboxes, speedometer cables, joints and couplings, aircraft and ship propellers, starters for automotive and aircraft engines, and machine tools of all types. Figure 11–14 illustrates some of the typical applications of torque and reaction-torque transducers.

In the role of *dynamometers* in conjunction with a mechanical load, torque transducers are widely used in testing of engines and other such primary power sources for mechanical power measurements. Dynamometers are classified as *absorbing dynamometers* when they provide an opposing load to a test specimen that provides output power, and as *motoring dynamometers* when they provide an output power to a test specimen that acts as the opposing load.

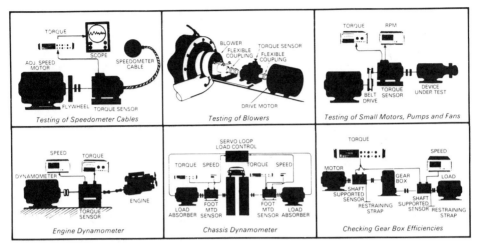

(a)

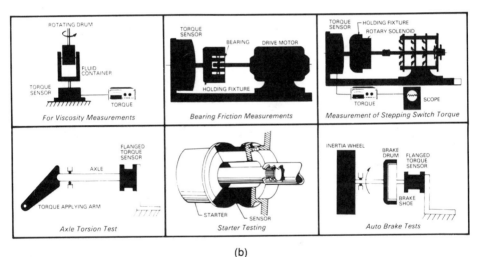

(b)

**FIGURE 11–14.** Selected typical applications of torque sensors: (a) some uses of in-line rotating-shaft torque sensors; (b) examples of uses of reaction-torque sensors. (Courtesy of Lebow Products, Eaton Corp.)

At least one application of torque transducers not related to rotating machinery is worth noting: the measurement of torques applied to fasteners while they are being tightened, usually with a socket wrench. The measurement is made by a transducer of special design mounted between the fastener and the wrench.

# 12

## Flowmeters

### 12.1  BASIC DEFINITIONS

*Flow* is the motion of a fluid.

*Flow rate* is the time rate of motion of a fluid, expressed as fluid quantity per unit time; the quantity is fluid mass for *mass flow rate* and fluid volume for *volumetric flow rate*.

*Total flow* is the flow rate integrated over a specific time interval.

A *flowmeter* is a flow-rate transducer.

*Laminar flow* (streamline flow) is the motion of fluid particles along lines that run parallel to the local direction of flow; it can be represented as layers of fluid sliding steadily over one another.

*Turbulent flow* is the motion of a fluid whose velocity at a point of observation fluctuates with time in a random manner.

The *Reynolds number* is a dimensionless number used to express the fluidity of a fluid in motion; it is given by the relationship $N_R = VD\rho/\mu$, where $V$ is the mean flow velocity, $D$ is the internal diameter of the pipe through which the fluid is flowing, $\rho$ is the density and $\mu$ the viscosity of the fluid, and $N_R$ is the Reynolds number. Reynolds numbers below about 2000 indicate laminar flow, whereas those above 4000 generally indicate turbulent flow.

(See also Section 15.1 for impact pressure and stagnation pressure, which are related to flow.)

## 12.2 UNITS OF MEASUREMENT

Flow rate is expressed either in units of volume per unit time (volumetric flow rate) or in units of mass per unit time (mass flow rate). The SI units are the $m^3/s$ for volumetric flow rate and the $kg/s$ for mass flow rate. The mass flow rate of gases is often expressed as equivalent volume flow at "standard conditions," i.e., 20°C and 1 standard atmosphere (see Section 15.2). When this is the case, the units are called standard $m^3/s$ (or standard other units that can be converted to m³/s).

Conversion factors

| To Convert From | to | Multiply by |
|---|---|---|
| foot³/minute | m³/s | $4.7195 \times 10^{-4}$ |
| inch³/minute | m³/s | $2.732 \times 10^{-7}$ |
| yard³/minute | m³/s | $1.2743 \times 10^{-2}$ |
| gallon/minute | m³/s | $6.309 \times 10^{-5}$ |
| liter/second | m³/s | $10^{-3}$ |
| pound-mass/second | kg/s | 0.4536 |
| pound-mass/minute | kg/s | $7.56 \times 10^{-3}$ |

## 12.3 SENSING METHODS

### 12.3.1 Differential-Pressure Flow Sensing

Flow is very commonly measured by forcing the fluid to flow through a restriction in a pipe so that its velocity changes at that point and a pressure difference proportional to flow is created. The fluid can also be caused to flow through curved pipe sections so that its velocity profile changes in the curvature region, causing a pressure differential to occur at a point in this region. Since the product of the cross-sectional area of the pipe and the flow velocity of the fluid remains the same when the fluid flows through a restriction, the velocity increases where the area is reduced because of the restriction and the pressure drops accordingly. The resulting pressure difference is then measured by a differential-pressure transducer (see Figure 12-1). Volumetric flow rate is roughly proportional to the square root of the differential pressure. The point at which the pressure is minimum in a pipe equipped with a restriction is known as the *vena contracta*.

Sensing elements used to measure flow by the differential-pressure ("pressure head") method are illustrated in Figure 12-2. The *orifice plate* (a) is typically inserted between two flanges. The opening in the plate can be circular and either concentric or eccentric relative to the centerline of the pipe, or it can be segmented (i.e., other than a full circle). The opening

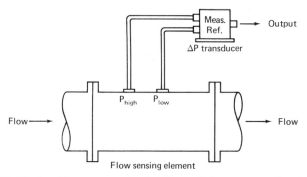

**FIGURE 12-1.** Measurement of differential pressure due to flow rate.

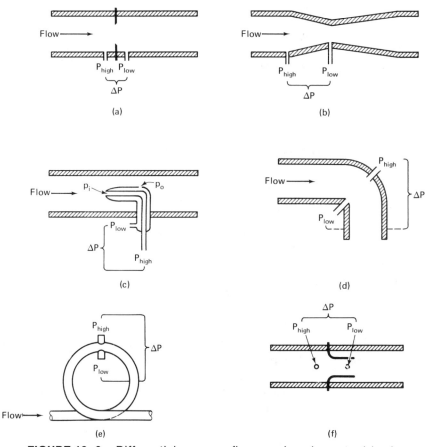

**FIGURE 12-2.** Differential-pressure flow-sensing elements: (a) orifice plate; (b) Venturi tube; (c) Pitot tube; (d) centrifugal section (elbow); (e) centrifugal section (loop); (f) nozzle.

can be square edged (i.e., drilled straight through the plate) or conical at the outlet side ("sharp edged"), the inlet side, or both sides. Pressure ports ("pressure taps") can be located in the flanges ("flange taps"), upstream of and directly at the vena contracta ("vena contracta taps"), or well upstream and downstream of the orifice ("pipe taps").

The *Venturi tube* (b), whose external configuration is often barrel shaped rather than following its internal configuration, usually has its pressure ports at the throat and just upstream of the entrance cone. The *nozzle* ("flow nozzle") (f) is typically clamped between two flanges. Its throat is shaped so as to cause the flow discharge to be parallel with the centerline of the downstream section of the pipe. In the *nozzle-Venturi* the cross section immediately downstream of the nozzle is configured as the exit cone of a Venturi tube.

The *Pitot-static tube* (c) is a right-angle-bent tube inserted into the fluid stream so that the stagnation pressure (the sum of the static and impact pressures, the latter denoted by $P_i$) is sensed by a port in its upstream-pointing tip, and the static pressure $P_o$ is sensed by a separately tubulated port normal to the flow direction. In the classical or basic *Pitot-tube* configuration the static-pressure port is located in the pipe wall instead, facing the point at which the stagnation-pressure port of the bent tube, in the flow stream, is located.

In the *centrifugal sections,* either the elbow (d) or the loop (e), the centrifugal force caused by the change in flow direction causes a pressure gradient along the radius such that the pressure is higher where the radius of curvature is larger.

### 12.3.2 Mechanical Flow Sensing

Mechanical elements have been designed to respond to fluid flow by displacing or deflecting, or by rotating at a speed proportional to flow rate. *Variable-area* flowmeters use a float in a vertical, tapered section of pipe; the float displaces upward with increasing flow rate so that the weight of the float balances the upward-acting force on it (Figure 12–3(a)). The *spring-restrained plug,* in conjunction with an appropriately shaped restriction, also operates on the variable-area principle (Figure 12–3(b)). Flow acting on a spring-loaded *hinged vane* (Figure 12–3(c)) or on a *cantilever beam* to which a disk-shaped body is attached (Figure 12–3(d)) respond to flow-rate changes with changes in deflection and strain in the beam, respectively. The latter principle, where the flow acts with impact on a disk ("target") at the end of a beam, is also used in *target flowmeters* operating in a force-balance mode. A related flow element is the *drag body,* supported in a pipe section so that its displacement due to flow causes deflection (or strain) in its supports or other mechanically linked members.

The mechanical flow-sensing elements just described are not used as

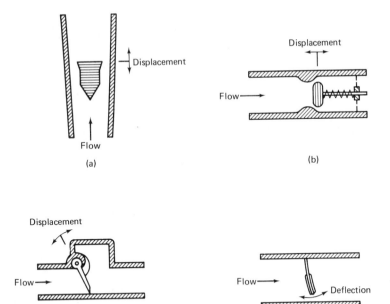

**FIGURE 12–3.** Mechanical-displacement flow-sensing elements: (a) float (variable area); (b) spring-restrained plug (variable area); (c) hinged vane; (d) cantilever vane.

frequently as the rotating types, employing design variations of a propeller (used for wind-speed and ocean current sensing) or, most notably, a turbine wheel or a rotating cup assembly (used in *anemometers,* wind-velocity sensing devices); these elements are illustrated in Figure 12–4.

### 12.3.3  Thermal Flow Sensing

The amount of heat transferred between two points in a moving fluid is proportional to mass flow rate. The first *thermal flowmeter* was described by C. C. Thomas in 1911, and later developments and design refinements did not substantially alter the basic components of this design (see Figure 12–5(a)): a heating element immersed in the fluid and two temperature sensors, one located upstream of the heater to sense the fluid temperature $t_1$ for reference purposes, the other located downstream of the heater to sense the additional heat transferred from the heater by the moving fluid. The heat input $Q_H$ and the mass flow rate $Q_M$ are related by $Q_H = c(t_2 - t1)Q_M$, where $c$ is the specific heat of the fluid. Thermocouples, originally used for temperature sensing, are now largely replaced by resistive temperature sensors.

Since the heater power required in this type of design increases dramatically with pipe diameter, the use of such flowmeters is limited to rela-

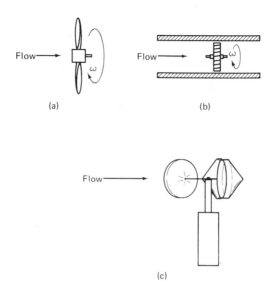

**FIGURE 12–4.**  Rotating mechanical flow-sensing elements: (a) propeller; (b) turbine; (c) cup assembly.

tively small pipe sizes. For larger pipe sizes, however, constant-ratio bypass sections of small diameter have been used to sample the fluid (typically gaseous) and sense its mass flow rate. Another way in which the heater power problem can be overcome is to inject the heat only into the very thin boundary layer of the fluid, adjacent to the pipe wall. In such a *boundary-layer flowmeter* the temperature sensors are installed flush with the inside pipe surface and the heater is embedded in a groove flush with the inside pipe wall but isolated from the fluid, or wound around the outside of the pipe (Figure 12–5(b)). The heat input varies with mass flow rate and temperature differential; however, the relationship is more complex than for the Thomas flowmeter.

The *hot-wire anemometer* (Figure 12–5(c)) consists of a thin heated wire or metal film whose cooling due to fluid flow is proportional to mass flow rate. The cooling is detected either by detecting the decrease in element resistance or by measuring the increase in current needed to maintain the same element temperature (in the *constant-temperature anemometer*). This sensing method applies primarily to air (the Greek word *anemos* means wind) and other gaseous fluids.

### 12.3.4  *Magnetic Flow Sensing*

A conductive (even mildly so) fluid flowing through a transverse magnetic field has an increasing electromotive force induced in it with increasing flow velocity (Figure 12–6). The magnetic field is typically created by an electromagnet that is excited by sinusoidal ac or pulsed dc current, and two electrodes are used to detect the voltage signal.

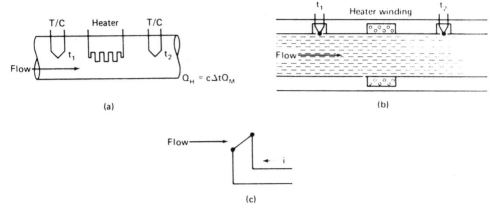

(a)                                    (b)

(c)

**FIGURE 12–5.** Thermal flow sensing: (a) Thomas flowmeter; (b) boundary-layer flowmeter; (c) hot-wire anemometer.

### 12.3.5 Oscillating-Fluid Flow Sensing

Devices can be introduced into a pipe or other conduit through which a fluid is flowing which will give the fluid an oscillatory motion proportional to volumetric flow rate. The two forms of oscillatory motion which are utilized for flow measurement can be classified as forced oscillation and natural oscillation. The type of forced oscillation used is that due to the generation of a vortex which causes the flow to become helical, with the velocity profile characterized by a higher velocity (lower pressure) along the centerline ("core") of the conduit, which is also the centerline of the sensing device. When the fluid encounters an area enlargement, the axis of rotation shifts from a straight-line path to a helical path. This path itself rotates (precesses). The frequency with which the precessing vortex core passes a given point is proportional to the volumetric flow rate for the vortex-precession method

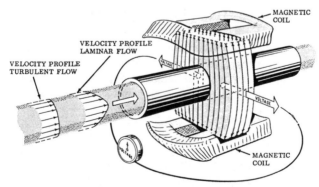

**FIGURE 12–6.** Magnetic flow sensing. (Courtesy of Fischer & Porter Co.)

of flow sensing (Figure 12–7(a)). This frequency is detected by a sensor with fast time response, such as a piezoelectric sensor, and a frequency output proportional to volumetric flow rate is produced.

Natural oscillations in the fluid are generated by immersing a non-streamlined *vortex-shedding* body into the flow stream. Vortices are produced (shedded) by this body due to surface frictional forces acting on the fluid boundary layer. The vortex pattern, a stable pattern of alternate clockwise and counterclockwise shedding, is known as a *von Kármán vortex street* pattern. The pattern will be different for different body shapes; that typical for a right circular cylinder is shown in Figure 12–7(b). The vortex street patterns produced by certain other shapes of shedding bodies were, however, found to be more suitable. The periodic fluctuations in vortex forces are detected by a device such as a strain-gage force sensor, and a frequency output proportional to flow rate is provided by it.

### 12.3.6 Ultrasonic Flow Sensing

Two methods are used to measure flow rate using ultrasound; both employ ultrasonic "transducers" which are either emitters or receivers of ultrasonic energy, or they can be both. In the *Doppler* method one transducer sends ultrasound, at a well-controlled frequency, through the measured fluid at a specific angle. Particles, bubbles, or turbulence in the fluid then reflect the ultrasonic energy, and a second transducer receives the reflected energy. The latter experiences a frequency shift due to the Doppler effect that is proportional to flow rate (and the injection angle, which is fixed and known). In the *transit-time* method, when wetted transducers (whose faces are flush with the inside pipe wall) are used in the direct-measurement mode, two transducers are mounted so that they face each other, as shown in Figure 12–8(a). Both can transmit as well as receive. A pulse is sent from one transducer to the other over the path length $S$ and reaches the other transducer after time $t_1$. A pulse is then sent from the second transducer to the first; its transit time is $t_2$. At zero flow, $t_1 = t_2$. On the other hand, when there is flow, $t_1$ decreases and $t_2$ increases, and the difference between the upstream and downstream transit time is directly proportional to the flow velocity $v$. When the inside area (based on diameter $D$) is introduced, the volumetric flow rate can be calculated. When clamp-on transducers are used (Figure 12–8(b)), they do not face each other because of refraction of the ultrasonic beam through the pipe wall (which must be sonically conductive), but they have to be the proper distance apart.

### 12.3.7 Special Mass-Flow Sensing Methods

Mass flow is usually determined from simultaneous volumetric flow and density measurements; electronic calculation can be used to provide real-time displays of mass flow rate. For gases primarily, mass flow can also be

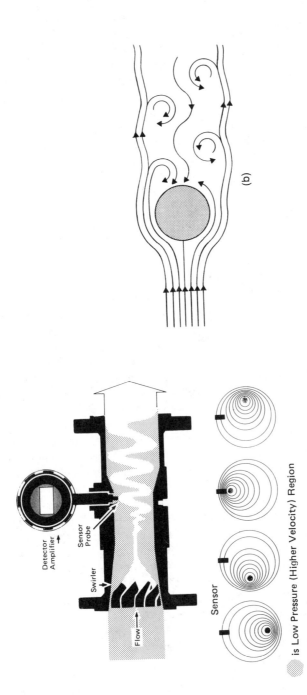

**FIGURE 12–7.** Oscillating-fluid flow sensing: (a) forced oscillation (vortex precession); (b) natural oscillation (vortex shedding). (Courtesy of Fischer & Porter Co.)

(b)

(a)

is Low Pressure (Higher Velocity) Region

Detector
Amplifier

Sensor
Probe

Swirler

Flow

Sensor

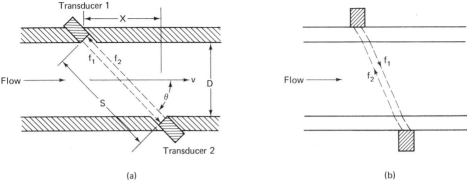

**FIGURE 12–8.** Ultrasonic flow sensing (transit-time method): (a) wetted; (b) clamp-on.

measured by thermal sensing methods. Additional specialized methods have also been employed. The *gyroscopic* mass-flow sensing method uses a circular loop of pipe (Figure 12–9) in a plane normal to the input line and output line. During flow of the measured liquid through the loop, the liquid develops angular momentum $H$ similar to that developed by the rotor of a gyroscope. The loop is mechanically vibrated through a small angle of constant amplitude about an axis in the plane of the loop. This vibration results in an alternating gyrocoupled torque $T$ about the orthogonal axis. The peak amplitude of this torque is directly proportional to mass flow rate and can be detected in terms of an alternating angular displacement about the torque axis.

Another method uses a U-shaped pipe which is vibrated at its natural frequency. The *Coriolis-type acceleration* imparted by the fluid causes an angular deflection of the pipe section which is proportional to mass flow rate. A *dual-rotor turbine flowmeter* employs two rotors with different blade angles, connected by an elastic member which acts as a torsion bar, thereby

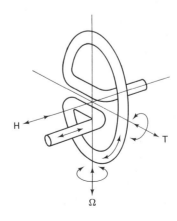

**FIGURE 12–9.** Gyroscopic (mass-)flow sensing.

permitting transduction of the torque developed in it. A related method uses an upstream rotor driven by a synchronous motor which imparts *angular momentum* to the measured fluid. The angular momentum is removed from the fluid by a downstream turbine rotor, and the torque (which is transduced) exerted on the turbine is linearly proportional to mass flow rate.

## 12.4  DESIGN AND OPERATION

### 12.4.1  Differential-Pressure Flowmeters

Differential-pressure flowmeters are really sensing systems comprising at least two major elements: (1) the orifice, Venturi tube, nozzle, Pitot tube, Pitot-static tube, or centrifugal section by which a differential pressure is developed; and (2) the differential-pressure transducer, which provides the output signal. Since flow rate is proportional to the square root of differential pressure, a "square-root extractor" is often added, in the form of electronic or electromechanical provisions. The pressure transducer can be separately installed and connected to the two pressure ports of the flow-sensing element by tubing, or it can be integrally packaged with the sensing device. Other elements, such as a temperature transducer and an absolute-pressure transducer, are added to refine the data presented by the output signal. (Both of these are needed when gas mass flow has to be determined.) The computation of volumetric or mass flow rate, on the basis of the outputs of these tranducers, is often performed by microprocessors ("flow computers").

The most common varieties of the sensing elements are described in Section 12.3.1. Additional design variations include a Pitot tube having a length essentially equal to the inside pipe diameter and provided with several appropriately spaced stagnation-pressure ports leading into a plenum tube in which a single port detects the average of the pressures.

### 12.4.2  Turbine Flowmeters

Flowmeters using a turbine-type rotor as the sensing element are widely used. The rotor is so designed that it converts the flow velocity into an equivalent angular velocity of the rotor (see Figure 12–10). Flow straighteners (flow conditioners) are always incorporated upstream of the rotor and often also downstream of it (see Figure 12–10(b)). The rotor is supported by one or more bearings, and various approaches have been used to minimize bearing friction. Most designs use electromagnetic transduction, an electromagnetic sensing coil in conjunction with ferromagnetic turbine blades (or blade tips). When electromagnetic transduction is used, a small amount of electromagnetic drag can be created which tends to retard rotor motion slightly. This minor problem does not occur when an RF-excited inductive

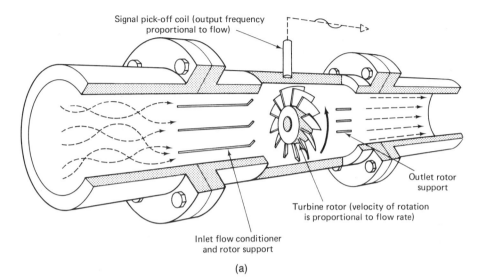

Signal pick-off coil (output frequency proportional to flow)

Outlet rotor support

Turbine rotor (velocity of rotation is proportional to flow rate)

Inlet flow conditioner and rotor support

(a)

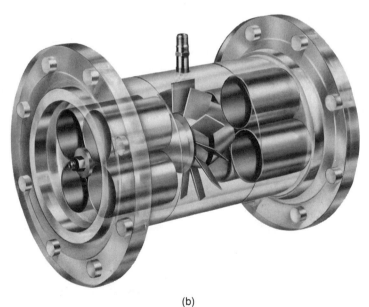

(b)

**FIGURE 12–10.** Basic elements of a turbine flowmeter: (a) operating principle; (b) typical flowmeter, with transparent case used to show elements. [(a) courtesy of Aqua Matic Inc., and (b) courtesy of ITT Barton Instruments Co.]

eddy-current sensing coil ("pick-off coil") is used instead, or when electro-optical transduction is employed. In an electro-optical version of a turbine flowmeter, the interruptions of a light beam by the turbine blades are sensed by a photodetector. This design lends itself to the use of optical fibers, with the light source and detector installed some distance away. Regardless of the transduction method used, viscous drag affects the rotor velocity at low speeds.

The output of a turbine flowmeter is a frequency that is proportional to volumetric flow rate; it can be sinusoidal or in the form of pulses of various shapes, depending on blade-tip geometry and on the transduction method employed. The sensitivity of a turbine flowmeter is expressed in terms of the *flowmeter coefficient* (or "flow coefficient") $K$, in hertz per liter (or per gallon, or per another volume unit). This coefficient is applicable under specified conditions of measured fluid density, viscosity, downstream temperature, back-pressure (downstream absolute pressure), and range of flow rates. The flowmeter coefficient can also be affected by the line (pipe) configuration; thus, the lengths of straight pipe runs upstream and downstream of the flowmeter, as well as the configuration and location of any additional flow straightener, should be known if the as-installed coefficient is to be compared with the coefficient obtained during a laboratory calibration.

Figure 12–11 illustrates two typical flowmeter designs. The large flowmeter is built for flange connections, whereas the small one is designed for threaded connections. Turbine flowmeter sizes generally range from about 1 cm to 25 cm (the size is given by the inside pipe diameter), with a few special designs to over 60 cm. Another flowmeter configuration is the insertable type, with an integral assembly consisting of the turbine and transduction element insertable into a pipe perpendicularly to the flow stream. The assembly is mounted and sealed to a mounting boss on the pipe.

### 12.4.3 Annular-Rotor Flowmeters

One flowmeter design is somewhat related to turbine flowmeters in that it uses a rotating mechanical element and produces a pulse-frequency output. The operating principle, however, differs substantially from that of the turbine flowmeter. As shown in Figure 12–12, the annular-rotor flowmeter contains a ring-shaped rotor supported by the fluid rather than by bearings. The measured liquid flows into the operating chamber through its circular periphery and forms tangential jets such that a spiral rotation is imparted to the liquid. This spiral rotation spins the rotor, stabilizing it in the middle of the chamber. The liquid makes a right-angle turn at the center of the chamber and exits from it. The rotor carries light-reflective marks. Fiber optics is used to carry a light beam to an optical window, located over these marks, and to carry back reflected light to a light sensor which produces an electrical pulse for each rotor mark passing the window. The light source and sensor

**FIGURE 12–11.** Typical large and small turbine flowmeters. (Courtesy of ITT Barton Instruments Co.)

are in a separate box which also contains the conditioning circuitry and which is connected to the flowmeter by the fiber-optics cable.

### 12.4.4 Rotating-Cup Anemometers

Rotating-cup anemometers (see Figure 12–13) are widely used to measure air-flow velocity (wind speed). They often incorporate a rotating vane assembly for the simultaneous measurement of wind direction; this then allows wind velocity (a vectorial quantity) to be determined. Electromagnetic or electro-optical transduction, which produces a pulse-frequency output proportional to wind speed, is most commonly employed. Other angular-speed transducers that produce an analog output have also been used.

### 12.4.5 Propeller-Type Flowmeters

Rotors having a shape similar to those of aircraft propellers are also used in some anemometers. Rotors whose shape resembles that of a ship's propeller are also used in some types of ocean current (or other water current) meters. Electromagnetic sensing coils or cam-operated switch contacts pro-

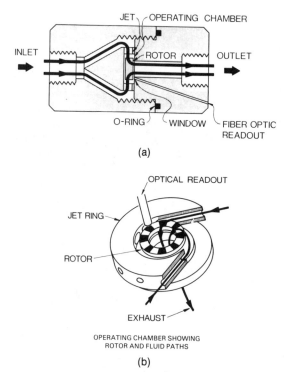

FIGURE 12–12. Annular-rotor flowmeter with electro-optical transduction: (a) cross section of typical flowmeter; (b) operating chamber, showing rotor and fluid paths. (Courtesy of Bearingless Flowmeter Co.)

vide a pulse-rate output that is proportional to water velocity. Propeller-
shaped rotors have also been used in pipeline flowmeters, wherein the propeller, shaft, and transduction element (e.g., magnetic-reed switch) form an
assembly that is inserted into the pipe through a port which protrudes from
the pipe wall at an angle. The assembly, with only the propeller end in the
fluid stream, is sealed to, but removable from, the port fitting.

### 12.4.6 Positive-Displacement Flowmeters

Positive-displacement flowmeters such as the sliding-vane, nutating-disk,
drum, or rotary-piston type employ rotary mechanical sensing elements
which are designed to trap sequential volumes of fluid so that the fluid is
"packetized" and then pass the "packets" to the exit port. The number of
"packets" per unit time is proportional to the volumetric flow rate. Such
flowmeters have been in use for many years, but normally have a mechanical
rather than electrical output. A dial- or counter-type indicator is attached
to the flowmeter and, driven by a shaft linked to the rotor, is used to provide

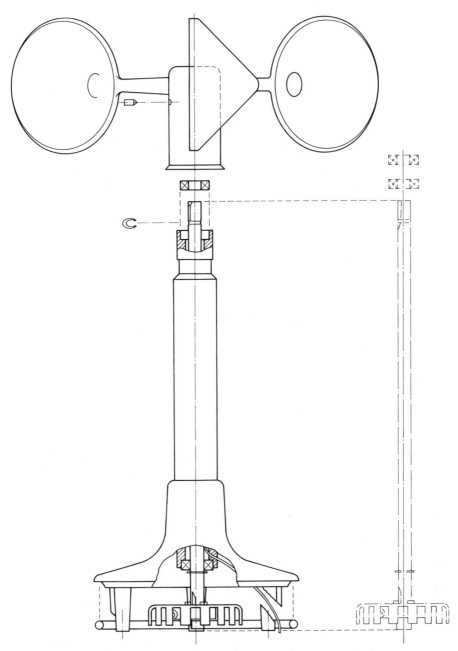

**FIGURE 12–13.** Rotating-cup anemometer (top portion). (Courtesy of Vaisala OY.)

a direct display of total flow (*flow totalizer*). However, positive-displacement flowmeters can also be provided with an angular-speed transducer (tachometer), which then provides an electrical output proportional to flow rate.

### 12.4.7   Target Flowmeters

Target flowmeters include those flowmeters containing a body, immersed in the flow stream and typically disk shaped, which is attached to a mechanical link such that the force on the "target" due to flow velocity can be transduced. The associated force transduction is either of the strain-gage or force-balance variety. It is important for the target to be positioned at the pipe centerline and oriented normally to the flow direction, within close tolerances. The force exerted on the target is theoretically proportional to the square of the fluid velocity through the annular region formed between the target and the pipe wall.

### 12.4.8   Thermal Flowmeters

Thermal flowmeters are categorized into two groups. (1) Energy-balance flowmeters add heat to a confined fluid stream and relate the resulting temperature rise in the fluid to mass flow rate (and the heat capacity of the fluid). (2) Thermal anemometers use a heated body immersed in the fluid stream and relate the heat transfer from the body to the fluid to mass flow rate (per unit area) or mass velocity.

**12.4.8.1   Energy-Balance Thermal Flowmeters.**   The *heated-fluid flowmeter* was first developed by C. C. Thomas in 1911. It uses an electrical heater to raise the temperature of the measured fluid and two temperature sensors, one upstream and the other downstream of the heater, to determine the temperature rise downstream of the heater as referred to the temperature upstream of the heater. The mass flow rate is related to the difference between the two temperatures by $Q_m = W/(c_p \, \Delta t)$, where $Q_m$ is the mass flow rate, $W$ the heater power, $c_p$ the specific heat of the fluid at constant pressure, and $\Delta t$ the temperature difference. This principle is illustrated in Figure 12–14(a). If flow rate is to be expressed in terms of moles per unit time instead of units of mass per unit time, $c_p$ is expressed in molar units instead of mass units. The primary application of thermal flowmeters is in measurements of mass flow rate (or "molar" flow rate) of gases.

Design variations of the *Thomas flowmeter* were developed subsequently, mainly to overcome the problem of high power consumption. One successful design was the boundary-layer flowmeter, in which only the "layer" of fluid closest to the inside pipe wall is heated and subjected to the two temperature determinations. Another design is the *heated-conduit*

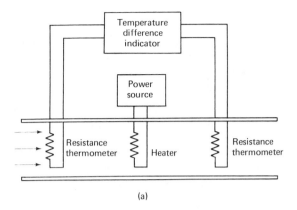

(a)

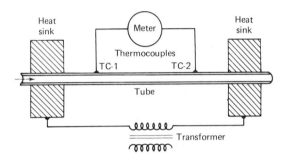

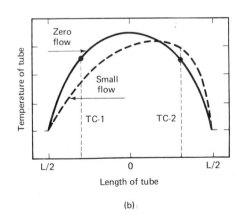

(b)

**FIGURE 12–14.** Heated-fluid thermal mass flowmeter: (a) basic operating principle (Thomas flowmeter); (b) simplified schematic of heated-conduit flowmeter with heat sinks at each end and associated temperature distribution. (Courtesy of Teledyne Hastings-Raydist.)

*flowmeter* [Figure 12–14(b)], in which heat is conducted along a conduit to heat sinks at each end. A differential thermocouple is arranged to measure the temperature distribution about the midpoint of the conduit. As the flow rate increases from zero, sensing junction TC-2 sees an increasingly higher temperature, whereas sensing junction TC-1 sees a decreasing temperature. The thermoelectric emf generated by the differential thermocouple is, therefore, proportional to the mass flow rate for a gas of given heat capacity, as indicated by the meter shown in the simplified schematic. (Additional signal-conditioning circuitry, including the means for eliminating the ac component in the output signal, is not shown; the full-scale thermoelectric potential for the linear portion of the range is about 8 mV.) A typical flowmeter of this type, with associated display device, is shown in Figure 12–15.

The measuring range of the heated-conduit flowmeter is limited; however, it can be increased by shunting the main flow pipe with a small-diameter bypass section and either placing viscous restrictions in both of the lines or placing an orifice in the main as well as the bypass line (in which the measurement is made) such that the ratio of orifice discharge coefficients is constant for variations in flow rate, temperature, and pressure (*constant-ratio bypass section*).

Related designs use resistance thermometers instead of a differential thermocouple, and they may employ different signal conditioning and display techniques, including providing a switch (discrete) output (flow/no flow).

**12.4.8.2 Thermal Anemometers.**  Hot-wire and hot-film anemometers are used primarily to measure flow velocities of air and other gaseous fluids. Some designs are also used for mass velocity (mass flow) measurements, and a few designs have been used in a limited number of liquids.

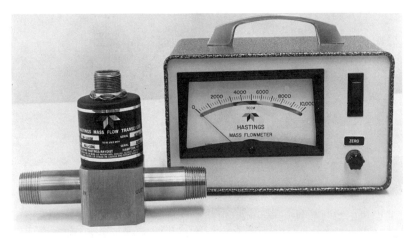

**FIGURE 12–15.** Thermal mass flowmeter for gases. (Courtesy of Teledyne Hastings-Raydist.)

Many types of anemometers are characterized by very small sensor size and a resulting fast response time. This makes them suitable for determinations of flow-velocity profiles in ducts, as well as for analyses of velocity micro-structures in turbulent contained or free-stream flows.

The resistive sensing element consists of a thin metal wire or deposited film supported by electrical connections in such a manner that heat conduction from the element to its supports is minimized. The sensor is connected into a Wheatstone-bridge circuit to which sufficient excitation is provided to cause self-heating in the element. When immersed in the measured fluid, the element loses heat by convection to the fluid stream. This convective heat loss varies approximately with the square root of the fluid velocity. Two operating modes are used for heated-element anemometers (see Figure 12–16). The bridge circuit of the *constant-current anemometer* obtains its excitation from a constant current supply (Figure 12–16(a)). The current is adjusted so that the sensor is heated to a temperature optimized for a given application. At increasing fluid velocities the sensor cools increasingly, and the resulting resistance change causes a bridge unbalance and commensurate changes in the bridge output voltage. The *constant-temperature anemometer* (see Figure 12–16(b)) has an unheated temperature-compensation sensor connected into the bridge arm adjacent to the arm constituted by the heated sensor. (The two sensors are in close physical proximity to each other.) As the sensor cools due to flow velocity, its resistance change causes a bridge unbalance. The bridge output voltage is fed to a control unit containing a high-gain feedback amplifier whose output is the bridge excitation. The amplifier sees any bridge-unbalance voltage as an error signal and causes its output (the bridge excitation) to change until the error signal is zero, under which condition the heated sensor is brought back to its initial temperature. Hence, the bridge-excitation changes also provide the output signal. The constant-temperature circuit is usually preferred for flow measurements.

Probe configurations have been developed for a variety of applications

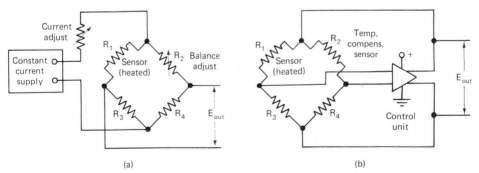

**FIGURE 12–16.** Heated-sensor anemometer circuits: (a) constant-current operation; (b) constant-temperature operation.

in anemometry. A probe consists of a sensor, the sensor support, the probe body, and the electrical connections. Typical cylindrical sensors are either a thin (on the order of 5 μm in diameter) wire or a slightly thicker nonconductive rod on which a metal film is deposited. Wire materials include platinum-plated tungsten and platinum-rhodium alloy. For film sensors, a metal with a high temperature coefficient of resistance, such as nickel, is deposited on a temperature-stable substrate (e.g., quartz), and the film sensor is then coated with a protective film, typically also of quartz. Coating thickness varies between about 0.5 μm (for use in air) and 2 μm (for use in water), while active lengths of the element range between 0.2 and about 2 mm. The sensors in the probe configurations shown in Figure 12–17 are supported by sturdy prongs. The prongs can be straight and either of equal length or of unequal length, the latter so as to place the sensor at an angle to the probe axis. They can be bent at a right angle so that the sensor is either parallel or perpendicular to the probe axis, and they can also be double bent so as to offset the sensor from the probe axis. Noncylindrical sensors are of the film type; they are used in wedge-shaped and conical probes (see Figure 12–18) as well as in flush-mounted and bondable probes used, for example, for skin-friction measurements. Two or three sensors at specified angles to each other can be incorporated at the tip of a single probe for measurements of flow direction and magnitude in a two-dimensional or three-dimensional flow field, respectively. The probe shown in Figure 12–19 contains three mutually orthogonal sensors.

Probes can have various body designs and are often protected by a sheath. Sheaths have been designed with a handle, for hand-held operation, or with mounting bosses, mounting chucks, or pressure fittings, for either fixed or rapidly removable installations.

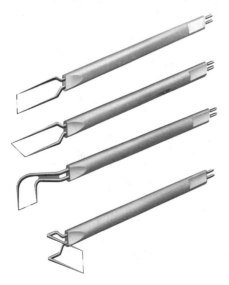

**FIGURE 12–17.** Hot-wire probe configurations. (Courtesy of Dantec Elektronik.)

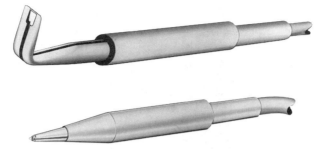

**FIGURE 12–18.** Hot-film anemometer probes. (Courtesy of Dantec Elektronik.)

A temperature-compensation sensor (platinum wire or film) is often incorporated in the probe tip, but with the velocity sensor (heated sensor) protruding beyond the temperature sensor. Figure 12–20 shows the sensor portion of a constant-temperature anemometer (CTA). It contains two platinum-wire windings: the mass velocity sensor winding $R_p$, which is kept at a constant temperature above the ambient (or measured-fluid) temperature, and the ambient-temperature sensor winding, $R_{tc}$. Both are wound on a ceramic mandrel and have a protective glass (or Teflon) coating. An immersion probe equipped with such a sensor is shown, installed in a large-diameter (up to 60 cm) pipe, in Figure 12–21. Other probe configurations are available, such as in-line multisensor arrays and hand-held probes that are typically used in conjunction with portable air velocity meters. The air velocity meter shown in Figure 12–22 displays either velocity or ambient temperature on its digital indicator. In-line multisensor arrays can be used to establish air velocity profiles in ducts or to provide velocity averaging across a large (in excess of 60 cm diameter) pipe or duct. Since the output-vs.-velocity relationship of anemometer systems is inherently nonlinear, the associated electronics usually provides linearization.

### 12.4.9  *Magnetic Flowmeters*

Magnetic flowmeters are used to measure the flow rate of conductive liquids; however, the conductivity need not be very high: most meter designs operate with liquids having conductivities down to 5 μS/cm (microsiemens per centimeter), and special designs accept conductivities down to 0.1 μS/cm. A

**FIGURE 12–19.** Triaxial triple-sensor hot-wire anemometer probe. (Courtesy of Dantec Elektronik.)

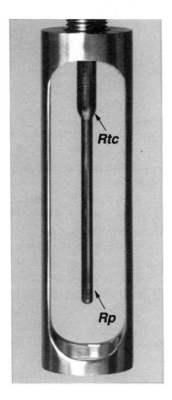

**FIGURE 12–20.** Sensor portion of CTA mass flowmeter. (Courtesy of Kurz Instruments, Inc.)

typical magnetic flowmeter consists of a steel body, an insulating liner through which the liquid flows, a set of electromagnetic coils, and a pair of electrodes in contact with the fluid (see Figure 12–23). The electrodes are essentially flush with the inside surface of the liner; hence, one of the advantages of this type of flowmeter is that it offers no obstruction to the moving liquid. The electrodes are located at right angles to both the flow and the magnetic field, as explained in Section 12.3.4. The output signal, the voltage generated between the two electrodes, is related to fluid velocity by $E = Bdv$, where $E$ is the output voltage, $B$ the magnetic flux density, $d$ the spacing between the electrode faces, and $v$ the average fluid velocity. $E$ is expressed in abvolts ($10^{-8}$ V) when $B$ is expressed in gauss, $d$ in cm, and $v$ in cm/s.

An ac voltage is usually applied to the magnetic coils, which can be connected in parallel or, when lower power consumption at a cost in output signal level is acceptable, in series. Series connection also reduces the heat input from the coils into the fluid. Pulsed dc has been used for powering the coils in lieu of sinusoidal ac; this reduces power consumption and assures the absence of a quadrature voltage in the output signal. The output is inherently linear with flow rate.

Different liner and electrode materials are available to fit the temper-

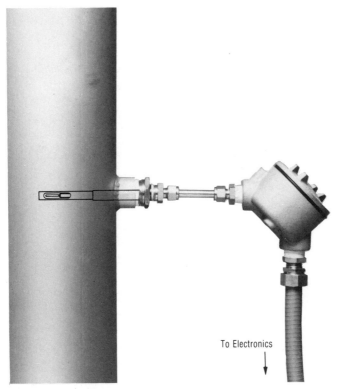

**FIGURE 12–21.** CTA mass-flowmeter immersion probe installed in large-diameter pipe. (Courtesy of Kurz Instruments, Inc.)

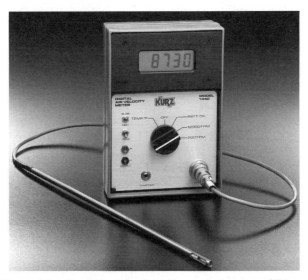

**FIGURE 12–22.** Portable air velocity meter using CTA sensor. (Courtesy of Kurz Instruments, Inc.)

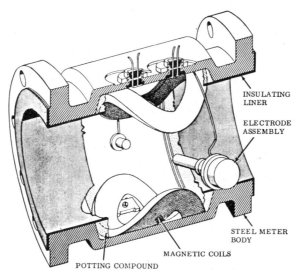

**FIGURE 12–23.** Typical basic layout of a magnetic flowmeter. (Courtesy of Fischer & Porter Co.)

ature and the physical (primarily abrasion) and chemical (notably corrosive) properties of the measured liquid. Glass or ceramic liners, for example, are usable at temperatures to 180°C, fiberglas to about 150°C, and polyurethane and synthetic rubbers to about 100°C. Various stainless steels are commonly used for electrodes, and other materials include nickel alloys, platinum, platinum-iridium, titanium, and zirconium.

The sizes of magnetic flowmeters (inside diameters) range from 0.25 cm to over 200 cm. For proper operation it is desirable that the fluid velocity be less than 10 m/s; for abrasive liquids it may have to be limited to about 2 to 3 m/s. Velocities somewhat above 2 m/s help to minimize coating of the electrode faces when the measured-liquid properties tend to cause such a coating problem.

The electromagnetic flow-sensing principle has also been applied to immersible water-current meters as well as to blood-flow sensors.

### 12.4.10 Oscillating-Fluid Flowmeters

There are two types of oscillating-fluid flowmeters: the *forced-oscillation* type, in which the fluid is forced into oscillatory motion, and the *natural-oscillation* type, exemplified by the vortex-shedding flowmeter. The forced-oscillation flowmeter, which employs *vortex precession* to generate a frequency output proportional to flow rate, is intended primarily for measurements of gaseous fluids. The vortex-shedding flowmeter is suitable for liquid fluids. The underlying principles are explained in Section 12.3.5.

The vortex-precession flowmeter (see Figure 12–24) uses a bladed entrance device to impart a swirling motion to the fluid, so that a vortex is generated. In the Venturi-shaped body this vortex is caused to precess: the axis around which the fluid is spinning changes from a straight-line path to a helical path. This oscillation is detected by a sensor located in the region where the oscillations occur. The swirling motion can then be removed by a deswirling device at the exit of the flowmeter. The oscillation can be sensed as temperature fluctuations, using a thermistor, or as pressure variations, using a fast-response sensor (e.g., piezoelectric); it can also be detected ultrasonically. Signal conditioning can be added to convert the frequency output into an analog voltage or current output. Output frequencies are in the range 10 to 1000 Hz.

In the vortex-shedding flowmeter a pattern of vortices is generated by immersing a nonstreamlined body into the flow stream. Various designs of vortex-shedding bodies have been used. Figure 12–25 shows one example, a torque tube with a sensing vane. The vortices shed by this element continuously change direction of rotation, applying an oscillatory force to the sensing vane on the torque tube. The resulting (small) oscillatory angular displacement of the sensor link can then be converted into an output signal by one of several types of transduction elements, as long as they have a

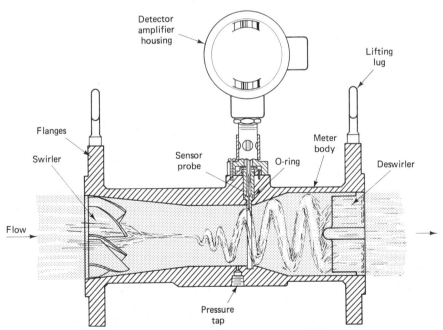

**FIGURE 12–24.** Oscillating-fluid flowmeter (vortex-precession type). (Courtesy of Fischer & Porter Co.)

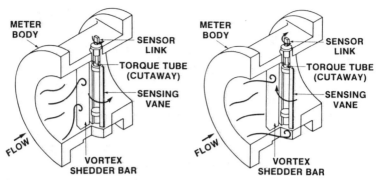

**FIGURE 12–25.** Example of vortex-shedding body in oscillating-fluid flowmeter. (Courtesy of Fischer & Porter Co.)

sufficiently high frequency response. A typical vortex-shedding flowmeter, together with its in-line installation, is shown in Figure 12–26. The frequency output of the transduction element can be converted, if desired, to other output forms, such as analog voltage or current.

### 12.4.11 *Ultrasonic Flowmeters*

Ultrasonic flowmeters are categorized according to their operating principle as either *transit-time flowmeters* or *Doppler flowmeters*. The operating principles are explained in Section 12.3.6. Both categories exist as wetted-transducer or clamp-on types.

Wetted-transducer ultrasonic flowmeters normally use two transducers; some Doppler flowmeters use only one transducer, and a few transit-time flowmeter designs use four transducers arranged in an X-shaped configuration. The transducers are mounted in a machined boss so that they protrude through the pipe wall, but are recessed so that their faces are flush with the inside pipe surface. The bosses are 180° apart on the pipe wall, and they are so machined as to position the transducers at the proper angle (typically 45°) and the proper distance apart (see Figure 12–8(a)). These flowmeters generally consist of a pipe section with transducers and, very often, also with an integrally mounted electronics unit and with connection provisions in the form of pipe threads or flanges.

Clamp-on ultrasonic flowmeters consist of two transducers with clamps, for mounting to the outside of a pipe, and a separate electronics unit which often also includes the display. The transducers are clamped to the pipe 180° apart. The pipe wall thickness and material must be known so that the refraction angle of the ultrasonic beam through the pipe wall can be established (see Figure 12–8(b)). The lateral distance between the transducers can then be adjusted appropriately. Some transducer designs include tracks (mounted by the clamps) along which the transducer can slide, to

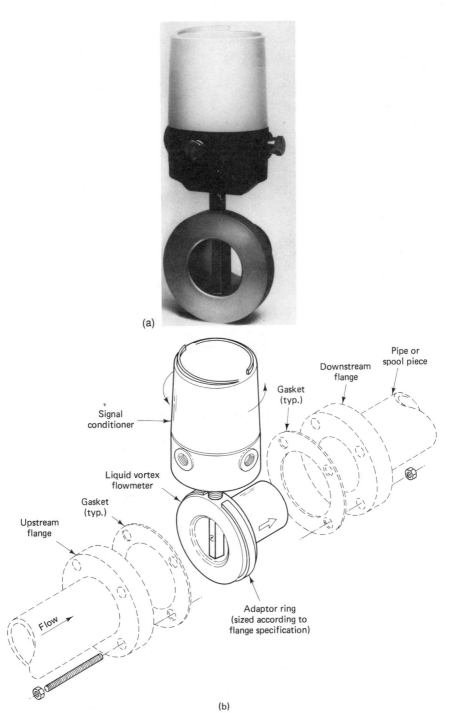

(a)

(b)

**FIGURE 12–26.** Oscillating-fluid flowmeter (vortex-shedding type):
(a) liquid-vortex flowmeter; (b) typical in-line installation. (Courtesy
of Fischer & Porter Co.)

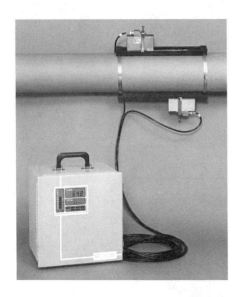

**FIGURE 12–27.** Clamp-on ultrasonic flowmeter. (Courtesy of Controlotron Corp.)

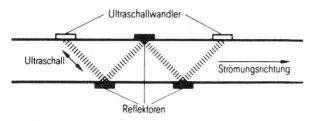

**FIGURE 12–28.** Ultrasonic flowmeter (with provisions for thermal energy measurement). (Note: *Ultraschall* = ultrasound; *-wandler* = transducer; *Reflektoren* = reflectors; *Strömungsrichtung* = flow direction.) (Courtesy of Siemens AG.)

facilitate this lateral adjustment. A clamp-on flowmeter system is shown in Figure 12–27. Clamp-on flowmeters have the advantage of being portable, with the transducers easily removed and reinstalled on another pipe.

Figure 12–28 shows a wetted-transducer ultrasonic flowmeter. The transducers are mounted at 45° angles, which is typical. This design also uses three reflectors to increase the path length and, hence, the transit time; this makes it easier to determine the upstream and downstream transit-time difference and thus improves resolution, especially at low flow rates. The flowmeter illustrated not only measures flow rate, but is intended for determinations of thermal energy used by central hot-water heating systems. For this purpose the flowmeter incorporates a platinum resistance thermometer, with a second such thermometer installed in the return line (lower pipe section in the photo). When fluid density and enthalpy are known, the temperature difference measured between the feed and return lines can be multiplied by the flow rate, and the result can be integrated over a period of time to yield the quantity of total heat energy, expressed in kW·h. The integral-electronics-and-display unit is powered by a battery that is reported to have a six-year life. The display can be switched from heat energy to flow rate as well as to each of the two measured temperatures.

In general, transit-time flowmeters work best with fluids having low concentrations of acoustic discontinuities such as gas bubbles or particles, whereas Doppler flowmeters work best with the exact opposite, fluids with high concentrations of sonic discontinuities.

## 12.5 FLOWMETER CHARACTERISTICS

Flowmeter characteristics, to be considered when selecting and specifying such a device, will depend largely on the type of flowmeter selected. Certain characteristics, however, are applicable to all in-line flowmeters to approximately the same extent. The size of the pipe into which a flowmeter is to be inserted generally dictates the size of the flowmeter; for some applications some types of flowmeters may be larger or, more often, smaller than the pipe. The physical and chemical characteristics of the measured fluid influence flowmeter selection and design characteristics very strongly. The fluid can be gaseous or liquid; if liquid, it may contain some amounts of gas, or solids in solution, undissolved solids, or abrasive particles. The specific gravity and the viscosity (over the complete temperature range) of the fluid must be known, and its conductivity and relative corrosiveness must be established. The maximum fluid pressure and the minimum and maximum fluid temperatures must also be known. Mechanical connections of the flowmeter can be by means of threads or flanges; dimensional standards exist for both. Specification drawings show the configuration and dimensions as well as materials of construction and the mass of the flowmeter. The location and type of external electrical connections must also be known or specified; codes or standards may apply to these and to the housing in which they are contained.

The sensing element of a flowmeter offers either no obstruction, a minor obstruction, or a major obstruction to the moving fluid, depending on its type and operating principle. The flowmeter may also cause a pressure drop in the line. These characteristics are quite important in the selection of a flowmeter for a given application. Provisions for checking or servicing a flowmeter are sometimes provided or required, and can be either mechanical (e.g., a spin port on a turbine flowmeter) or electrical in nature. Some designs permit servicing or replacement of the sensing assembly without removal of the flowmeter body.

Performance characteristics include the range of flow rates to be measured and the portion of this range over which the measurand-vs.-output relationship is linear within a specified tolerance. Where "accuracy" is shown, this term usually applies to repeatability. Sensitivity (such as the $K$ factor of a turbine flowmeter) is expressed in terms of hertz per unit of volume ($m^3$, L, gallon, etc.) when the flowmeter provides a frequency output; when the output is in analog form, sensitivity is shown in units of voltage or current per unit of flow rate. For dc analog outputs the amount of ripple or other ac component must be covered by tolerances. Frequency outputs should be shown not only in range of output frequency over the measuring range, but also in terms of their wave shape (and total harmonic distortion if sinusoidal). Dynamic characteristics of a flowmeter are sometimes shown in terms of frequency response, but more commonly in terms of time constant or response time.

## 12.6  APPLICATION NOTES

### 12.6.1  Installation Precautions

Most flowmeters require a minimum length of straight pipe upstream, and many of them further require such a minimum length downstream. The main reason for these requirements is the prevention of random fluid swirls at the sensing element. For turbine flowmeters the recommended lengths are 20 pipe diameters upstream and 5 diameters downstream. For transit-time flowmeters the requirement is typically for 10 diameters upstream only. For other types these numbers can be different; the manufacturer should be consulted. Some installations may benefit from a strainer that is installed upstream.

Flowmeters are usually supported only by the pipe in which they are installed. Therefore, they can be subject to case stresses introduced by bending, thermal deformation, vibration, or improper mounting. Following the manufacturer's recommendation for mounting usually overcomes such problems. The direction of flow is marked on the flowmeter housing (unless the flowmeter is bidirectional), usually in form of an arrow; this information must be heeded during the installation. Flowmeters with rotating mechanical elements, such as turbines, have a maximum-speed rating. If this rating is exceeded (for example, by the presence of substantial amounts of gas in a supposedly liquid-only fluid), the rotor can be damaged; as a worst-case result, parts of the rotor may break off and cause damage to downstream pumps and other equipment.

### 12.6.2  Data System Considerations

The output signal amplitude of some types of flowmeters is inherently low. Accordingly, unless an amplifier is incorporated into the flowmeter, the usual precautions associated with low-level signal sources must be taken: shielding, twisting the conductors, limiting the length of the cable run, installing a floating ground at the flowmeter end, and moisture-proofing the connectors. When noise is present in the frequency output of a flowmeter, a lowpass or band-pass filter can be added to eliminate spurious signals at frequencies above the maximum flowmeter output frequency.

Signal-conditioning equipment can include frequency-to-analog converters, voltage-to-current converters, and analog-to-digital converters. Hot-wire anemometer systems can include electronics that permits detailed analysis of the output wave shapes. The electronics associated with the display often provides an additional *totalizer* function, in which case total mass or volumetric flow over a specified time interval can then also be displayed. If a flowmeter is used as part of a closed-loop control system, such a totalizer can be used for batching. When the outputs of a volumetric flowmeter and a densitometer (at or close to the flowmeter location) are combined in a small computer (or microprocessor), mass flow readings can be obtained.

### 12.6.3  Typical Applications

Flowmeters are used in virtually all industries and sciences to measure the flow of gases, liquids, and slurry-like fluids. In some industries, flow measurements are made literally from origin to end use. For example, in the petroleum industry the measurements begin at the oil well, continue through a pipeline or the loading and unloading of a transfer vehicle, continue further at many points in the refinery and at intermediate loading and unloading points, and end just upstream of the nozzle of a gasoline pump or at a tank truck from which fuel oil is fed to a residential heating system. Flowmeter installations in long pipelines, often at points that are far away from any human habitation and that may be difficult to reach, are a prime example of the need for long-term, unattended, reliable operation of a transducer.

The reliability of flowmeters in long pipelines has additional importance in terms of cost: if a flowmeter has to be replaced, the oil flow has to be shut off in most cases, causing a substantial loss of revenue to the oil producer.

There are innumerable applications of flowmeters in the chemical and energy-producing industries. In many cases, an industrial or private customer is billed on the basis of a flowmeter reading; the requirements for accuracy and stability in such instances are self-evident. Flowmeters are also essential to pollution control: they measure the flows of smoke, industrial effluent, sewage, and wastewater. Ocean current meters are used in oceanography, and rotating-cup anemometers are used in meteorology. Hot-wire anemometers measure air flow in heating and air-conditioning systems and are also used for research on turbulent flows, making them an important tool in the design of ships, aircraft, automobile bodies, and fixed structures that are exposed to wind loads. In medicine, examples of flowmeter applications are in external and internal blood flow, respiratory diagnostics, and anesthesiology.

In some applications the use of flowmeters that have sensing elements immersed in the measured fluid is undesirable. In such cases the "nonintrusive" types of flowmeters may well be usable instead. These include magnetic flowmeters (as long as the fluid is electrically even slightly conductive) and ultrasonic flowmeters of the transit-time or Doppler type as well as some thermal flowmeter designs, all of which are also characterized by a very small pressure drop.

# 13

# *Humidity and Moisture Sensors*

## 13.1 BASIC DEFINITIONS

*Humidity* is a measure of the water vapor present in a gas.

*Absolute humidity* is the mass of water vapor present in a unit volume.

*Relative humidity* is the ratio of the water-vapor pressure actually present to the water-vapor pressure required for saturation, at a given temperature. This ratio is expressed in percent and is temperature dependent; it is the most commonly used quantity in humidity measurement, including in weather reports.

*Specific humidity* is the ratio of the mass of water vapor in a sample of gas to the total mass of the sample.

The *humidity mixing ratio* is the mass of water vapor per unit mass of the dry constituents.

*Moisture* is the amount of water (unless another liquid is specified) adsorbed or absorbed by a solid, or in a liquid, or chemically bound to a liquid.

The *dew point* is the temperature at which the saturation water-vapor pressure is equal to the partial pressure of the water vapor (in the atmosphere). Any cooling of the atmosphere below the dew point would produce water condensation. The term has also been defined as the temperature at which the actual quantity of water vapor in the atmosphere is sufficient to saturate this atmosphere with water vapor. The relative humidity at the dew point is 100%.

## 13.2 UNITS OF MEASUREMENT

*Relative humidity* is expressed in *percent* ("%RH").

*Absolute humidity* is usually expressed in units of mass per unit volume ($kg/m^3$, or, more commonly, $g/m^3$).

*Moisture* is expressed in percent by volume or percent by "weight" (of either the total or the dry "weight").

The *dew point* is expressed in units of temperature, typically, °C.

## 13.3 SENSING METHODS

Four types of methods are used for humidity and moisture sensing. Those employed in hygrometers produce an output that is calibrated directly in %RH. Those employed by psychrometers produce two temperature readings, and a chart has to be used to correlate these readings with %RH or with moisture (by volume or by weight). Those employed by dew-point sensors provide a temperature reading from which humidity (see Section 13.3.3) can be inferred from a table, unless the dew point (or frost point) itself is the quantity to be indicated. Finally, those employed in remote sensing systems provide a reading of moisture (by mass or by volume).

### 13.3.1 Hygrometric Sensing Methods

The earliest hygrometer elements were mechanical elements which lengthen or deflect with increasing humidity. Human hair, certain animal membranes, and some plastic materials have this property. Mechanical elements are obsolete for modern humidity and moisture sensors, but are still widely used in simple, e.g., residential, humidity indicators.

*Capacitive* hygrometer elements operate on the fixed-electrode/variable-dielectric principle. In the example shown in Figure 13–1(a), a thin hygroscopic film is the dielectric between the two electrodes on the bottom and the electrode on top. The top electrode is porous so that it can admit water to the film. The film changes its dielectric characteristics and hence varies the capacitance between one bottom electrode and the top electrode, and between the top electrode and the other bottom electrode. This results in a net change in capacitance between the two bottom electrodes.

*Resistive* hygrometer elements are widely used, both in a wafer and a cylindrical (probe) form. The first successful resistive hygrometer was one using a hygroscopic film consisting of a 2 to 5% aqueous solution of *lithium chloride* (LiCl) and provided with two electrodes so that the resistance change of the film, due to a change in humidity, could be measured. Such an element was first developed by F. W. Dunmore in 1938 at the National Bureau of Standards in Washington, and, although the design has been re-

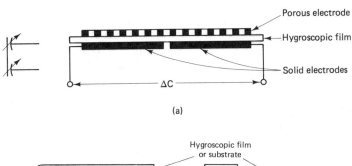

(a)

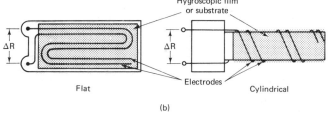

(b)

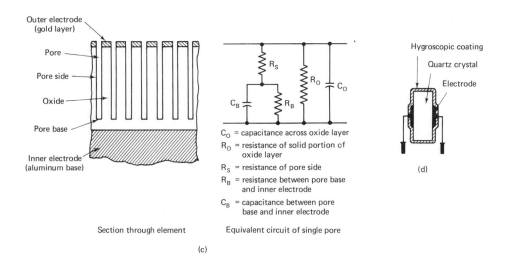

$C_O$ = capacitance across oxide layer
$R_O$ = resistance of solid portion of oxide layer
$R_S$ = resistance of pore side
$R_B$ = resistance between pore base and inner electrode
$C_B$ = capacitance between pore base and inner electrode

Section through element      Equivalent circuit of single pore

(c)

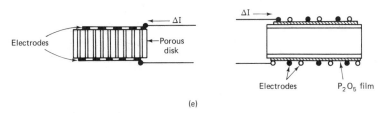

(e)

**FIGURE 13–1.** Hygrometer sensing elements: (a) capacitive; (b) resistive; (c) capacitive/resistive (aluminum oxide); (d) oscillating-crystal; (e) electrolytic.

fined since then, it is still often referred to as a "Dunmore element" or "Dunmore hygrometer." Bifilar wire electrodes, wound around an insulating (typically polystyrene) mandrel, or foil electrodes in a zigzag pattern on a substrate of similar material, are coated with the hygroscopic-salt film (see Figure 13–1(b)). Similar hygrometer elements use changes in the surface resistivity of the substrate itself without employing a hygroscopic film. An example is polystyrene which is treated with sulfuric acid to obtain the desired surface-resistivity characteristics; such *sulfonated polystyrene* elements are sometimes referred to as "'Pope elements" (or "Pope cells"), after their developer.

*Aluminum-oxide* elements exhibit a change in resistance as well as capacitance, and hence a change in impedance, with a change in humidity. This type of element consists essentially of aluminum whose surface is anodized so that a thin layer of aluminum oxide is formed. The structure of such a film has been determined to consist of a multitude of fibrous pores (see Figure 13–1(c)). A very thin film of gold, vacuum-deposited on the outside surface of the aluminum oxide layer, acts as one electrode (the film is thin enough to be porous), and the aluminum substrate acts as the other electrode. The equivalent circuit of a single pore is included in the figure. Water vapor is transported through the gold layer and equilibrates on the pore walls in a manner functionally related to the vapor pressure of water in the ambient atmosphere. The number of water molecules absorbed on the oxide structure determines the change in impedance of the element.

*Oscillating-crystal* elements (Figure 13–1(d)) consist of a quartz crystal with a hygroscopic coating. The crystal is connected as the frequency-controlling element in an oscillator circuit. The mass of the crystal changes with the amount of water sorption on the coating. This results in changes in the frequency at which the crystal oscillates, and a frequency output proportional to humidity is produced by the circuit. Hygroscopic polymers appear to be the most suitable coating materials.

*Electrolytic* hygrometer elements (see Figure 13–1(e)) are supplied with current sufficient to electrolyze water vapor into hydrogen and oxygen. The water vapor is usually absorbed by a desiccant such as a thin film of phosphorous pentoxide ($P_2O_5$) on which the bifilar electrodes are wound. Another design uses a porous glass disk with electrodes on both its surfaces; water vapor sorption occurs on the walls of the pores. The amount of current required for electrolysis varies as a function of water vapor absorbed, and hence of humidity, and the current itself provides the sensor output indicative of humidity.

### 13.3.2 Psychrometric Sensing Methods

The sensing elements of psychrometric sensors are temperature-sensing elements (see Chapter 19). Two separate elements are used: one, (the "dry bulb") measures ambient temperature, while the other (the "wet bulb") is enclosed by a wick which is saturated with distilled water (see Figure 13–2). The air whose humidity or moisture is to be measured is forced to blow over the wick so that water in the wick evaporates. This evaporation cools the wick and the sensing element underneath it; the latter then senses a temperature below ambient. The evaporation is dependent on the vapor pressure or the moisture content of the air (or other measured gas or gas mixture). Humidity (or moisture) can then be determined from the two temperature readings by referring to a *psychrometric chart* (see Section 13.4.2).

The most common way to obtain proper ventilation of the wet bulb is to manually swirl the entire sensing assembly with a circular motion, as in the *sling psychrometer*. Another way is to use a small blower (*aspirated psychrometer*). The wick is usually made of cotton, but can also be a porous ceramic sleeve fitted over the temperature-sensing element. The temperature sensors which are used most frequently are of the resistive type.

### 13.3.3 Dew-Point Sensing Methods

The *dew point* is that temperature at which the liquid and vapor phases of a fluid are in equilibrium. (The temperature at which the vapor and solid phases are in equilibrium is usually called the *frost point*). At the dew point, only one value of saturation (water) vapor pressure exists. Hence, absolute humidity can be determined from this temperature as long as the pressure is also known. The most commonly used method of measuring the dew point is to cool the surface whose temperature is being measured until dew (or frost) first condenses on it. As soon as this point in the cooling cycle is reached, the temperature of the surface is read out. The sensing element used for this method of dew-point measurement must provide two functions: it must sense the temperature of the surface that is being artificially cooled, and it must sense the change from the vapor to the liquid (or solid) phase. Resistive or thermoelectric elements (see Section 19.3) are commonly used for the temperature measurement. Sensing the instant of condensation can be performed by various methods (see Figure 13–3). All methods require a condensing surface, which is typically a thin disk or plate in close thermal coupling with a cooling device, usually a thermoelectric (Peltier-effect) cooler. The photoelectric method (Figure13–3(a)) is used most frequently.

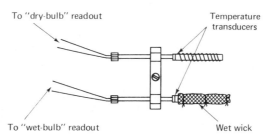

**FIGURE 13–2.** Psychrometric sensing element.

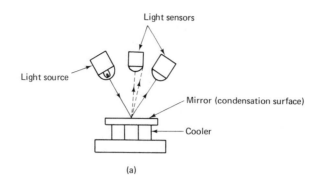

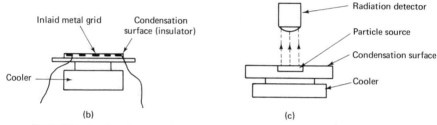

**FIGURE 13–3.** Dew-point condensation detection elements: (a) photoelectric; (b) resistive; (c) nucleonic.

The condensation surface is polished to mirror-quality reflectivity, and a light beam is aimed at the mirror. One or more light sensors receive the light reflected from the mirror, and condensation is accompanied by an abrupt change in the amount of light reflected to the sensor(s). A resistive type of condensation detector employs a surface of insulating material with an inlaid metal-electrode pattern; a change in surface resistance occurs when condensation forms (see Figure 13–3(b)). In nucleonic condensation detectors, an alpha- or beta-particle radiation source is located flush with the condensation surface and a radiation detector above the surface senses the drop in particle flux when condensation forms over the radiation source (Figure 13–3(c)).

A different method of dew-point sensing is employed in the *heated saturated-salt-solution* sensor; it has been referred to as an "energy-balance" method. Since the salt most commonly used is lithium chloride, such sensors are also known as *saturated heated lithium chloride* dew-point sensors. The sensor (see Figure 13–4) consists of a thin-walled metal tube covered by a fabric sleeve which is impregnated with a lithium chloride solution. A bifilar winding around the impregnated sleeve is used for heating the sensor. A temperature transducer inside the tube, and in good thermal contact with it, is used to provide a dew-point reading. The sensor is heated until the vapor pressure of the LiCl solution is in equilibrium with the vapor

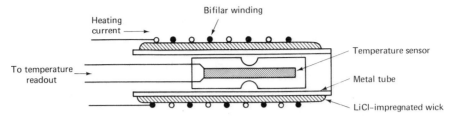

**FIGURE 13–4.** Saturated-solution dew-point sensing element.

pressure of the fluid (e.g., ambient air) whose dew point is to be determined. The resistance of the salt solution is indicative of the point at which the equilibrium temperature is reached. The output of the temperature sensor is displayed and related to the corresponding (lower) water-vapor dew point. (The vapor pressures for saturated LiCl are well established.)

### 13.3.4 Remote Moisture-Sensing Methods

Remote moisture-sensing methods employ electromagnetic techniques in regions of the spectrum extending from radio frequencies to ultraviolet and including especially microwaves and infrared. Devices employing such methods can be calibrated to display the moisture content of gases, liquids, and solids in terms of volume ratio (percent) or mass ratio (parts per million). Some devices simply use the absorption of microwaves in a sample for the indication of moisture content; the absorption increases with moisture content and results in attenuation of microwave energy between a source and a detector.

More commonly used devices are spectroscopic in nature. Their operation is based on the partial and selective absorption of radiation, due to moisture content, at specific wavelengths. These wavelengths are located in many portions of the spectrum extending from ultraviolet (UV) through infrared (IR) and including visible light. Commercially available moisture-sensing systems are based primarily on the use of infrared absorption spectra. In some, the IR absorption at a specific wavelength (characteristic of $H_2O$) is measured for a sample volume of the measured fluid and for a volume of a reference fluid with known moisture content, and the two readings are compared. Another system looks at two specific wavelengths in the measured fluid and compares the attenuation (in IR energy incident on a photodetector) at the two "dips" in the spectral curve, at only one of which significant changes due to absorption occur, but with the "dip" at the other wavelength used as reference. Absorption bands characteristic of moisture content can also be observed at microwave and submillimeter-wave frequencies. Instruments using these frequency bands and capable of distinguishing specific wavelengths (*multispectral microwave radiometers*) are used for remote sensing of atmospheric moisture content from satellites.

## 13.4  DESIGN AND OPERATION

### *13.4.1  Hygrometers*

The production of electronic hygrometers seems to have been spurred on primarily by the humidity-sensing requirements of meteorological *radiosondes*, small battery-powered telemetry packages carried aloft by balloons and tracked by radar. Sensors in the package measure temperature, pressure, and humidity, and the tracking data provide information about the location of the balloon and the time rate of changes of location, from which wind speed and direction can be inferred. Designers of radiosondes prefer that all sensors in the package have the same transduction principle so that they can operate in a standardized telemetry circuit.

In Finland, the meteorological radiosonde sensors must all be capacitive. The *capacitive* hygrometer designed for this application consists of a glass substrate on which the two lower electrodes are formed by etching of a metallized surface. A thin (about one micrometer thick) polymer film is coated over these electrodes. The upper electrode, permeable to water vapor, is then vacuum-deposited on the film. The changes in dielectric due to water absorption on the film make the sensor into a variable capacitor which is then connected into a high-frequency ac circuit. This sensor, known by its trade name HUMICAP (registered trademark of Vaisala OY, Helsinki, Finland), is 6 × 4 × 0.2 mm in size and is used in many applications besides radiosondes. A variety of housings have been designed to contain the basic sensor, for fixed installations as well as portable applications. The packaged sensor is connected by a cable to its electronics unit, which may also contain the display.

In the U.S.A., the radiosondes of the late 1930s (and used for many years thereafter) required all sensors to be resistive. Since the hygroscopic-salt resistive humidity sensor had just recently been developed by Dunmore (see Section 13.3.1), it was chosen for radiosonde use. The telemetry circuit was so designed that a resistance change from 5 megohms to 5 kilohms was needed for a humidity range of 15 to 100 %RH. This range of resistances remained typical for the lithium-chloride humidity sensor for a long time. The applications of these sensors soon extended from radiosondes into other fields, and manufacturers developed improvements to and variations of the sensor design. It was soon determined, for example, that many applications did not call for measuring such a wide range of humidities and that the performance of hygroscopic-salt sensors improved if they were designed for narrower portions of the 15 to 100 %RH range. When such improved performance was desired for the entire range, several sensors, each for a narrow portion of the range, could be employed together. As the development of electronic humidity sensors progressed, types other than lithium-chloride sensors became available for radiosonde use.

*Hygroscopic-salt* humidity sensors have been fabricated in the form of rectangular wafers as well as cylindrical elements. The wafer is typically made of a plastic material such as polystyrene. Electrodes are printed on both sides of the wafer, and a humidity-sensitive coating consisting of an aqueous solution of the hygroscopic salt in a plastic binder is applied to both surfaces. The printed electrodes have a zigzag pattern to optimize detection of the resistance changes. Probe-type elements have been constructed with a bifilar winding of palladium wire over a polystyrene bobbin, with the hygroscopic-salt coating then applied over the winding. Hygrometric sensors using the resistance changes, due to ion exchange, of the surface of a sulfonated polystyrene substrate can also be wafers with zigzag electrodes printed on one or both surfaces, or they can be cylindrical. The probe shown in Figure 13–5 uses such an element, which incorporates a temperature-compensating thermistor. A platinum resistance thermometer is integrally packaged with the humidity-sensing element so that simultaneous measurements of temperature and relative humidity can be obtained. A perforated stainless-steel cover protects both sensing elements.

A *carbon-film* resistive hygrometer has been used for radiosonde applications. The wafer-shaped or cylindrical element is made from acrylic plastic, provided with metallized electrodes and coated with a carbon-powder suspension in a gelatinous cellulose carrier. The resistance of such sensors increases with increasing relative humidity, whereas it decreases in hygroscopic-salt sensors. However, the carbon-film sensors are still capable of meeting the same kilohms-to-megohms resistance change required for radiosonde use, but with the lowest resistance (about 15 k$\Omega$) at 10% RH. Development was also started on other types of resistive hygrometer elements, such as lead iodide on glass, polyelectrolyte combinations with ion-

**FIGURE 13–5.** Relative-humidity/temperature sensor probe; sulfonated-polystyrene hygrometer element is mounted on the rear of the element assembly, which shows the platinum resistance thermometer mounted to its front side. (Courtesy of General Eastern Corp.)

exchange resins, and cerium titanate, but these developments have apparently not resulted in satisfactory production designs.

The *aluminum-oxide* (resistive-capacitive) hygrometer element has seen additional development since its original conception prior to 1960. Typical designs now use very small wafer-shaped elements (see Figure 13–6). Associated signal-conditioning and readout equipment can indicate dew point/frost point or moisture content in parts per million by weight. The sensor is unaffected by variations in pressure over wide ranges and can be used for in situ measurements in gases as well as liquids.

*Electrolytic* hygrometers are used for measurements of moisture content of a variety of gases. Those using a phosphorous pentoxide ($P_2O_5$) sensor for electrolyzing the water vapor in a gas sample, such as the unit shown in Figure 13–7, can be used for most gases except corrosive gases, alcohols, ammonia, and unsaturated hydrocarbons (i.e., fluids that combine or interact with $P_2O_5$). The unit illustrated has gas inlet and outlet ports in its rear panel and a flow-control valve on its front panel. The sensor, internal to the unit, is connected to the ports, valve, and an internal differential-pressure regulator by plumbing. The sensor consists of a bifilar winding of inert electrodes on a fluorinated hydrocarbon capillary coated with a thin film of partially hydrated $P_2O_5$. Direct current is applied to the electrodes so that the water absorbed by the $P_2O_5$ is dissociated into hydrogen and oxygen; this current is directly proportional to the number of water molecules elec-

(a)    (b)

**FIGURE 13–6.** Aluminum-oxide hygrometer: (a) probe assembly with sintered stainless-steel end cap protecting the sensor assembly; (b) sensing element, with lead wires, on its mechanical mount. (Courtesy of Panametrics, Inc.)

**FIGURE 13–7.** Electrolytic hygrometer system with internal $P_2O_5$ sensor. (Courtesy of General Eastern Corp.)

trolyzed. The current can then be displayed as parts per million by volume, at a specified sample-gas flow rate.

*Oscillating-crystal* humidity sensors rely on changes in the frequency of oscillation of a quartz crystal coated with a hygroscopic material for their operation. The crystal is connected into an oscillator circuit, normally mounted together with it. The crystal case can be equipped with inlet and outlet ports for sample-gas flow. The oscillator frequency is either read out directly or mixed with that of a reference crystal, and the beat frequency is then read out.

### 13.4.2  Psychrometers

Since many types of direct-indicating electronic hygrometers and dew-point sensors are now available, psychrometers have become less and less popular; however, some are still in use, and replacement units are available for those who have gotten accustomed to their use. Compared to the direct-indicating sensors, psychrometers have two disadvantages: the need for the use of a chart and the need for ventilation. Once the two temperature readings (from the "dry bulb" and the "wet bulb" thermometer) have been obtained, a *psychrometric chart* has to be consulted in order to determine humidity or moisture. Such a chart is shown in Figure 13–8. The term "bulb" is a carryover from the days when liquid-filled thermometer systems were used in psychrometers. The dry-bulb thermometer senses the ambient temperature. The wet-bulb thermometer is covered with a water-saturated wick and measures a temperature lower than ambient, due to evaporative cooling. The difference between the two temperatures is called *wet-bulb depression*, and relative humidity, at 1 standard atmosphere barometric pressure, can also be determined from a *psychrometric table* which, for given dry-bulb temperatures, shows %RH as a function of wet-bulb depression. At other

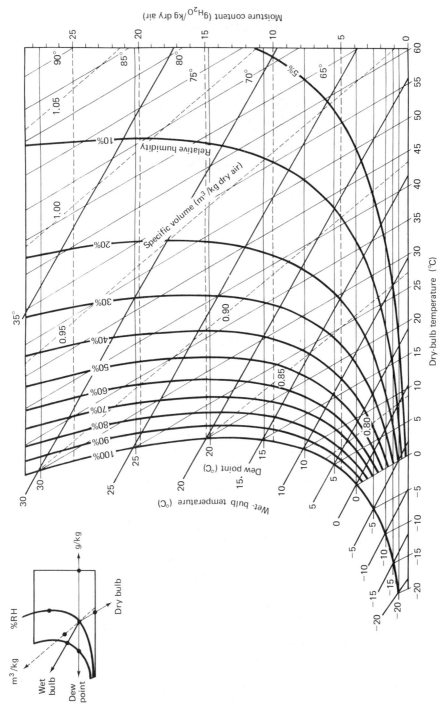

**FIGURE 13–8.** Psychrometric chart (at 101.32 kPa barometric pressure).

barometric pressures, corrections to the humidity values are required or alternative tables can be consulted.

To obtain a meaningful wet-bulb temperature reading, some means of forced ventilation must be used. For the *sling psychrometer*, which dates back to the late nineteenth century in its original form employing mercury-in-glass thermometers, this is done simply by holding the assembly by an attached hand sling and swinging it around. The ventilation rate should be on the order of 300 m/min. For electronic psychrometers this function is often performed by a blower or fan. When such a device is integrally packaged with the sensing assembly, the package is referred to as an "aspiration" or "aspirated" psychrometer.

In most electronic psychrometers the wet-bulb and dry-bulb temperatures are obtained by resistive temperature sensors (platinum-wire, nickel-wire, or thermistors). One design uses a resistive element for dry-bulb temperature in conjunction with a differential thermopile to measure wet-bulb depression; one set of thermopile junctions is covered with a wick, while the other set is placed close to the dry-bulb resistive element. A small water reservoir is sometimes included in electronic psychrometers for wetting the wick.

### 13.4.3 Dew-Point Sensors

Most electronic dew-point sensors are either of the cooled-condensation-surface type or of the saturated-lithium-chloride-solution type. The concept underlying the former is based on techniques developed during the nineteenth century involving polished silver condensation surfaces, visual observation of the condensation, and surface temperature measurements with mercury-in-glass thermometers. Modern devices based on this concept not only can detect very accurately the surface temperature at which condensation first occurs upon cooling (the dew point), but they can also track the dew point, as it varies, by using closed-loop cooler control. The measured fluid can be any of a large variety of gases at pressures to about 2000 kPa.

The operation of such dew-point sensors is illustrated in Figure 13–9. The thermoelectrically cooled (Peltier effect) mirror reflects light from light-emitting diodes (LEDs) to light sensors (phototransistors) when the mirror surface is dry. The mirror is cooled until condensate forms on it. The appearance of condensate on the surface causes light scattering. An optical sensing bridge detects the resulting change in light level and produces a signal that is used for closed-loop proportional control of cooler temperature. The mirror temperature is stablilized at the dew point, and variations in dew point are continuously tracked. The temperature of the mirror is sensed by a platinum-wire resistance thermometer embedded in it. The output of the temperature sensor, with appropriate signal conditioning, provides a continuous display of the dew point. Peltier-effect coolers offer the advantage

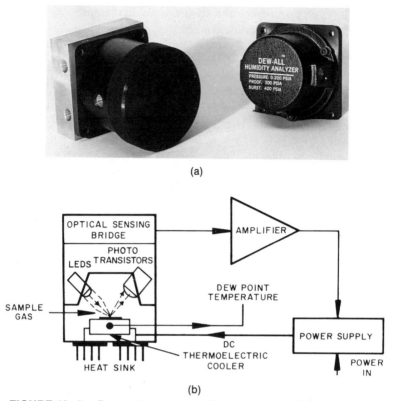

(a)

(b)

**FIGURE 13–9.** Dew-point sensor with electro-optical condensation detection: (a) sensor configurations; (b) sensing-system block diagram. (Courtesy of E G & G Environmental Equipment Div.)

of also being usable as heaters by simply reversing the polarity of the dc current supplied to them. Other, related designs use a separate resistive heating element to provide this function. The two sensor configurations shown each have inlet and outlet ports for the sample gas. The sensor at the left, which is provided with a threaded spin-off cover, is equipped with a coolant jacket (whose two ports are visible at the extreme left), an optional accessory. The power supply and control circuitry are contained in the electronics/display unit. The system can be augmented by an ambient-temperature sensor assembly and a microprocessor used to compute %RH, which can be displayed additionally and separately on the panel of the electronics unit.

The electro-optical method of chilled-mirror condensation detection is the most popular one. Other designs have used an inlaid metal grid in a mirror of insulating material; condensation is detected by monitoring the resistance across the grid. A mildly radioactive mirror surface (or integral portion thereof), in conjunction with semiconductor radiation detectors, has also been used for condensation detection. Partial absorption of the (alpha or beta) radiation occurs in the condensate, resulting in a drop in output counts of the detector. This output change can be used for closed-loop Peltier-cooler control and dew-point tracking.

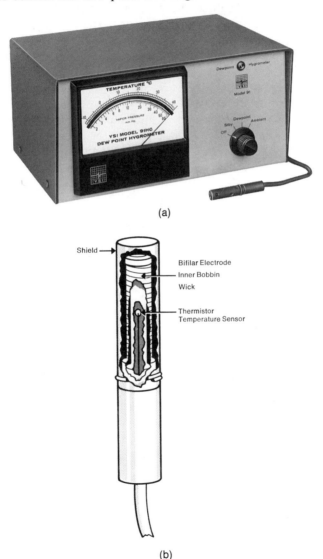

(a)

(b)

**FIGURE 13–10.**  Dew-point sensing system using saturated-lithium-chloride-solution sensor: (a) electronics/display unit with probe; (b) probe construction. (Courtesy of Yellow Springs Instrument Co., Inc.)

Figure 13–10 illustrates an example of a dew-point sensor of the saturated-lithium-chloride-solution type. Bifilar electrodes are wound on a wick covering a hollow bobbin. The wick is impregnated with a saturated solution of lithium chloride. When a current is caused to flow through the electrodes, and flows through the solution, the wick is heated by $I^2R$ heating. As moisture evaporates due to heating, the resistance of the wick increases and evaporation is reduced since heating is reduced. The wick (and bobbin) then start to cool and take on water again, and the heating increases. The cycle repeats until a heat/moisture equilibrium condition is reached. This equilibrium temperature can be directly related to the dew point. A thermistor thermometer inside the bobbin senses this temperature, and its resistance changes are electronically converted to output variations that are displayed in terms of the dew-point temperature. Other such designs can use a platinum resistance thermometer instead of a thermistor thermometer. Thermal shielding of the bobbin from flow-induced heat loads is an important design characteristic. An additional (sometimes optional) measurement of ambient temperature is often provided by such sensing systems.

## 13.5  TRANSDUCER CHARACTERISTICS

The essential characteristics of moisture and humidity sensors vary somewhat, depending on whether a hygrometer, a psychrometer, a dew-point sensor, or a remote moisture sensor is to be specified. *Mechanical characteristics* that should be shown include the sensor configuration and overall dimensions (and the same for any associated separate electronics). For fixed-mounted (as opposed to portable) sensors, all pertinent overall and mounting dimensions, including, for duct- or pipe-mounted probe-type sensors, the dimensions of the pressure-tight fitting should be shown. For measured fluids other than open air, fluid pressures and sensor compatibility with the fluids must be established. *Electrical characteristics* such as element-resistance range (and nominal resistance at a given humidity and temperature), power supply requirements, warm-up time, and output characteristics must be shown for sensors that are intended to be connected into the user's signal-conditioning and excitation circuitry. More commonly, the sensors are procured in conjunction with circuitry and often also complete excitation/conditioning/display equipment, furnished by the manufacturer. In this case only line-voltage power requirements are shown (or type of battery if the sensor is only battery operated). However, any significant limits on electrostatic and electromagnetic environmental conditions (conducted or radiated) should be stated.

*Performance characteristics* comprise primarily range, measured-fluid temperatures, and output and accuracy characteristics. For hygrometric sensors the range is usually shown in %RH, for psychrometric sensors in terms of wet-bulb and dry-bulb temperatures, and for dew-point sensors in terms of the dew point (temperature). For many dew-point sensors and some hygrometers, as well as for moisture sensors of other types (e.g., infrared, microwave), the range may be shown in terms of moisture content, in "ppm" (parts per million) or units equivalent to grams of water per kilogram of dry air. For psychrometric sensors, the wet-bulb depression range (dry-bulb minus wet-bulb temperatures) is also shown, for a given operating temperature range.

The various quantities used for range are interrelated as shown in the *psychrometric chart* of Figure 13–8. The chart, applicable at 1 standard atmosphere barometric pressure, shows how other quantities can be determined from the knowledge of two measured quantities. For examples, if the dry-bulb temperature (measured-fluid temperature) was 35°C and the dew point was measured as 15°C, then from the point where the two lines intersect the humidity can be read off as 30% RH. Had a wet-bulb temperature been taken instead of the dew point, that temperature would have been approximately 21.5°C. (Psychrometric charts in actual use are more finely graduated than the example shown, but some interpolation is still required for them.) It can further be seen that the moisture content, based on the two measured quantities, would be 10.8 g/kg. The dew point (at the stated barometric pressure) relates directly to moisture content. For example, following the dew-point line horizontally across the chart, a dew point temperature of $-10°C$ translates into a moisture content of 1.6 g/kg (or 1600 ppm). Further, for this dew point and a dry-bulb temperature of approximately $+8°C$, the specific volume can be read off as 0.8 m³/kg for air at the stated barometric pressure. Measured frost points have to be converted to dew points.

The output of the sensor can be shown in terms of the intrinsic transduction characteristic (i.e., resistance of the element, or impedance at a stated carrier frequency, or current, or voltage) for a stated humidity or moisture range. For most sensors the range limits, to which end points apply, are not at 0% and 100% RH (or equivalent values in other units). For some types of hygrometers, ranges such as 5 or 10% to 90 or 95% are more typical. Those are, of course, the ranges over which stated accuracy characteristics apply; for the same sensor, different accuracy characteristics may apply to portions of the measuring range above and below those limits. Since the output-vs.-measurand characteristics of most humidity sensors are not linear, accuracy characteristics are limited to repeatability and hysteresis.

Dynamic performance characteristics are shown in terms of a time constant, sometimes rise time or response time. This time is typically different for increasing vs. decreasing humidity or moisture. It is also temperature dependent and often strongly dependent on flow (or ventilation) rate; all applicable conditions should be stated. Measured-fluid temperature ranges are always shown, and limits of temperature ambient to the head of an immersion probe (immersed in a pressurized fluid) should also be shown, as should other conditions that may affect probe performance. Storage conditions for hygrometers usually involve keeping the sensor in its packaged condition together with a desiccant such as silica gel.

## 13.6   APPLICATION NOTES

Most humidity sensors are used to measure the relative humidity of air. The two fields in which humidity sensors are used most widely are meteorology (outdoor air) and heating and air conditioning (indoor air). In meteorological applications, humidity sensors (primarily hygrometers) are used in weather stations located in fixed positions on the ground, on ocean buoys, and in balloon-borne radiosondes; some aircraft are also specially equipped for weather monitoring. Many ground weather stations now operate autonomously and transmit their data via telemetry just like ocean buoys and radiosondes.

In heating and air-conditioning systems, humidity measurements are made partly for comfort control; much more frequently, however, they serve to control humidity in specific areas where humidity (and temperature) must be kept either below a specified level, above a specified level, or within a specified band. Examples of such areas are museums (especially art museums), libraries, computer rooms, warehouses, greenhouses, testing and calibration laboratories, and manufacturing facilities of many types, notably electronics assembly, paint application, plastic-film wrapping, printing, textile production, drying, pharmaceuticals processing, and "clean rooms" in general.

Moisture measurements are made primarily in industrial processes where the moisture content of gases, nonaqueous liquids, and solids must be known and controlled. Such measurements are often done on a sampling basis, but are increasingly made on-stream on a continuous basis. Remote moisture-sensing systems are particularly useful for continuous moisture-content measurements in which physical contact of a sensor with the measured material is undesirable—for example, material carried on conveyors, webs in paper and textile mills, and streams of certain process liquids. Among the many other areas where moisture content needs to be measured are soil mechanics, agriculture, and horticulture, where in situ as well as remote measurements are employed.

# 14

## Liquid Level Sensors

### 14.1 BASIC DEFINITIONS

*Liquid level* is the height of the surface of a liquid or a quasi-liquid (e.g., slurry or a powdered or granular solid) relative to a reference point.

*Continuous level sensing* refers to stepless sensing over a specified height range.

*Point level sensing* is a discrete measurement of the presence or absence of liquid at a specified height or location. A given installation may, however, employ two or more point level sensors.

The *interface* between two fluids, sometimes a liquid and a gas, but more often two nonmixing liquids, is another parameter detectable by level sensors.

Pressure methods of level sensing are based on a measurement of *head,* the height of a liquid column at the base of which a pressure is developed (see Section 14.3.1).

### 14.2 UNITS OF MEASUREMENT

Liquid level is usually expressed as a length dimension, the height of the liquid surface relative to a reference point.

Related measurements can be inferred from liquid level measurements by calculations of a type that can be handled easily by a microprocessor. Thus, when the geometry and dimensions of a tank are also known, the *volume* of the liquid can be determined. If, additionally, the density of the liquid is known, the *mass* of the liquid can be established.

## 14.3  SENSING METHODS

### 14.3.1  *Pressure Methods of Level Sensing*

A very common method of measuring liquid level is in terms of *head,* the height of a liquid column at the base of which a pressure is developed (see Figure 14–1). When the specific weight $w$ of the liquid is known, the level $h$ above the point at which the pressure $p_L$ is measured relative to the pressure above the liquid's surface $p_H$ is given by

$$h = \frac{p_L - p_H}{w}$$

When the tank is closed, the pressure difference $p_D$ must be measured by a differential-pressure transducer, whose ports are connected with plumbing to the top and bottom of the tank. For a liquid of a given specific weight, the output of the transducer is then directly proportional to the level of the liquid. When the gas above the liquid is pressurized (so as to force the liquid to flow from a port at the bottom of the tank), the gas pressure is known as *ullage pressure*.

A number of techniques of varying complexity are illustrated in Figure 14–2 for level sensing by differential-pressure measurement. The simplest measurement is that which can be made in a stationary open tank by flush-mounting a gage-pressure transducer close to the bottom of the tank (Figure 14–2(a)), so that the sensing element of the transducer is always wetted by

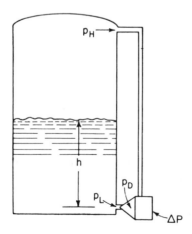

**FIGURE 14–1.**  Principle of pressure methods used for level sensing.

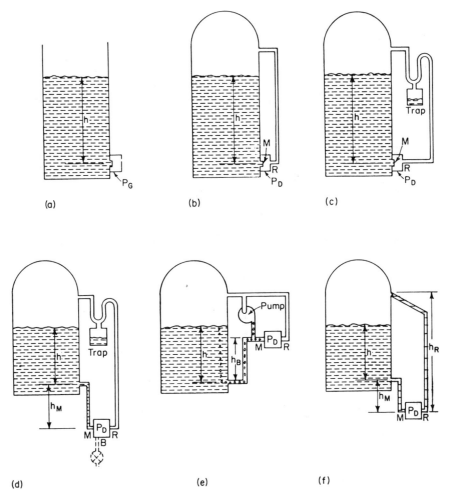

**FIGURE 14–2.** Level sensing by differential-pressure measurement: (a) open tank; (b) flush-mounted transducer; (c) moisture trap
in reference line; (d) lines to both ports; (e) bubbler; (f) transfer fluid
in reference line.

the liquid. The pressure $P_G$ sensed in this manner is directly equal to $hw$.
In the other examples illustrated, differential pressure $P_D$ is measured by a
transducer whose measurand port is identified by $M$ and whose reference
port is identified by $R$. The head of any gas in either line is assumed to have
a negligible effect on the accuracy of measurement. The specific weight
of the liquid is assumed to be much larger than that of the gas; when
this assumption cannot be made, the gas head must be accounted for in the
total measurement. The configuration shown in Figure 14–2(b), in which the
measurand port is flush mounted to the tank, can be used where there

is no possibility of moisture condensation in the reference-pressure line. When vapor in the ullage gas may enter and condense in the reference line, a moisture trap can be inserted in this line (Figure 14–2(c)) to prevent a liquid head of unknown height from building up in the line.

In many installations the pressure transducer cannot be flush mounted to the tank; instead, both of its ports must be connected to the appropriate points in the tank by means of plumbing. A simple scheme for accomplishing this is shown in Figure 14–2(d). The transducer is located below the level of the lower measuring point (*tap*) in the tank and is often equipped with a *bleed port* (*B*) so that any air or gas in the line can be bled off and only liquid then fills the line. The differential pressure sensed by the transducer is now $(h + h_m)w$, where $h_m$ is the (fixed) head of liquid in the measurand line, and the level measurement must account for this fixed head. *Bubbler* systems are useful when the transducer is located above the measurand tap in the tank, or when its sensing element should not come in contact with the liquid. For open tanks, purge air is bubbled at low flow into the tank through a standpipe; the (gage) pressure transducer is connected to the top of the standpipe and senses a back-pressure proportional to level (at constant density). For closed tanks, a pump (Figure 14–2(e)) can be used to bubble ullage gas through the measurand line, or a separate gas supply can be used for this purpose. The differential pressure is then equal to $hw - h_Bw_B$, where $w_B$ is the specific weight of the gas.

When it is undesirable to have the ullage gas come in contact with the reference port of the transducer (e.g., when the gas is corrosive or contaminating), the reference line can be filled with a *transfer fluid* having a specific weight (density) equal to or greater than that of the liquid in the (closed) tank. An elastic membrane is located at the reference tap to isolate the transfer fluid. Since the head $h_R$ in the reference line is always greater than the head seen by the measurand port of the transducer (when the latter is installed below the bottom tap of the tank), a negative differential pressure will be sensed, varying from zero when the liquid is at the reference tap to a maximum value when it has decreased to the measurand tap (Figure 14–2(f)). In all applications where a differential-pressure transducer is used and the ullage pressure can vary, the *reference-pressure error* of the transducer must be negligible (or known and accounted for in level determinations).

### 14.3.2 Level Sensing by Weighing

By weighing a tank of known geometry, subtracting the weight of the empty tank (*tare*), and allowing for the specific weight of the liquid, the level in the tank can be determined (see Figure 14–3). Such mass determinations are usually made by load cells. The tank can also be mounted in a weighing arrangement in which its tare weight is balanced by a ballast mass equivalent to the tare weight of the tank.

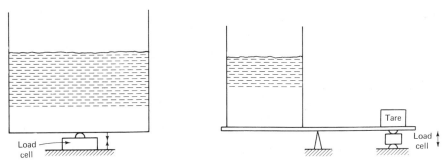

**FIGURE 14–3.** Level determination by weighing.

### 14.3.3 Buoyancy Level Sensing

Archimedes's principle that a body submerged wholly or partially in a fluid is buoyed up by a force proportional to the mass of the fluid displaced is put to use in level sensors whose sensing element is a float, either hollow or made of a material lighter than the measured fluid (Figure 14–4). The up/down motion of the float relative to the (fixed) case of the sensor is converted into an output, indicative of level, by a transduction element in the case. Potentiometric or reluctive transduction is typically used in continuous-level sensors. A magnetic reed switch in combination with a permanent magnet, which can be embedded in the float, has been popular in point-level sensors. When the transducer case can be wetted by the measured liquid, the case as well as the actuation mechanism must be hermetically sealed. In a related sensing method, the float (*displacer*) does not actually move; rather, the force acting on it due to buoyancy is transduced by an appropriate transduction element, typically of the strain-gage or force-balance type.

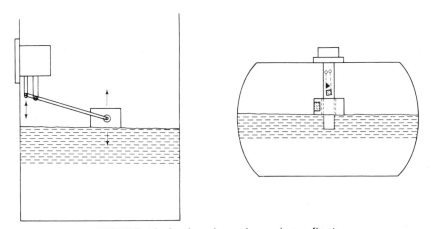

**FIGURE 14–4.** Level sensing using a float.

### 14.3.4   *Conductivity Level Sensing*

The level of an electrically conductive liquid can be sensed by two electrodes in contact with the liquid by monitoring the change in resistance between the electrodes (Figure 14–5). The conductivity of the liquid can be used for continuous-level as well as for discrete-level indications. The tank wall, if metallic, can be used as one of the two electrodes.

### 14.3.5   *Capacitive Level Sensing*

The dielectric constant of a liquid is usually different from that of the air or any other gas above it. When one or more pairs of electrodes (Figure 14–6) are immersed in a liquid, the variation in dielectric due to rising or falling liquid level will cause a change in capacitance between electrode pairs. The tank wall, if metallic, can be used as one electrode of a pair. This principle is applicable to continuous-level as well as point-level sensing. The sensing element can be configured as two or four coaxial tubes, with alternate tubes ganged together when more than one pair of electrodes is used. A four-arm ac bridge network is typically used, with the level-sensing capacitive element constituting one arm of the bridge. Accuracy can be improved by placing a second capacitive element, which remains submerged, below the level-sensing element to compensate for changes in characteristics of the liquid. As indicated in the illustration, this reference capacitive element $C_R$ sees a fixed head $h_R$ which constitutes a small fraction of the head, or capacitive-element height $h$, sensed by capacitive element $C$. The tank level is measured between the top of $C_R$ and the top of $C$, and the level is then determined by $h/h_R = \Delta C/\Delta C_R$.

### 14.3.6   *Heat-Transfer Level Sensing*

The rate of heat transfer from a heated element is generally larger to a liquid than to a gas. This principle is applied in several point-level sensor designs (Figure 14–7). Resistive elements are typically used for the purpose (ther-

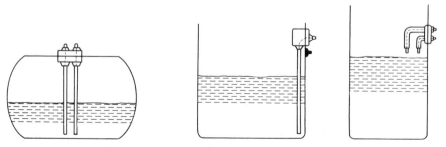

**FIGURE 14–5.**   Conductivity level sensing.

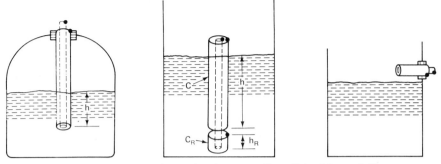

**FIGURE 14–6.** Level sensing by capacitor dielectric variation.

mistors are popular for such applications), with a current through them sufficiently high to cause some self-heating. When the level rises so that the liquid comes in contact with the warm element, the element will be cooled. The resulting step change in the resistance of the element is used for a point-level indication. Another method employs a thermocouple to sense the change in temperature of the element.

### 14.3.7 *Photoelectric Level Sensing*

Photoelectric sensing methods are used for point-level sensing in two different modes (Figure 14–8). In the transmittance mode, the light beam from a source to a sensor, installed either in the opposite tank wall or immediately below the source, is attenuated when the level of the liquid rises into the optical path. In the reflection mode, an optical prism is so arranged that less light is reflected back to a light sensor when the prism is immersed in liquid, due to the change in the index of refraction, than when it is immersed in gas.

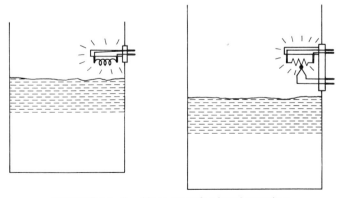

**FIGURE 14–7.** Heat-transfer level sensing.

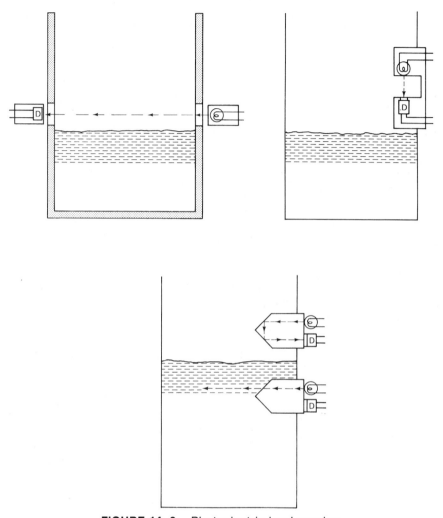

**FIGURE 14–8.**   Photoelectric level sensing.

### 14.3.8   Damped-Oscillation Level Sensing

The change in damping of an oscillating element when its ambient fluid changes from gas to liquid is used in two types of point-level sensors (Figure 14–9). One type uses a vibrating paddle whose oscillation amplitude is reduced, due to increased viscous damping, when the paddle is submerged in liquid. The amplitude changes are detected by a (typically reluctive) transduction element. The other type utilizes a piezoelectric or magnetostrictive element so designed that it oscillates in a gaseous medium but stops oscillating, due to acoustic damping, when the medium changes to liquid. Os-

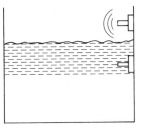 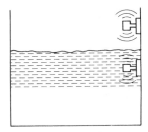

**FIGURE 14–9.** Level sensing by oscillation damping.

cillation frequencies are chosen in the ultrasonic range for the latter type, and in the low audio-frequency range for the former type.

### 14.3.9 Ultrasonic Level Sensing

Two categories of ultrasonic sensing methods are employed for level sensing: cavity-resonance sensing and sonic-path sensing (Figure 14–10). The damped-oscillation ultrasonic sensors described in the previous section could be considered a third category, but deserve separate treatment.

The cavity-resonance method (Figure 14–10(a)) is a volume-sensing technique from which level can be derived. Electromagnetic oscillations at ultrasonic or radio frequencies are excited within the cavity bounded by the tank walls and liquid surface from a coupling element at the top of the tank. As the liquid rises, the cavity volume shrinks and its resonant frequency changes accordingly. When the resonant frequency of the empty tank is known and a scaling factor is applied, the volume or level of the liquid can be determined. Variable-frequency oscillators can be used for the resonant-frequency search. The radio-frequency method can be used with dielectric liquids.

Sonic-path methods can be used for continuous-level as well as discrete-level sensing; the reflectance mode is commonly used for the former and the transmittance mode for the latter (see Figure 14–10(b)). For continuous-level sensing, either a separate transmitting and receiving element or a single element, operating alternately in the transmitting and receiving mode, can be employed. Pulsed ultrasonic energy is directed at the liquid/gas interface, and the travel time of the pulse reflected back by this interface is measured. When the velocity of sound in the fluid through which the pulse travels is known, the distance between the transmitting/receiving element(s) and the interface, and hence the liquid level, can be determined. Point-level sensors normally use a transmitter and a receiver. When liquid enters the sonic path between the two, the amount of sound energy at the receiver is attenuated significantly. (Circuit elements can even be adjusted so that the receiver output drops to zero.) With the exception of the *gap* type of point-level sensor, transmitting and receiving elements can be of either the wetted

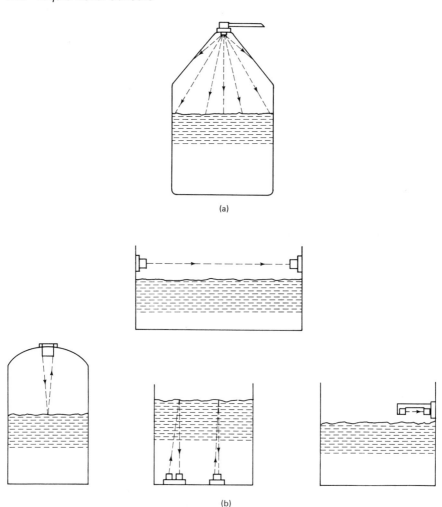

**FIGURE 14–10.** Ultrasonic level sensing: (a) cavity-resonance method; (b) sonic-path methods.

or externally mounted type; the latter, however, may not be feasible in some installations, or its use may be precluded by characteristics of the measured fluid.

### 14.3.10  Nucleonic Level Sensing

Radiation emanating from a radioactive source at a constant rate will reach one or more detectors located at the opposite wall of the tank to a lesser degree when the path is through liquid than when it is through gas. Figure 14–11 shows typical sensing configurations, in which $S$ indicates the source

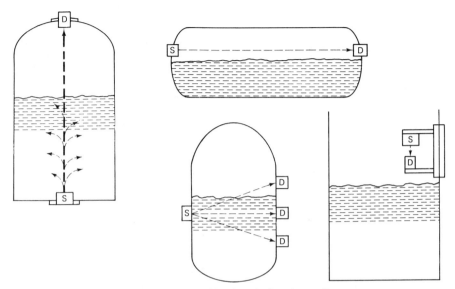

**FIGURE 14–11.**  Nucleonic level sensing.

and *D* indicates a detector. Gamma radiation from a source such as $^{137}$Cs, $^{60}$Co, or $^{226}$Ra is most commonly used. Its attenuation by liquid is caused mainly by absorption. Single or multiple point-level sensing is accomplished by one source and one or more detectors located at horizontally opposite tank walls. Continuous-level sensing can be achieved by using a vertical radiation path and monitoring the detector output, which will decrease with rising level. Other means of continuous-level sensing involve either the use of long vertical sources and detectors in the form of strips or an electro-mechanical servo control system which raises and lowers the source and detector simultaneously so that they follow the liquid/gas interface. In most installations nucleonic level-detection systems can be installed or placed at the outside of the tank (or other vessel); this is their major advantage.

### 14.3.11  Microwave Level Sensing

The changes in transmittance or reflection of microwave energy when the fluid between a transmitter and a receiver changes from gas to liquid (or quasi-liquid) have been used for point-level sensing (Figure 14–12). When the container wall material is relatively transparent to microwaves, the trans-mitter and receiver can be mounted external to the container walls. The attenuation of the microwave energy, used for level indication, is dependent on the interaction between instrument characteristics (angle of incidence, wavelength, polarization) and material characteristics (conductivity, per-mittivity, permeability, surface quality). Pulsed energy at a frequency around 10 GHz is typically employed.

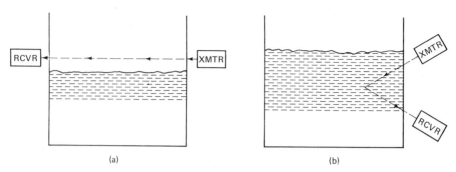

**FIGURE 14–12.** Microwave level sensing: (a) transmittance; (b) reflectance.

## 14.3.12    Superconductive Cryogenic Level Sensing

Certain metals become superconductive at low cryogenic temperatures; that is, their resistance drops to zero. This principle has been applied to level determinations of liquid helium. Point-level sensing has been achieved with, for example, a niobium film on a quartz substrate. The resistance of the film becomes zero when it is in contact with helium at a temperature slightly above its boiling point; hence, the niobium film is superconductive when wetted by liquid helium. Use of this principle for continuous-level indications has also been demonstrated, using a niobium–titanium wire connected to a low-capacity constantan-wire heater which is located in the region in which helium is expected to remain in its vapor phase. A portion of the vertically suspended wire is thus sufficiently heated to be nonsuperconductive ("normal"), whereas the portion submerged in liquid helium dissipates this small amount of heat rapidly enough to become superconductive. As the level of the liquid helium rises, an increasing portion of the wire becomes superconductive and the resulting change in the wire's end-to-end resistance can be converted into an *IR* change (i.e., can be read out as a varying voltage).

## 14.3.13    Variable-Coupling Liquid-Metal Level Sensing

The variation in mutual inductance between two windings of a transformer due to the relative presence of a conductive liquid has been used for sensing levels of liquid metals. An example of a device employing this technique is a ceramic mandrel, intended for vertical immersion and continuous-level sensing, wound bifilarly with primary and secondary windings of nickel-alloy wire. The wound element is protected by a metallic sheath or thermowell. When excited by an ac current, the primary winding causes eddy currents in everything that is adjacent and metallic, including the liquid metal that is wetting the sheath or thermowell. These eddy currents increase as the level of liquid metal rises and wets more of the thermowell. As a result, the amount of current induced by coupling into the secondary winding decreases. The output of the secondary winding is, therefore, inversely proportional to the liquid level. Levels of metals having boiling points up to about 700°C have been measured using this technique.

## 14.4  DESIGN AND OPERATION

This section covers designs of the more commonly used level sensors. Pressure transducers used for level sensing employing pressure methods are covered in Section 15.4.

### 14.4.1  Float-Type Level Sensors

Most of the sensor designs that use the buoyancy force acting on a float are point-level sensors. A design concept that is often employed in such a sensor uses a permanent magnet, embedded in the float, and a magnetically actuated transduction element (e.g., magnetic reed switch or Hall-effect switch) in the portion attached to the tank, vessel, or duct. Thus, the actuated element and its wires are inherently sealed. Designs using mechanical coupling between the float and the transduction element need some sort of seal, such as a bellows, to isolate the element and wiring from the measured liquid. Top-mounted sensors generally use a toroidal float, whereas side-mounted sensors typically use a spherical or cylindrical float.

The actual motion of the float is rarely equal to the full range of liquid-surface changes, simply because there is no need for such a large float motion. Instead, the float typically moves only a few millimeters from its non-actuating to its actuating position. However, when two or more magnetic elements need to be actuated, so as to indicate various levels, the float travel must be appropriately longer.

The relatively few continuous-level sensor designs available either use a full-travel float to actuate a potentiometric or reluctive transduction element, or sense the buoyancy force acting on an essentially motionless float and then employ strain-gage or force-balance transduction to provide an analog output signal proportional to level.

A derivative of the float-type designs is the electromechanical plumb-line type of level sensor (Figure 14–13). A plumb line wound on a drum is unwound until a float (for liquids) or weight (for granular solids) touches the surface. An angular-displacement transducer, e.g., an incremental encoder (see Section 5.4), provides an output signal indicative of the number of turns, and fraction of a turn, of the drum and, hence, of the distance between the drum (whose location is known) and the float or weight that has come to rest (whose position is thus being measured). As soon as the float (or weight)

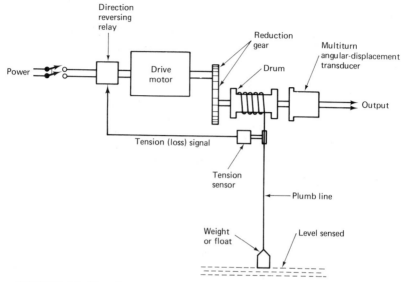

**FIGURE 14–13.** Operating principle of plumb-line type of electro-mechanical level sensor.

touches the surface, tension in the plumb line is lost. A tension sensor detects this and sends a signal to a polarity-reversing relay which then causes the drum drive motor to reverse its rotation so that the plumb line is wound back onto the drum. The motor stops when the plumb line is wound up completely, and the system is then ready for the next level measurement. The resolution of the system depends primarily on the number of output pulses generated per turn of the drum.

Flexible steel tape is often used as the plumb line, and a number of design variations of the system exist, including a design in which a disc-shaped float is balanced by a counterweight (float and counterweight both hang down).

### 14.4.2 *Conductivity Level Sensors*

Conductance is the inverse of resistance. Many liquids are more or less electrically conductive, thus allowing the use of conductivity probes for point-level sensing. A typical probe consists of a metal rod (usually stainless steel or titanium) coated through most of its length with an insulating material (e.g., glass, ceramic, plastic) so that only its tip comes into electrical contact with the measured liquid. The rod is attached to, but electrically insulated from, a threaded or flanged mounting head which also provides the external electrical connections.

In most installations the tank wall is electrically conductive; the conductance path can then simply be formed between the probe tip and the tank wall. If the inside surface of the tank wall is electrically nonconductive, a second probe must be used. Dual-level probes contain two rods next to each other, one short (for "high" level sensing), the other long (for "low" level sensing). Conductivity probes are most frequently installed vertically, from a mounting boss in the top of the tank.

The conductivity probe is typically connected electrically as one leg of an ac-excited Wheatstone bridge network. The bridge can be adjusted to allow for the conductivity of the liquid, or even for a range of conductivities. The current through the liquid should be kept as small as possible, to minimize electrolysis and to prevent a possible explosion hazard. Conductivity probes should be designed for ease of removal and reinstallation, since they require relatively frequent cleaning.

### 14.4.3 Capacitive Level Sensors

Continuous-level and point sensors whose capacitance changes as the dielectric between one or more pairs of electrodes changes exist in a number of different configurations. Capacitance increases with rising level of the liquid or quasi-liquid, whose dielectric constant is greater than the air or other gas it replaces. A single probe, in the form of a metal (typically steel) rod insulated from its mounting head, can be immersed vertically downward into a tank and act as one electrode of a capacitor when the tank walls are conductive and the measured fluid has a reasonably high dielectric constant. When the tank walls are nonconductive, or when the measured fluid has a low dielectric constant, the metal rod can be mounted within a coaxial metallic cylinder which acts as the second electrode. This cylinder is partly perforated or slotted to permit free flow of liquid and gas. Two or four concentric cylinders can be used for liquids having very low dielectric constants (e.g., some liquified gases). When severe foaming or sloshing is expected, a stillwell can be built into the tank and the probe can be immersed into the stillwell.

Single-electrode probes can also be dagger shaped or provided with rectangular extension plates to increase their surface area when a greater capacitance change is needed. Probes used for point-level sensing are often covered with an antifouling coating for a portion of their length. For very long immersion depths the electrode can take the form of a weight at the end of a cable ("cable probe" or "rope probe"). Point-level probes are best side mounted so that the entire probe length becomes the active electrode sensing the capacitance change when the probe becomes wetted by the measured liquid. Probe design, material, and finish should minimize any adherence of liquid when the level falls. Plate- and blade-type probes are often more suitable for this reason than rod-probes when the measured material is a granular solid. Completely coated insulated probes can be used in conductive liquids; the liquid then acts as the grounded electrode whose height, relative to the probe, increases with rising level. In such cases the changes in capacitance as well as liquid resistance can be measured as changes in impedance (or admittance).

The capacitive element formed by the probe, or by the probe and tank wall, is usually connected into one arm of an impedance-bridge circuit excited by low-voltage ac at frequencies between about 400 Hz and over 10 kHz. The output due to bridge unbalance is then amplified and conditioned to either actuate a relay, provide a gated pulse, or furnish an analog or digital continuous display. The cable between probe and electronics should be low-capacitance coaxial shielded cable which should also be constrained as to its length. Many capacitive level sensors overcome the cable problem by having their bridge circuit and amplifier, and sometimes further signal conditioning, packaged into the mounting head of the probe.

### 14.4.4  Heat-Transfer Level Sensors

The principle of heat transfer to the measured fluid from a heated element in a probe is used in some point-level sensor designs. Heat transfer is more rapid when the fluid is a liquid than when it is a gas. Thermistors (see Section 19.3.3), operated at sufficient current to produce some self-heating when they are in air (or another gas), are quite suitable for such applications. As the thermistor comes in contact with liquid, its resistance undergoes a step change due to cooling. This resistance change can be converted into a voltage change by such means as a voltage-divider or bridge circuit.

Various heat-balance designs are also used for point-level sensing. One design employs a heated thermocouple sensing junction at the probe tip, with its reference junction (see Section 19.3.1) located a small distance away from it but still within the probe envelope. When immersed in liquid, the sensing and reference junctions are at approximately the same temperature. When immersed in gas, the sensing-junction temperature rises above that of the reference junction; as a result, a measurable thermal emf is produced and can be amplified for display or control purposes. Another heat-balance design uses the thermal coefficient of expansion to cause a relay-type contact to be held open when a heated metal rod, provided with a good heat conduction path to the measured fluid, remains heated because the probe is in air. When the probe is in liquid, the metal rod cools due to the increased heat transfer into the liquid, and subsequent contraction causes the contact to close.

Besides thermistors, wire-wound resistive elements which are excited to produce self-heating when in gas ("hot-wire probes") have been used for liquid level sensing, notably for cryogenic liquids. Exposed-element probes, either wound around a hollow mandrel or wound, as planar winding, in the form of a grid supported by small insulating studs, have been used in such applications. The amount of self-heating must be carefully controlled to cause only *nucleate boiling,* that is, the formation of vapor bubbles at various points of the element, which leave the element and rise to the surface (usually recondensing into liquid before they reach the surface). With further heating the *film boiling* phase is reached, in which an unstable film of vapor is formed along the wire surface, greatly reducing heat transfer from the probe to the liquid since the film forms an insulating layer. Hot-wire probes can be connected into a Wheatstone-bridge circuit, or they can be excited from a constant-current source so that the voltage across their terminals changes when their resistance changes.

### 14.4.5  *Photoelectric Level Sensors*

Photoelectric sensors, in the form of a light source installed in one wall and a light sensor installed in the opposite wall, have been used for the detection of fill levels in hoppers and other vessels, mostly with granular solids. When the material reaches the level at which the light beam crosses the vessel the *transmittance* is either interrupted or greatly reduced, and this causes a drop in the output of the photosensor. The resulting output changes can then be used for level indication or control.

Point-level sensors using the index of refraction of the measured liquid to change *reflectance* have the advantage of requiring only one mounting boss. This sensing method is employed in the sensor shown in Figure 14–14. The operating principle is illustrated in Figure 14–15. An optical prism is so designed that a light beam, from light source to prism, is reflected back to a detector in the sensor when immersed in gas. When immersed in liquid, however, the index of refraction of the liquid in contact with the prism causes most of the light energy to be directed into the liquid, so that virtually no light is reflected back to the light detector (see Figure 14–15(a)). The mounting head of the sensor contains the LED light source and the photodiode light detector and its preamplifier. Fiber-optic cables conduct the light beam to and from the prism; hence, the prism and probe stem conduct only light and are nonelectrical in nature, an advantage when the measured liquid is potentially flammable or explosive. An electronics unit connected to the sensor by cabling (see Figure 14–15(b)) generates a pulsed current to the LED and receives the pulsed signal from the photodiode preamplifier. The two pulse streams are of different amplitudes so as to avoid erroneous indications. The signal from the sensor is then conditioned into an actuating relay which, in the illustration shown, turns off a pump. In other applications it might, for example, turn on an indicator light.

### 14.4.6  *Vibrating-Element Point-Level Sensors*

Viscous damping will cause the oscillation amplitude of a vibrating mechanical element to be reduced when the ambient fluid around the element changes from gas to liquid. A simplified typical design of a sensor using this principle is shown in Figure 14–16.

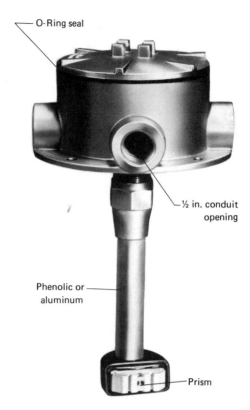

**FIGURE 14–14.** Photoelectric point-level sensor using light refraction. (Courtesy of MICRO SWITCH, a Division of Honeywell.)

The vibrating element is a paddle supported by two thin rods held by seals at their node points (points of minimum oscillation amplitude). One rod is driven so that it causes the paddle to oscillate. The other rod leads to the (reluctive) transduction element, which detects reductions in paddle vibration amplitude when the paddle sees liquid and then provides an appropriate output signal. The oscillation frequency is typically in the low audio-frequency range and can be at or near the line-voltage frequency. Damped-oscillation level sensors operating at ultrasonic frequencies are discussed next.

### 14.4.7 Ultrasonic Level Sensors

Ultrasonic level sensors employ one of three basic sensing methods: cavity-resonance, damped-oscillation, and sonic-path sensing. The sensors share one characteristic: their operating frequency is in the ultrasonic range, that

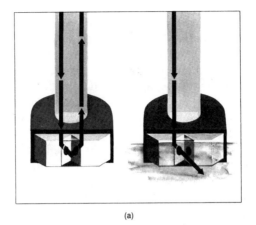

(a)

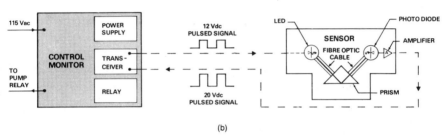

(b)

**FIGURE 14–15.** Operation of light-refraction-type photoelectric point-level sensor: (a) operating principle; (b) block diagram. (Courtesy of MICRO SWITCH, a Division of Honeywell.)

is, above the audio-frequency range (and generally well into the radio-frequency, or RF, range). Sensors specifically designed for the cavity-resonance method (acoustic resonance) are relatively rare and have been used only for specialized applications.

*Damped-oscillation* ultrasonic point-level sensors can be of the pie-

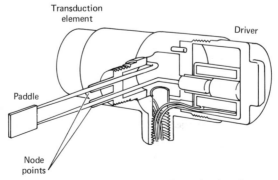

**FIGURE 14–16.** Vibrating-paddle point-level sensor.

zoelectric or magnetostrictive type. *Piezoelectric* sensors employ a quartz or ceramic crystal mounted in the tip of a hermetically sealed probe sheath. The crystal, driven by an oscillator circuit, resonates at normal amplitude when immersed in gas and at a significantly reduced amplitude when immersed in liquid. The amplitude change is detected, and the detection elements of the circuit provide the discrete output signal. *Magnetostrictive* sensors use an assembly consisting of a drive coil and a feedback coil, both wound on the same ferromagnetic rod, whose tip is in physical contact with the inside surface of the tip of a hermetically sealed probe sheath. The rod (and probe tip, as in piezoelectric sensors) are set into vibration by an oscillator connected to the drive coil. Two effects are employed in the magnetostrictive assembly with its associated circuitry to maintain the rod in longitudinal elastic vibration at a frequency controlled by the characteristics of the rod (typically around 40 kHz): (1) when an iron rod is subjected to a longitudinal magnetic field, it increases slightly in length, whereas a nickel rod would decrease in length under these conditions (*Joule effect*); (2) a change of magnetic induction occurs within the ferromagnetic rod under longitudinal stress (*Villari effect*). The drive and feedback coils are connected in a current-driven feedback oscillator, with signal levels so adjusted as to maintain oscillation in the rod only when the probe tip is exposed to a compressible fluid such as air, froth, or foam. As soon as the probe tip encounters a noncompressible fluid (a liquid), the vibration is damped and the oscillation stops; output-conditioning circuitry then provides a resulting discrete output signal. The circuitry can be contained in a separate box, with interconnecting cables kept short to minimize their capacitance, or it can be packaged into the probe head. The liquid sensed by such point-level sensors should not contain any material that may remain on the probe tip and dry and then harden.

Sonic-path ultrasonic sensors are used for continuous-level sensing in the reflectance mode and for point-level sensing in the transmittance mode. For both types of sensing, either a single element (usually a piezoelectric crystal), acting alternatingly as transmitter and receiver, or a pair of elements, one acting as transmitter the other as receiver, can be employed. Continuous-level sensors generally see the surface of the liquid from the top of the tank, but sometimes (particularly when the level of an interface between two dissimilar liquids is to be sensed) from the bottom of the tank. Typically, the transmitter is excited by repetitive pulses of ultrasonic energy which are then reflected back to the receiver, and the time elapsed between pulse transmission and reception is a measure of the distance between the sensors and the surface and, hence, of level.

Point-level ultrasonic sensors can be installed from the top of a tank (when the surface rises close enough to the top that the probe need not be too long) or from the side. Many such sensors are of the gap type; that is, a single assembly contains both receiver and transmitter separated by a relatively small distance, and a discrete output signal is generated when liquid (or granular solids) fill the gap. A sensor of this type, for use with liquids, is shown in Figure 14–17. The gap is sized by the thin metal rod between the probe body and probe tip. Body and tip contain the transmitter and receiver, respectively, facing each other across the gap. All associated excitation and signal conditioning, including the relay actuated by output-conditioning circuitry, is contained in an assembly mounted under the cover. The electronics assembly operates directly off the line voltage. (A power supply is also included in the package.) Terminals are provided for a normally open as well as a normally closed relay contact.

Sensor pairs installed separately and facing each other across the inside of a pipe, hopper, tank, or open duct are used in many applications. Long gap-type probes can be equipped with two gaps, with a crystal pair associated with each gap, where one gap corresponds to a "low" level and the other to a "high" level. Special sensor designs are used for point-level detection of the interface between two liquids. Some applications permit the use of externally mounted (cemented or clamp-on) sensors.

### 14.4.8  Nucleonic Level Sensors

Nucleonic level sensing systems consist of three major elements: a source of nuclear radiation, usually gamma rays; one or more radiation detectors; and the electronics associated with the detector. For point-level sensing, a source emitting a narrow conical beam is typically used, mounted opposite the radiation detector. For (multi-) point-level and continuous-level sensing, a source emitting a fan-shaped beam is used in conjunction with either a long detector (see Figure 14–18) or two or more individual detectors.

The *radiation source* contains the radioactive isotope that emits the gamma rays. Americium ($^{241}$Am, energy 0.066 MeV, half-life 455 years) and radium ($^{226}$Ra, energy 1.5 MeV, half-life 1,620 years) have been used in some sources; however, cobalt ($^{60}$Co, energy 1.25 MeV, half-life 5.5 years) and, especially, cesium ($^{137}$Cs, energy 0.66 MeV, half-life 30 years) are most commonly used. The source is always enclosed in a *source holder,* which not only provides shielding of the activated source in all directions except the viewing direction and must meet rigid radiation safety standards, but also provides for a (usually mechanical) "shutter" or positioning device that either rotates the source away from its viewing port or places a radiation shield between it and the viewing port. A source holder with the source in the "off" position must be capable of meeting radiation safety standards in all directions, so that it is completely safe during shipping, handling, installation, and maintenance.

The *detector* is usually a Geiger-Mueller tube (Geiger counter) or an ionization chamber (see Chapter 23). The *electronics* comprises those items necessary to provide excitation potentials to the detector and condition the

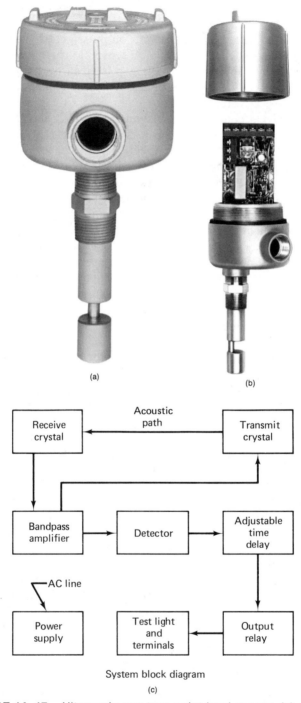

(a)

(b)

Acoustic path

Receive crystal → Transmit crystal

Bandpass amplifier → Detector → Adjustable time delay

AC line

Power supply

Test light and terminals ← Output relay

System block diagram

(c)

**FIGURE 14–17.** Ultrasonic gap-type point-level sensor: (a) assembled unit; (b) shown with cover open; (c) system block diagram. (Courtesy of Xertex Corp./Sensall.)

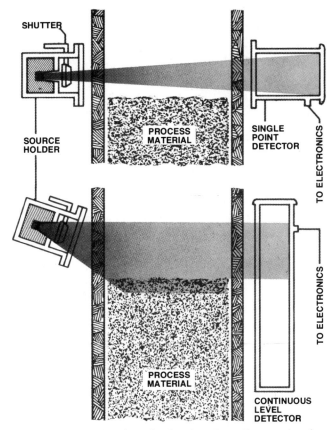

**FIGURE 14–18.** Nucleonic point-level and continuous-level sensor systems, schematic representation. (Courtesy of Kay–Ray, Inc.)

output of the detector. Ion chambers, with their associated electronics, have been used to form the "long" detector needed for continuous-level monitoring. Level sensing is based on the amount of radiation absorbed in the path between source and detector. The thickness and density of the material in this path are usually expresed in terms of *half-value thickness* (inversely proportional to material density), or "half-value layer," the thickness of a material necessary to reduce incident radiation by 50%. Hence, two half-value thicknesses reduce the radiation by 75%; for example, for a $^{137}$Cs source, the half-value of thickness of steel is 1.5 cm, and two such thicknesses (or layers), i.e., 3 cm of steel, would reduce an incident radiation of 10 milliroentgens per hour (mR/h) to 2.5 mr/h. The safety of the installation can be determined from the half-value thickness (the radiation penetrates two wall thicknesses and no additional material when only air or gas is in the transmission path). The required radiation at the source, in millicuries, also depends on the container half-value thickness, as well as on the measured material, the detector threshold (dosage required at the detector), a dosage constant given by the type of radioactive material used, and the square of the distance between the source and the detector. When more than one point-level detector is used, the difference in transmission path length due to angle must be considered in sizing the source and adjusting the gain of each detector.

## 14.5   TRANSDUCER CHARACTERISTICS

Mechanical characteristics of level sensors are given by their specification drawing supplemented by statements of such items as operating and proof-pressure rating, materials in contact with the measured fluid, recommended installation method, and rating of the probe and head in terms of applicable industrial or governmental codes (e.g., whether it is explosion-proof, how it responds to a corrosive atmosphere, and its safety as regards radiation). The latter also applies to any separately mounted electronics unit. Outline and mounting dimensions and details of mounting provisions are shown for electronics boxes as well as probe assemblies; additionally, for the latter, locating dimensions of the sensing element relative to the level sensed, as well as any plumbing connections, should be stated. The physical and chemical characteristics of the measured fluid must be known, of course, since they are critical to the selection of a particular type of sensor.

Electrical characteristics include type of line voltage and power, any limitations on cable length and protection, insulation resistance of the probe, and output characteristics, including, for continuous-level systems, the output impedance. Intrinsic safety in accordance with applicable codes, if provided, should also be stated. All external connections must be identified.

Performance characteristics for point-level sensors are essentially limited to level-sensing accuracy (expressed in terms of a unit of height, e.g., "± 0.5 cm") and time constant or (typically 98%) response time; if the latter is different for rising and falling level, both values should be shown. Repeatability is often shown, with a tolerance smaller than the overall accuracy tolerance. For continuous-level sensors (and their associated electronics) the measuring range and output range are shown, together with tolerances on sensitivity or end points and on linearity, hysteresis, and repeatability.

Environmental characteristics primarily comprise the operating temperature range (the fluid temperature range is one of the measured-fluid characteristics, of course) of the probe head and the electronics unit, as well as cabling; secondarily, they should include all other significant environmental conditions, such as atmospheric conditions, vibration, shock, noise level, magnetic fields, electromagnetic interference, and nuclear-radiation levels if such conditions are expected to occur and to have potentially deleterious effects on sensor system operation.

## 14.6 APPLICATION NOTES

Level sensors are used in virtually all industries for monitoring as well as automatic control applications. Examples of such applications are tank metering; high- and low-level warning; liquid leak detection; filling and emptying of tanks, hoppers, bins, silos, ship containers, and railway cars; and propellant utilization in vehicles ranging from automobiles to rockets and spacecraft.

Examples of liquids whose level is sensed are beverages (both alcoholic and nonalcoholic); syrups; soups; fresh and salt water; wastewater; petroleum and refinery products; chemicals of all types; pharmaceuticals; cryogenic liquids (liquified gases); wood pulp; solid suspensions; and slurries. Applications also extend to normally solid materials in their liquid state, e.g., molten glass and metals, ready-mixed concrete, liquid rubber, and plastics. When two immiscible liquids are present in one container, level sensors are used to locate the interface between them.

Level sensors are also used for powdered and granular solids, e.g., grains, granular food products, sand and gravel, ceramic and other mineral powders, plastic and wood chips, tobacco, coal, ore, and solid waste.

Of the two types of level sensors, continuous-level sensors predominate over point-level sensors. Continuous-level sensors are used mostly in storage vessels such as tanks. Point-level sensors are also used in storage vessels, but they are also used in pipes and open ducts, in hoppers and filling nozzles, and in vessels where such process operations as agitation, blending, and centrifuging are carried on. Continuous-level systems are used primarily for monitoring and information, whereas point-level sensors are used not only for level indications and warnings, but also in a variety of automatic control systems.

Considerations for selection of a level sensor are based primarily on feasibility, reliability (with emphasis on durability in an installation before removal for cleaning, refurbishing, or replacement), and cost. Feasibility determinations include such factors as the physical and chemical nature of the material (e.g., its conductivity, dielectric constant, index of refraction, density, viscosity, stickiness, and pH), the nature of the container (e.g., wall thickness, whether the wall material is conductive or nonconductive, and location), and accessibility to the potential location of the sensor. Finally, safety to personnel, to the facility, and to the process is an important consideration.

# 15

# *Pressure Transducers*

## 15.1  BASIC DEFINITIONS

*Pressure* is a multidirectionally uniform type of stress, a force acting on a unit area; it is measured as force per unit area exerted at a given point.

*Absolute pressure* is measured relative to zero pressure (a "perfect vacuum").

*Gage pressure* is measured relative to ambient pressure.

*Differential pressure* is the pressure difference between two points of measurement, measured relative to a reference pressure or a range of reference pressures.

*Partial pressure* is the pressure exerted by one constituent of a mixture of gases not reacting chemically with each other. According to *Dalton's law*, the total pressure in the mixture equals the sum of the partial pressures.

*Static pressure* is the pressure of a fluid exerted normal to the surface along which the fluid flows.

*Impact pressure* is the pressure in a moving fluid exerted parallel to the direction of flow (due to flow velocity).

*Stagnation pressure* (also called *total pressure*) is the sum of the static pressure and the impact pressure. According to *Bernoulli's equation* for horizontal flow, $p_s = p_0 + \frac{1}{2}\rho V_0^2$, where $p_s$ is the stagnation pressure, $p_0$ the static pressure, $\rho$ the density of the fluid, and $V_0$ the flow velocity upstream from the stagnation point. The flow velocity can thus be determined as $V_0 = \sqrt{2/\rho\,(p_s - p_0)}$.

*Head* is the height of a liquid column at the base of which a given pressure would be developed, for a liquid having a given *specific weight*

**Table 15–1. ALTITUDE (ABOVE MEAN SEA LEVEL) VS. ATMOSPHERIC PRESSURE\***

| Altitude (km) | Pressure (kPa) | Altitude (km) | Pressure (kPa) |
|---|---|---|---|
| 0 | 101.32 | 16 | 10.50 |
| 1 | 89.74 | 17 | 8.91 |
| 2 | 79.92 | 18 | 7.55 |
| 3 | 70.10 | 19 | 6.50 |
| 4 | 61.63 | 20 | 5.52 |
| 5 | 54.18 | 21 | 4.77 |
| 6 | 47.41 | 22 | 4.06 |
| 7 | 41.31 | 23 | 3.49 |
| 8 | 35.90 | 24 | 2.96 |
| 9 | 31.15 | 25 | 2.56 |
| 10 | 26.75 | 26 | 2.22 |
| 11 | 23.03 | 27 | 1.90 |
| 12 | 19.64 | 28 | 1.64 |
| 13 | 16.80 | 29 | 1.40 |
| 14 | 14.39 | 30 | 1.20 |
| 15 | 12.22 | | |

\* Pressure values shown are approximate and rounded off; officially accepted values can be obtained from tables furnished by government aviation agencies.

(density multiplied by the acceleration due to gravity—see Section 18.1); the pressure $p_L$ at a depth $h$ below the surface of a liquid is given by $p_L = wh$, where $w$ is the specific weight of the liquid.

The ambient pressure decreases with altitude (above sea level) and increases with depth (below sea level). Thus, pressure measurements are widely used to determine altitude, especially on aircraft and other airborne vehicles. Standard pressure-vs.-altitude relationships are used for the correlation of such measurements; typical values are shown in Table 15–1.

Similarly, pressure measurements are used to determine depth underwater, below sea level, and below the surface of freshwater bodies. Pressure-depth readings are affected to varying degrees by water temperature, density, and salinity. When corrections for these parameters are not, or need not be, considered, the average *seawater* depth can be calculated from pressure measurements on the basis of *10.066 kPa/m* (1.46 psi/m), or very nearly *10 kPa/m*; the average freshwater depth can be calculated on the basis of *9.817 kPa/m* (1.424 psi/m).

## 15.2 UNITS OF MEASUREMENT

*Pressure* is expressed in units of force per unit of area, e.g., $N/m^2$; in the special case of pressure the $N/m^2$ is called the *pascal* (*Pa*), and the Pa is now the SI unit of pressure. A large variety of different units have been used to express pressure (see the subsequent table of conversion factors); relatively low pressures, for example, have been expressed in terms of liquid head (column height), such as millimeters of mercury (at 0°C)—a unit also called *torr* in vacuum work—or centimeters of water (at 4°C). A frequently used unit, the *normal atmosphere*, representing the ambient atmospheric pressure at earth sea level, was defined as the pressure indicated by a 760-mm-high column of mercury (at 0°C, at a density of 13.595 g/cm$^3$, and at an acceleration due to gravity of 980.665 cm/s$^2$). The most commonly used unit of pressure in the United States is the *pound force per square inch (psi)*, which has been in use in more specific forms denoting what pressure the expressed pressure is referenced to: *psia* (for *absolute*), *psig* (for *gage*), and *psid* (for *differential*) pressure; such a differentiation is often very important, and no equivalent suffixes have been assigned to the Pa nor are any expected. It is, therefore, important to remember that Pa is a unit of absolute pressure unless otherwise stated (e.g., "tires are typically inflated to a *gage* pressure of 200 kPa," or "resulting in a *differential* pressure of 2.4 MPa"). Since the Pa is a very small unit of pressure (about $1.45 \times 10^{-4}$ psi), its acceptance will meet considerable resistance, except for low-pressure measurements, especially in the United States. However, decimal multiples of the Pa offer more reasonable conversion magnitudes: 1 psi = 6.894 757 kPa or, where a very coarse approximation is adequate, 7 kPa; 1 kPa = 0.1 450 377 psi, or very close to 0.145 psi. Megapascals (MPa) can be used for higher and very high pressures; for example, 1 MPa $\approx$ 145 psi. Table 15-2 gives quick conversions from psi to kPa and MPa. Finally, in weather reports, atmospheric pressure is still very often expressed in millibars (*mbar*), where 1 mbar = 100 Pa = 1 hPa (*hectopascal*). Some countries have already adopted the hectopascal as a unit for atmospheric pressure and are using this unit daily in their official weather reports.

The following table gives conversion factors from various units of pressure to pascals:

| To Convert From | to | Multiply by |
|---|---|---|
| atmosphere (normal) | Pa | $1.013\ 25 \times 10^5$ |
| bar | Pa | $10^5$ |
| dyne/centimeter$^2$ | Pa | $10^{-1}$ |
| foot of water (39.2°F) | Pa | $2.989 \times 10^3$ |
| inch of mercury (0°C) | Pa | $3.3864 \times 10^3$ |
| inch of water (4°C) | Pa | $2.491 \times 10^2$ |
| kilogram-force/cm$^2$ | Pa | $9.806\ 65 \times 10^4$ |
| kilogram-force/meter$^2$ | Pa | 9.806 65 |
| millibar | Pa | 100 |
| millimeter Hg (0°C) | Pa | 133.32 |
| pound-force/foot$^2$ | Pa | 47.88 |
| pound-force/inch$^2$ | Pa | $6.895 \times 10^3$ |
| torr | Pa | 133.32 |

**Table 15–2. PRESSURE UNIT
CONVERSIONS, psi TO kPa
AND MPA**

| psi | kPa |
|---|---|
| (1.000 | 6.895) |
| 1.00 | 6.9 |
| 2.00 | 13.8 |
| 3.00 | 20.7 |
| 4.00 | 27.6 |
| 5.00 | 34.5 |
| 6.00 | 41.4 |
| 7.00 | 48.3 |
| 8.00 | 55.2 |
| 9.00 | 62.1 |
| 10.0 | 69 |
| 20.0 | 138 |
| 30.0 | 207 |
| 40.0 | 276 |
| 50.0 | 345 |
| 60.0 | 414 |
| 70.0 | 483 |
| 80.0 | 552 |
| 90.0 | 621 |
| 100 | 690 |

| | MPa |
|---|---|
| 100 | 0.69 |
| 145 | 1.00 |
| 200 | 1.38 |
| 300 | 2.07 |
| 400 | 2.76 |
| 500 | 3.45 |
| 600 | 4.14 |
| 700 | 4.83 |
| 800 | 5.52 |
| 900 | 6.21 |
| 1000 | 6.90 |
| 1450 | 10.00 |
| 2000 | 13.79 |
| 5000 | 34.47 |
| 10 000 | 68.95 |
| 20 000 | 137.9 |

## 15.3 SENSING METHODS

### 15.3.1 *Pressure Sensing and Referencing*

Pressure is sensed by elastic mechanical elements, such as plates, shells, and tubes, which offer the *force* a surface (an *area*) to act upon. (Pressure is measured as force per unit area, e.g., $N/m^2$). The element will deflect whenever this force is not balanced by an equal force acting upon its opposite surface. Depending on the type of transduction used, the transduction element then sees this deflection usually as either displacement or strain. The most commonly used pressure-sensing elements are illustrated in Figure 15–1 and are explained in more detail in Sections 15.3.2 through 15.3.6. The displacement of a diaphragm is relatively small, whereas the displacement of bellows and multiturn Bourdon tubes is relatively large. This explains why certain types of pressure transducers tend to employ one or the other type of sensing element.

Although all pressure-sensing elements respond to a change in the differential pressure across them, transducers can be designed to measure either differential, gage, or absolute pressure, depending on the *reference pressure* admitted to, or maintained at, the reference side of the element. The basic pressure reference configurations are shown in Figure 15–2, with a diaphragm used to exemplify the sensing element. The reference side of an *absolute-pressure* sensing element is evacuated and sealed. *Gage pressure*, on the other hand, is measured when the reference side is vented to ambient pressure. A *differential-pressure* sensing element deflects with an increasing difference between two pressures, both of which may vary. The normally lower or less varying pressure is called the *reference pressure*, and the other pressure is called the *measured pressure*. When the measured pressure is always higher than the reference pressure, the transducer has a *unidirectional* range (e.g., 0 to 300 kPa differential). When the measured pressure can be either lower or higher than the reference pressure, the transducer has a *bidirectional range* (e.g., ±200 kPa differential). In a special version of the differential-pressure configuration (Figure 15–2(d)) a fixed pressure greater than zero is maintained permanently (by sealing it in) on the reference side. Note that in the case of hollow sensing elements the measured pressure is not always admitted into the element; instead, the inside of the element can be part of the reference side, and the measured pressure is admitted into the cavity surrounding the hollow element.

### 15.3.2 *Diaphragms*

A *diaphragm* is a thin circular plate fastened continuously around its edge. Two basic types of diaphragm are used in pressure transducers: the flat diaphragm (Figure 15–1(a)) and the corrugated diaphragm (Figure 15–1(b)).

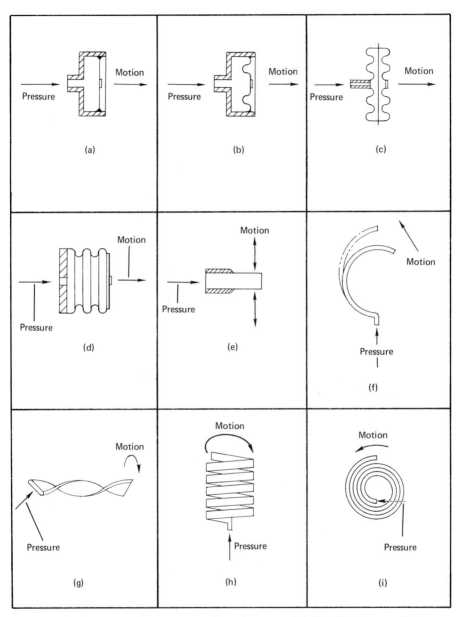

**FIGURE 15–1.** Pressure-sensing elements: (a) flat diaphragm; (b) corrugated diaphragm; (c) capsule: (d) bellows; (e) straight tube; (f) C-shaped Bourdon tube; (g) twisted Bourdon tube: (h) helical Bourdon tube; (i) spiral Bourdon tube.

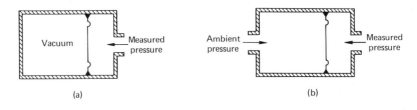

**FIGURE 15–2.** Pressure reference configurations: (a) absolute pressure; (b) gage pressure; (c) differential pressure; (d) differential pressure, sealed reference.

The proper design of a diaphragm requires specialized knowledge and exact calculations.

*Flat diaphragms* (Figure 15–3) deflect in accordance with laws generally applicable to circular plates under conditions of symmetrical loading. The basic flat diaphragm is an uninterrupted straight circular web supported at its edge. Diaphragms are either machined from stock that includes their edge support as well as additional portions of the transducer, or they are formed separately and then welded (sometimes brazed) to their support. In the *spherical diaphragm* the web is slightly concave. The *catenary diaphragm* (Latin: *catena*, chain) is additionally supported by the edge of an inner ring or tube concentric with the structure to which the web is fastened. *Drumhead diaphragms* are bent around and fastened to the outside of their supporting structure while radial tension is applied to stretch the diaphragm. An *annular diaphragm*, which can be flush or recessed, is a diaphragm with a central reinforcement ("boss") to facilitate the translation of its deflection into a secondary mechanical displacement.

*Corrugated diaphragms* (Figures 15–1(b) and 15–4) contain a number of concentric corrugations that increase the stiffness as well as the effective area of the diaphragm, thus providing a larger useful deflection than that of a flat diaphragm. The corrugations become progressively shallower from the periphery toward the center because bending is maximum near the periphery and minimum at the center.

The deflection of a diaphragm varies inversely to the 1.2 to 1.6 power of its thickness and approximately the fourth power of its diameter. Within

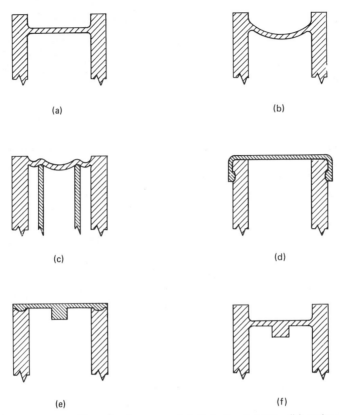

(a)

(b)

(c)

(d)

(e)

(f)

**FIGURE 15–3.** Flat diaphragms: (a) flat diaphragm; (b) spherical diaphragm; (c) catenary diaphragm; (d) drumhead diaphragm; (e) annular diaphragm (flush); (f) annular diaphragm (recessed).

the limits of usable deflection, deflection is proportional to pressure; it is also influenced by the design of the corrugations (if any), the material and its preparation and treatment, the manner of attachment of the diaphragm to its peripheral supporting wall (and its fillet radius at the wall interface, if machined) and the diameter of the central reinforcement (if any).

Materials used for diaphragms are elastic metal alloys such as brass, bronze, phosphor bronze, beryllium copper, stainless steel (a predominant material), and such proprietary alloys as Monel, Inconel-X, and Ni-Span-C

**FIGURE 15–4.** Corrugated diaphragm.

(a ferrous nickel alloy with good thermal properties). The choice of diaphragm material is strongly influenced by the chemical properties of the measured fluid that comes in contact with the diaphragm. Heat treating and pressure cycling help reduce elastic aftereffects (drift) and hysteresis in diaphragms.

### 15.3.3 Capsules

A *capsule* (sometimes called "aneroid") consists of two annular corrugated diaphragms formed into shells of opposite curvature and sealed together at their peripheries (Figures 15–1(c) and 15–5). In a single capsule, one diaphragm is provided with a pressure port, while the other has a boss from which the mechanical displacement originates. Alternatively, one diaphragm is provided with an internal boss to which a pushrod is attached, and the pushrod leads through a port in the opposite diaphragm at which (usually reference) pressure is admitted into the capsule. The use of two diaphragms in the form of a capsule nearly doubles the deflection obtained from a single diaphragm. Additional multiplication of deflection can be obtained by ganging two or more capsules together (Figure 15–5).

### 15.3.4 Bellows

*Bellows* (Figures 15–1(d) and 15–6) are typically made from thin-walled tubing formed into deep convolutions and sealed at one end, which displaces axially when pressure is applied to a port at the opposite end. The number of convolutions can vary from less than 10 to over 20, depending on pressure range and displacement (*stroke*) requirements and on outside diameter. Since inside diameters of bellows range between 50 and 90% of outside diameters, the effective area of one convolution is substantially less than that of a capsule. Bellows have been used mainly for low-pressure ranges

**FIGURE 15–5.** Capsules (left: triple; right: single)

**FIGURE 15–6.** Bellows.

and when no significant vibration exists in their environment. They have also been used for isolation, in the same manner as previously described for membranes, incorporated within a transducer. Additionally, they have found use as *expansion bellows*, sealed at both ends and containing an inert gas at low pressure, to compensate for changes in the volume of damping oil, due to temperature changes, in viscous-damped transducers.

### 15.3.5  Straight Tubes

The *straight-tube* sensing element has been used in a limited number of transducer designs. The tube, of circular cross section, is sealed at one end and expands and contracts with changes in the pressure admitted into its open end. These minute displacements are transduced either as strain or as changes in the resonant frequency of the tube (in a vibrating-element transducer).

### 15.3.6  Bourdon Tubes

The *Bourdon tube* (Figures 15–1(f)–(i) and 15–7) is a curved or twisted tube, oval or elliptical in cross section, which is sealed at one end (the *tip*). When pressure is applied into its open end, the tube tends to straighten. This results in an angular tip deflection in a twisted tube and in a curvilinear tip deflection (*tip travel*) in a curved tube. The Bourdon tube is named after the French inventor Eugène Bourdon, who patented it in 1849, although it was reportedly built and used by a German locomotive engineer prior to that date.

The *C-shaped Bourdon tube* (Figure 15–1(f)) has an angle of curvature between 180 and 270°, and tip travel is outward with increasing pressure. A related element is the *U-shaped Bourdon tube*, which has its pressure port at the center of its about 270° curve; both tips are sealed and travel outward and away from each other with increasing pressure. The *helical Bourdon tube* (Figure 15–7(a)) is similar in deflection behavior to a C-shaped tube. Since it is coiled into a multiturn helix, with a total angle of curvature of 1800 to 3600°, its tip travel is proportionately greater. The *spiral Bourdon tube* (Figure 15–7(b)) also amplifies tip travel due to its multiturn configuration (typically between four and eight turns). The *twisted Bourdon tube*

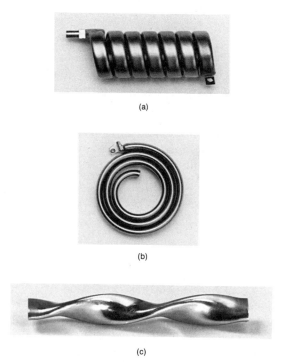

(a)

(b)

(c)

**FIGURE 15-7.**   Bourdon tubes: (a) helical; (b) spiral; (c) twisted.

(Figure 15-7(c)) is a flattened tube, twisted along its length (about two to five twists), with the centerline of the tube straight throughout its length.

The deflection of a Bourdon tube varies with the ratio of its major to minor cross-sectional axes, the tube length, the radius of curvature and total angle (rate of twist of a twisted tube), and applied pressure; it also varies inversely to tube wall thickness and to the modulus of elasticity of the tube material after processing (including heat treating). Materials are similar to those used (and described) for diaphragms; additionally, quartz has been used in at least one design. Metal tubes can be machined from bar stock or drawn, flattened, and then formed into the desired configuration. Great care is taken to achieve a leakproof seal at the tip and at the pressure port. Pressure cycling as well as temperature cycling is needed to provide long-term stability and a repeatable thermal behavior. Multiturn Bourdon tubes tend to be vibration sensitive and, when their environment will include vibration, are often damped using viscous damping provided by a fluid such as silicone oil. The oil fills a cavity containing, and just slightly larger than, the tube. Thermal effects on damping and apparent changes in reference pressure can be counteracted by the use of expansion bellows.

## 15.4 DESIGN AND OPERATION

### *15.4.1 Capacitive Pressure Transducers*

The capacitive transduction principle is utilized in pressure transducers in either of the following two designs:

1. *Single-stator:* pressure is applied to a diaphram which moves to and from a stationary electrode *(stator)*.
2. *Dual-stator:* pressure is applied to a diaphragm supported between two stationary electrodes.

In single-stator designs the diaphragm can be either the grounded or the ungrounded electrode. In the design shown in Figure 15–8 the diaphragm is integrally machined with its support member, moving toward the stator electrode on an insulating substrate. Full-scale diaphragm deflection is about 0.1 mm, and a lead connects the stator to an external terminal; the case acts as the other terminal. The internal cavity of the absolute-pressure transducer shown is evacuated and then sealed. The diaphragm is of the (patented) ''free-edge'' type: the free edge acts as a hinge when pressure ($P_s$) is applied to the diaphragm, reducing stress levels by a factor of about five over typical prestressed flat diaphragms; this reduces hysteresis and nonrepeatability.

Insulated stators, or, as used in other designs, insulated diaphragms, are now frequently made of quartz or ceramic with the electrode vacuum-deposited or sputtered onto the substrate.

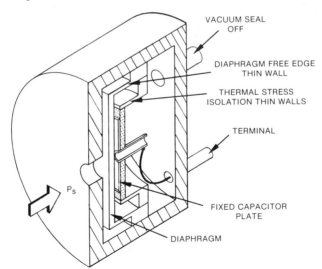

**FIGURE 15–8.** Single-stator capacitive absolute-pressure transducer. (Courtesy of Rosemount Inc.)

Dual-stator designs offer the advantage of a multiplication of the capacitance change, since, as the diaphragm deflects, its capacitance to one stator increases while it simultaneously decreases to the other stator. This effect is best utilized by connecting the two halves of the sensor as two arms in an ac bridge. In the transducer shown in Figure 15–9, the diaphragm is grounded by a welded seal to the case and the two metallized electrodes on ceramic substrates have external lead connections. This design also uses two isolating diaphragms (isolating membranes) and a transfer fluid to isolate the sensing cavity from the measured fluid, as well as reducing shock and vibration effects and providing a fluid with a known and invariable dielectric constant between the capacitor electrodes. Use of appropriate metal alloys for the isolation diaphragms also enables the transducer to be used for corrosive measured fluids. Figure 15–10 shows the dual-stator sensor in a rugged housing intended for industrial applications. The cylindrical box on top of the transducer contains the excitation- and signal-conditioning circuitry; the latter can provide the frequently required two-wire, 4-to-20-mA or 10-to-50-mA dc output signal.

A somewhat different design approach is used in a pressure transducer in which two stators are formed by metallizing both surfaces of an insulating substrate and two moving electrodes are incorporated by bonding a corrugated diaphragm to each side of the substrate. The resulting aneroid, with the electrode-carrying substrate sandwiched in its middle, is evacuated. The applied pressure acts on both diaphragms, making the two capacitance changes additive. Many capacitive pressure transducer designs, including the type just described, are used for barometric measurements, measuring either barometric pressure at a fixed location or pressure-altitude in an airborne vehicle. Designs are available for differential-pressure measurement for low and medium ranges. Although capacitive pressure transducers are

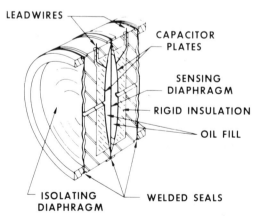

**FIGURE 15–9.** Dual-stator capacitive differential-pressure transducer. (Courtesy of Rosemount Inc.)

**FIGURE 15–10.** Capacitive differential-pressure transducer in industrial-type configuration. (Courtesy of Rosemount Inc.)

most frequently employed for low-pressure measurements, and special designs are used for vacuum measurements (see Section 17.4.1), some general-purpose models are available for absolute and gage pressure measurements up to 70 MPa.

The basic circuit into which capacitive pressure transducers are connected is the ac-excited impedance bridge (see Figure 15–11), although other circuits are used as well. Most designs incorporate circuitry to demodulate the ac output signal into a dc voltage and to provide for temperature compensation as well as zero, gain, and linearity adjustments. Some designs employ digital techniques for compensation and adjustment and can furnish an output in digital form.

### 15.4.2 Inductive Pressure Transducers

In an inductive pressure transducer (Figure 15–12) the self-inductance of a single coil is varied by pressure-induced changes in displacement of a metallic diaphragm in close proximity to the coil. Past designs used a diaphragm of magnetic material and its motion to and from the ferric core around which the coil is wound, or actuation of a movable ferromagnetic core within a coil, to obtain the inductance changes. Some more recent designs use a metallic diaphragm and a coil excited by ac current at RF frequencies to use the changes in eddy currents in the diaphragm for obtaining the changes in self-inductance. A second (reference) coil is often included in the same housing; it remains unaffected by pressure variations and provides compensation for temperature changes.

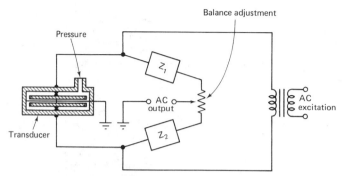

**FIGURE 15–11.**   Basic bridge connection schematic for dual-stator capacitive pressure transducer.

### 15.4.3  *Reluctive Pressure Transducers*

Reluctive pressure transducers include two major types of reluctive transduction elements used in pressure transducers: the differential transformer
(usually the LVDT) and the two-coil inductance bridge. The former uses
capsules, bellows, or Bourdon tubes as sensing elements, and the latter uses
diaphragms or Bourdon tubes as sensing elements.

Inductance-bridge reluctive pressure transducers use a magnetically
permeable member to increase the inductance of one coil while decreasing
the inductance in the second coil. The coils are connected in a bridge circuit
so that the increase and decrease in inductance of the two coils are additive
in the resulting bridge-output voltage change. When a diaphragm is used
(Figure 15–13), the diaphragm itself is the magnetically permeable induct-
ance-changing member. When a Bourdon tube is used as the sensing element,
as in the transducer illustrated in Figure 15–14, it is usually a twisted Bourdon tube to whose tip a strip of magnetically permeable material (an *armature*) is attached. The armature is located between the two coils, and their
inductances change simultaneously in opposite directions as the flux gaps
of the two coils are changed by the armature's moving away from one coil
and toward the other coil.

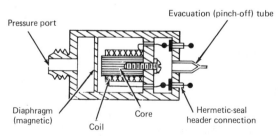

**FIGURE 15–12.**   Inductive pressure transducer.

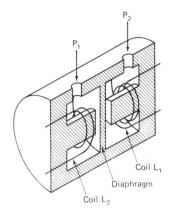

**FIGURE 15–13.** Reluctance
change by diaphragm deflection.

Many pressure transducers of the inductance-bridge type (as well as of the LVDT type, described in the next paragraph) incorporate circuitry that enables the transducers to be powered by dc and provide a dc output voltage. Typical circuitry (Figure 15–15) includes a regulator (so that fairly large variations in power-supply voltage can be accepted); an oscillator, which provides the required ac voltage for the transduction coils (sensor) and can also feed a small transformer-rectifier module that supplies regulated voltages to internal elements; a demodulator for converting the bridge-output voltage to dc; and an operational (differential) amplifier that boosts the dc output signal to the required level (e.g., 0 to 5 or 0 to 10 V dc). Circuitry can also be included for transducer operation on a two-wire system where its output is in the form of a change in current (e.g., 4 to 20 mA). Figure 15–14 illustrates a complete inductance-bridge transducer design with an integrally packaged electronics module containing the circuitry described. The twisted-Bourdon-tube sensing element is usable for pressure ranges between 0 and 350 kPa (50 psi) and 0 and 35 MPa (5000 psi) when used in conjunction with the armature and coil assembly designed for this transducer model. Resistive temperature-compensation components are connected to the electronics module.

In LVDT-type pressure transducers an assembly of three coils is wound on a hollow mandrel in which a core of magnetic material moves axially. One coil acts as the primary winding, while the other two act as the differentially connected secondary windings of a transformer. As illustrated by the operating principles shown in Figure 15–16, the moving core is actuated by the displacement of a Bourdon tube or capsule. As the core moves from a center (null) position toward either of the secondary windings, the coupling between that winding and the primary winding is increased (because the reluctance path is decreased). Various winding configurations can be used; in the most common one, core motion toward either secondary produces an output change that increases in amplitude equally for both directions of

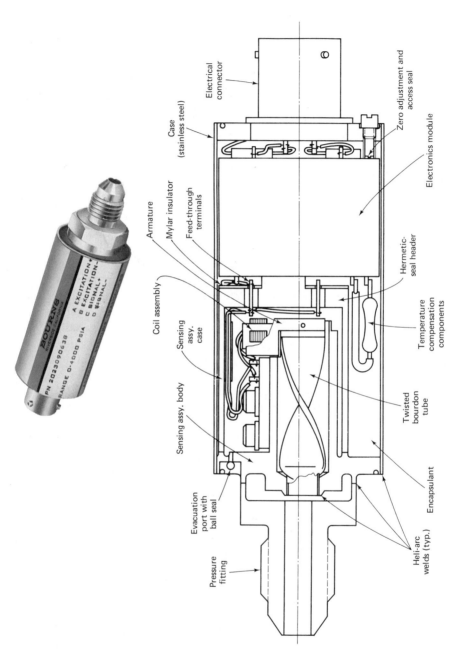

**FIGURE 15–14.** Reluctive dc output pressure transducer with twisted-Bourdon-tube sensing element (Courtesy of Bourns Instruments, Inc.)

Electrical connector

Case (stainless steel)

Zero adjustment and access seal

Electronics module

Armature

Mylar insulator

Feed-through terminals

Hermetic-seal header

Coil assembly

Sensing assy. case

Temperature compensation components

Sensing assy. body

Twisted bourdon tube

Evacuation port with ball seal

Encapsulant

Pressure fitting

Heli-arc welds (typ.)

BOURNS

PN 202309063B

RANGE 0–4000 PSIA

A EXCITATION+
B EXCITATION-
C SIGNAL+
D SIGNAL-

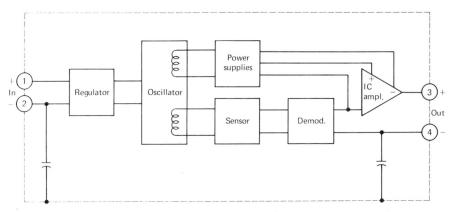

**FIGURE 15–15.** Typical dc-to-dc reluctive pressure transducer block diagram. (Courtesy of Bourns Instruments, Inc.)

motion but differs in phase. The amplitude and phase changes of the (ac) output signal can then be converted into a unidirectionally varying dc voltage by appropriate circuitry.

Most LVDT pressure transducers incorporate electronic components which, operating from a dc power supply, produce the ac excitation for the primary winding (typically at a frequency much higher than the 50 or 60 Hz line voltage has) and provide the conversion of the ac output signal into a dc voltage (e.g., 0 to 5 Vdc). Capsules are most commonly used as sensing elements.

In addition to the widely produced designs just described, a different approach has also been employed in pressure transducers, involving the change in permeability due to a change in stress of a straight-tube sensing element. This element is used as a fixed (but variable-permeability) core in conjunction with a differential-transformer winding configuration.

### 15.4.4 *Potentiometric Pressure Transducers*

Potentiometric transduction was used in some of the earliest pressure transducers developed. Over the many intervening years a large variety of designs have been produced for a multitude of applications. Since their output varies between 0 and 100% of the applied excitation voltage, they are inherently high-level-output devices needing no amplification.

Single or multiple capsules are used for relatively low-pressure ranges, and Bourdon tubes (helical, spiral, or twisted) are used for high-pressure ranges; the region of overlap (where either type of sensing element can be used) is around 3.5 MPa (500 psi). Although units with a range below 7 kPa (1 psi) have been produced, the ranges of the more commonly used production models extend from 100 kPa (15 psi) to 70 MPa (10 000 psi). A high-pressure transducer is shown in Figure 15–17. This configuration mounts

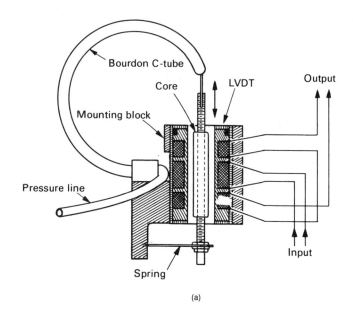

(a)

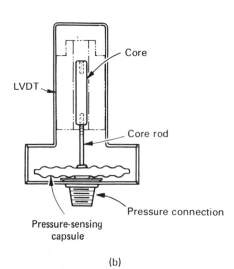

(b)

**FIGURE 15–16.** Operating principles of LVDT (reluctive) pressure transducers: (a) with Bourdon-tube sensing element; (b) with capsule sensing element. (Courtesy of Schaevitz Engineering.)

directly into a boss at the point of measurement (with a compression-type washer or O-ring used as seal). A helical Bourdon tube is used as sensing element in this transducer. The tube tip is attached to a rotary support which rides on ball bearings. The (electrically insulated) wiper arm is attached to this support and slides over an exposed strip on the (otherwise insulated-wire) resistance element. The entire cavity containing the Bourdon tube and transduction element is usually filled with silicone oil for purposes of vibration damping.

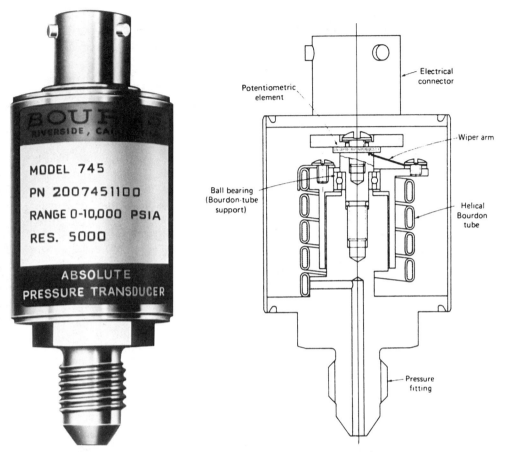

**FIGURE 15–17.** Potentiometric high-pressure transducer and its internal layout. (Courtesy of Bourns Instruments, Inc.)

High-pressure transducers using multiturn or multitwist Bourdon tubes usually need no amplification linkage to drive the wiper arm over the potentiometric element, which varies in length between about 5 and 20 mm. Low-pressure transducers, however, often require some mechanical amplification, since the displacement obtained from a single capsule is only on the order of 1 mm. Lever or flexure linkages have been used for this purpose, and jewel pivots as well as ball bearings have been employed in such mechanisms. Mechanical overpressure stops are often included to limit the displacement of the sensing element and keep the wiper arm from traveling beyond the end of the resistance element. Some overrange can be handled by winding the element a little longer than its nominal resistance requires and then shorting a number of turns at the "high" end with solder to form a short bus bar.

Vibration damping has been effected by other means besides filling the close-fitting cavity with silicone oil (or similar viscous damping liquid); counterbalance weights and pneumatic dashpots have also been used for this purpose.

The potentiometric resistance elements can be wire wound, or they can be made of conductive plastic; conductive ceramics (cermets), carbon-film, and metal-film elements have also been used, but have been found to be generally less producible or satisfactory. Wire-wound elements are typically made by winding between 300 and 600 turns of precious-metal or nickel-alloy wire, 0.008 to 0.05 mm in diameter (platinum alloy is often used), around a flat, oval, or round mandrel which should be of a material that matches the wire in temperature coefficient of expansion. Epoxy or ceramic cements are sometimes used to fix the winding in position on the mandrel. Use of bare wire requires very precise spacing between turns to avoid turns shorting to each other. Wire coated with a thin insulation is frequently used. After completion of all processes on the element, the insulation is carefully abraded so that a thin strip of bare wire contacts are provided for the wiper arm to slide over. Wipers are usually stamped from spring alloy and grooved or dimpled to provide a narrow contact surface. Other wiper designs include a slim spring-alloy arm to the bottom of which a precious-metal wire is attached to provide the required narrow contact surface. Wiper contact pressure must be carefully controlled: it must be high enough to prevent lift-off or noise yet low enough to minimize friction error that can be caused by wiper drag, and, generally, it must minimize friction, which reduces wiper or element life.

Potentiometric pressure transducers can be furnished with certain optional features. (1) One sensing element can drive two wipers over two identical resistance elements (dual output). (2) Rather than the two elements being identical, one of the elements can be linear and the other nonlinear (following a specific transfer function), or the other element can be so constructed that it provides outputs at one or more discrete levels of pressure (*switch outputs*). (3) The transducer can be designed to provide an output that is linear with altitude (*pressure-altitude*; see Table 15–1). (4) A differential-pressure transducer can be designed to provide an output that is linear with *airspeed* (with the pressure ports connected to a Pitot-static tube). Note that other types of pressure transducers, notably capacitive and servo transducers, are similarly available for altitude and airspeed measurements.

### 15.4.5   *Resistive Pressure Transducers*

Some early transducer designs used material that undergoes changes in resistance when pressure or force acts on it. Carbon is one of these materials. It was used, in powder form, in the earliest microphones. Stacked carbon discs were also used as transducers, with a bellows as force-summing member. Indeed, even carbon resistors have been used as pressure transducers.

The only material that is used in commercially available sensors, however, is *manganin*, an alloy of approximately 84% Cu, 12% Mn, and 4% Ni. This material is used in the form of wire, loosely coiled or shaped in a grid pattern (similar to a metal-wire strain gage), or as foil (photoetched in a pattern resembling a metal-foil strain gage) which can be purchased as such and installed by the user or as incorporated in a complete transducer assembly; in either form they are often referred to as *manganin gages*. The pressure sensitivity of manganin wire and foil is usually between 2.1 and 2.8 microhms/ohm/kPa (0.0021 to 0.0028 ohms/ohm/kbar). Manganin gages are probably the most suitable sensors for very high pressures. Pressures to about 1400 MPa (14 kbar) are commonly measured with manganin gages, and transient pressures (e.g., in high-pressure shock-wave studies) to over 40 TPa (400 kbar) have been measured experimentally.

Research has also been reported on high-pressure sensors using resistance changes in monocrystalline tellurium and indium antimonide (performance improved when the two materials were connected in series and both were exposed to the pressure), and current or voltage changes in planar transistors on which the pressure is caused to act directly.

### 15.4.6 Strain-Gage Pressure Transducers

The conversion of pressure changes into resistance changes, due to strain, in two or, much more commonly, four arms of a Wheatstone bridge has been used in pressure transducers for many years. Unbonded wire strain gages were used in some of the earlier models. Bonded metal-foil gages are used in many current designs, typically mounted to a beam to which a bending force is applied by a pushrod attached to a diaphragm, but sometimes mounted to the back side of the diaphragm itself. Thin-film techniques have been applied to bonded-gage designs, with the gages sputtered or vacuum-deposited onto either a beam or the back of the diaphragm. Diaphragms are used in the majority of designs; however, straight-tube sensing elements have proved very satisfactory in transducers that measure pressures in the higher ranges (10 MPa and up). Figure 15–18 shows a transducer with a straight-tube element; strain gages, first wound around and then bonded to the tube, respond to the hoop stress induced in the tube with increasing pressure.

Figure 15–19 illustrates a strain-gage pressure transducer which typifies many features of such transducers in general, although this particular design has thin-film gages sputtered onto the beam that is part of the strain sensing assembly (Figure 15–19(b)). As pressure is applied to the diaphragm (Figure 15–19(a)) the pushrod pushes against the beam so that the gages nearest the pushrod experience compression, whereas the gages near the ends of the beam experience tension. The gages are connected as four arms of a Wheatstone bridge. The gages undergoing compression decrease their resistance whereas the gages undergoing tension increase their resistance, and their connection into the bridge circuit is such the resistance changes

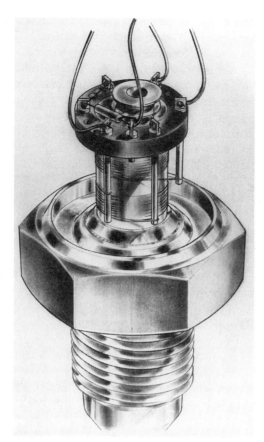

**FIGURE 15–18.** High-pressure transducer with strain gages bonded to straight-tube sensing element. (Courtesy of Paine Corporation.)

are additive, thus producing a larger bridge output voltage. This action is typical for most strain-gage transducers for various measurands. A discussion of strain-gage bridges and their adjustment and compensation is presented in Section 9.6. Note that the bridge balance resistors are located on the sensing assembly, whereas the compensation resistors are located in the transducer case, between the sensing assembly and the electrical connector. In absolute-pressure transducers the cavity behind the diaphragm must be evacuated and sealed. A ball seal is used for this purpose in this design. The stress isolation cutouts isolate the diaphragm and sensing assembly from any case-induced stresses. Some other designs have a longer transducer case to allow excitation conditioning and signal conditioning, including amplification circuitry to be accommodated between the sensing portion and the electrical connector.

Advances in semiconductor technology led first to the use of bonded semiconductor gages and later to integrally diffused gages. Both of these provide a much greater sensitivity than metal gages and, hence, a higher bridge-output signal level; however, they are more temperature sensitive, and temperature compensation is more critical and somewhat more complex. Bonded gages offer the advantage of being able to be bonded to metal diaphragms that can withstand fluids that would not be compatible with silicon.

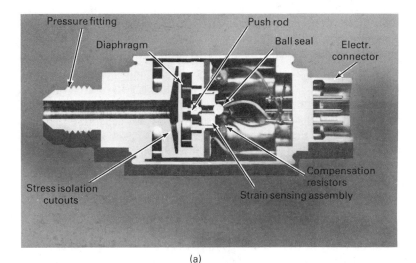

(a)

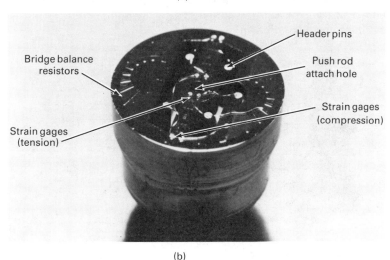

(b)

**FIGURE 15–19.** Pressure transducer with sputtered thin-film strain gages: (a) transducer; (b) strain sensing assembly. (Courtesy of CEC Instruments Div., Transamerica Delaval Inc.)

An alternative to this arrangement is to use an integrally diffused force-sensing beam deflected by a pushrod connected to a metal diaphragm.

Integrally diffused diaphragms offer the advantage of being producible fairly quickly and in large quantities, once the initial design and fabrication problems have been resolved. As a result, such pressure sensors are often available at a low cost. The techniques employed in diffusing a four-arm strain gage into a silicon diaphragm are closely related to integrated-circuit chip production methods. A typical wafer consists of a circular slice of *n*-type silicon into which strain-sensitive *p*-type areas have been diffused, using an appropriate dopant. Interconnecting strips and tabs for lead attachment are vacuum-deposited or also diffused. Photolithographic techniques allow the production of very small sensors; some are only 0.75 mm in diameter. In some designs a fair degree of bridge adjustment and temperature compensation is obtained by diffusing appropriate resistive areas directly into the diaphragm and interconnecting them. In other designs producibility is enhanced by diffusing two complete strain-gage bridges into one diaphragm, with either one or the other used in the final sensor on the basis of test results. Figure 15–20 shows a pressure transducer with an integrally diffused diaphragm. The unit consists of the basic four-arm-bridge sensor encapsulated in a glass-filled nylon package. No additional circuitry is incorporated; however, the manufacturer supplies suggested circuit diagrams for temperature compensation, constant-current excitation, and amplification (all included in the schematic diagram shown in Figure 15–20(d). There are many specialized configurations for integrally diffused pressure sensors, such as a 1-mm-thick sensor for use on airfoils and turbine blades, and sensors having a diameter of 1 mm or less and mounted near the tip of a hypodermic needle, for use in biomedical applications.

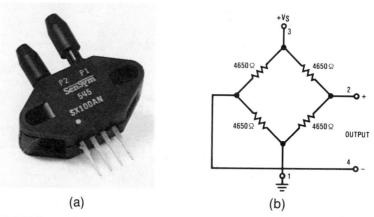

(a)                                              (b)

**FIGURE 15–20.** Integrally diffused semiconductor strain-gage pressure transducer: (a) sensor package; (b) physical construction of basic sensor; (c) equivalent circuit; (d) typical application schematic. (Courtesy of Sensym, Inc.)

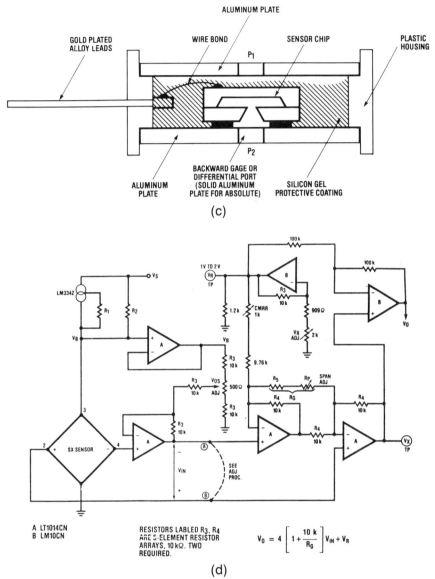

(c)

(d)

**FIGURE 15–20.** *(continued)*

### 15.4.7 *Servo-Type Pressure Transducers*

Servo-type pressure transducers generally provide very good accuracy, but at the cost of greater complexity. The three types of servo loops that have been employed are illustrated in the simplified block diagrams of Figure 15–21. It should be noted at the outset that the first two types, with shaft output,

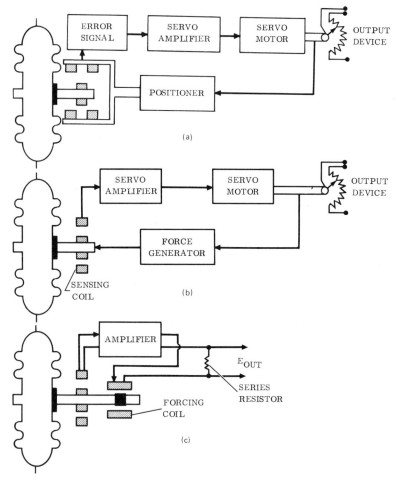

**FIGURE 15–21.** Basic block diagrams of servo-type pressure transducers: (a) null-balance pressure transducer; (b) force-balance pressure transducer (with shaft output); (c) force-balance pressure transducer (with direct voltage output).

are rarely manufactured anymore, now being considered "specialty items." Only the (less complex) force-balance type of transducer is readily available commercially, and even this is not being built by as many manufacturers as in the past, mainly because improved techniques have been applied to simpler types of pressure transducers to enhance their accuracy.

In the *null-balance* transducer (Figure 15–21(a)) the sensing element (capsule or bellows usually) is allowed to deflect freely. The displacement is detected by a null-type transduction element such as a differential transformer. Any unbalanced output due to the displacement (e.g., of a pushrod attached to the sensing element) becomes an error signal in the servo loop. The signal is amplified and applied to a servo motor, which drives the coil system to its new null position while simultaneously driving an output device. The latter can be a rotary potentiometer, a synchro or other reluctive device, or a shaft-angle encoder.

A motor-driven output device is also used in the *force-balance* transducer illustrated in Figure 15–21(b). However, the sensing element is not allowed to deflect freely, but is restrained by a force generator. When the sensing element tries to deflect in response to applied pressure, a transduction element detects the beginning of displacement and produces an error signal. The transduction element can be inductive (as implied by "sensing coil"), reluctive, or capacitive. The error signal is amplified and applied to a motor which drives an output device while simultaneously causing a (mechanically linked) force generator to apply sufficient force to the sensing element to restore a condition of balance. The transducer output (from the output device; at least one such system uses a shaft-angle encoder) is proportional to the force required to restrain the sensing element from displacing.

In the force-balance transducer shown in Figure 15–21(c) the additional electromechanical devices (servo motor, force generator, shaft-driven output device) are omitted, which makes this type of servo pressure transducer less complex than the other two types. The error signal that is produced by the transduction element is fed to an amplifier, and the output of the amplifier is applied to a forcing coil through a series resistor. The forcing coil operates electromagnetically to prevent any deflection of the sensing element; therefore, the current through the forcing coil/series resistor circuit is proportional to the applied pressure. The drop in *IR* across the resistor becomes the transducer output signal, which can be amplified or conditioned as required.

### 15.4.8 *Piezoelectric Pressure Transducers*

Piezoelectric pressure transducers are widely used for pressure measurements where a very high frequency response (up to 500 kHz in some designs) is required or where an equivalently short response time is required. Piezoelectric crystals are made of quartz or of a variety of proprietary ceramic mixtures; ceramic crystals acquire their piezoelectric characteristics by exposure to an orienting electric field during cooling after first being heated.

A typical piezoelectric pressure transducer using quartz crystals is shown in Figure 15–22. The pressure-sensing diaphragm acts against a stack of quartz disks which produce the output signal. The crystals are mechanically preloaded. The design shown also includes a small seismic mass with an associated quartz crystal which senses acceleration and produces a signal that is used for compensating the pressure-generated signal for simultaneously experienced acceleration. The housing may contain an (optional) IC

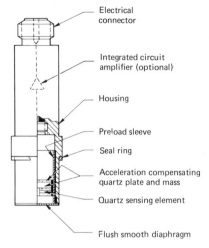

Electrical connector

Integrated circuit amplifier (optional)

Housing

Preload sleeve

Seal ring

Acceleration compensating quartz plate and mass

Quartz sensing element

Flush smooth diaphragm

**FIGURE 15–22.** Piezoelectric pressure transducer. (Courtesy of PCB Piezotronics, Inc.)

amplifier, a device very useful in providing a low-impedance output; the output impedance of the quartz crystal is very high.

The crystal materials originally developed for use in transducers, such as barium titanate and lead zirconate, are no longer used in production designs, and the single quartz crystal formerly used is now almost invariably replaced by stacked crystals. Quartz crystals are usable to temperatures up to about 350°C. Some recently developed types of ceramic crystals can withstand higher temperatures, while others have lower temperature limits. The limiting temperature of ceramic crystals is the *Curie point*: when heated above this point they lose their piezoelectric characteristics. Pressure ranges extend up to 150 MPa, although most transducers capable of measuring such a high range may only be calibrated for a portion of the range. Ranges to 50 or 100 MPa are more common; on the other hand, experimental sensors using a lithium niobate crystal have been reported to be usable for very short pressure transients up to 1.7 TPa.

Piezoelectric pressure transducers exist in many general-purpose as well as specialized configurations; the latter include designs that can replace a spark plug in an engine; those that fit into a small cavity in a fuel-injection line, also for automotive testing; and those that fit against a cartridge in a gun chamber, for ballistic measurements. Water-cooling adaptors are available for many configurations.

### 15.4.9  Vibrating-Element Pressure Transducers

Pressure transducers using the change in the resonant frequency of a vibrating mechanical member, due to pressure changes, are capable of providing extremely close repeatability. They also produce a frequency output or frequency-modulated output (frequency deviation from a center frequency) which lends itself to digitization without conversion error (or direct display on a frequency counter). In some FM/FM telemetry systems such transducer outputs were fed directly into the system as one of the subcarrier channels. A significant number of vibrating-wire pressure transducers were built and used primarily in aerospace and oceanography. However, they were found to be difficult to produce. A vibrating-diaphragm transducer was designed and used experimentally, and vibrating-cylinder transducers were produced in several countries but suffered from high temperature sensitivity as well as producibility problems. These problems were finally overcome, and the vibrating-cylinder pressure transducer is now commercially available. Further development may bring similar results for other types of vibrating-element transducers.

The vibrating-wire pressure transducer uses a pressure-sensing diaphragm to the center of which a very thin wire (typically tungsten) is attached. The other end of the wire is mechanically anchored and electrically insulated (see Figure 15–23). The wire is located in a magnetic field usually obtained from a permanent magnet. When a current is passed through the wire, the wire moves within this field sufficiently to have a current induced in it. The emf due to this induced current is amplified and fed back to the wire to sustain its oscillation. The output of this oscillator circuit is then further amplified, with the output amplifier also acting as buffer.

The transducer is normally so designed that increasing pressure results in decreasing wire tension and hence in a reduction of its frequency of oscillation. Vibrating-wire transducers are inherently nonlinear; however, they are capable of repeating points on their calibration curve within very close tolerances. The minute deflection of the stiff sensing diaphragm poses a

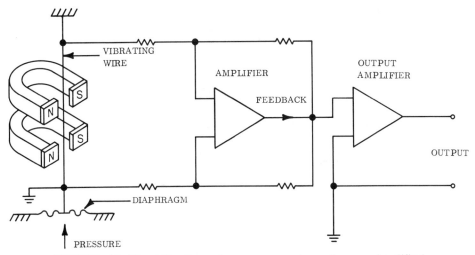

**FIGURE 15–23.** Vibrating-wire pressure transducer: simplified schematic.

requirement for very close tolerances in the diaphragm/wire/magnet assembly as well as for near-perfect matching of thermal coefficients of expansion to minimize thermal effects on performance. Thermal effects can be reduced further by enclosing the sensing assembly in a proportionally controlled heater jacket.

The vibrating-cylinder pressure transducer (Figure 15–24) uses a sensing element of the straight-tube type; the cylindrical wall of this tube undergoes changes in hoop stress as pressure is applied to it. When the cylinder is set into oscillation, its frequency of oscillation (natural frequency) will increase with increasing hoop stress due to increasing pressure. The cylinder is maintained in oscillation by a feedback amplifier/limiter combination which is connected to the driving coil and receives its input from the sensing coil. The output of this amplifier/limiter is also fed to another amplifier followed by circuitry that converts the sinusoidal signal into a square-wave output signal (pulse-frequency output) which can then be converted into a digital output by conventional means.

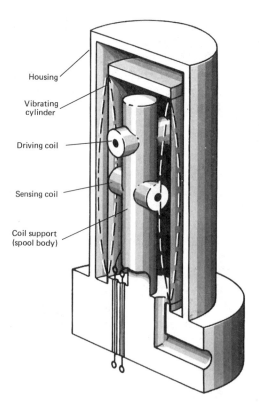

**FIGURE 15–24.** Vibrating-cylinder pressure transducer. (Courtesy of Hamilton Standard.)

The cavity between the cylinder and the housing is evacuated so that absolute pressure is measured. The mode of oscillation is given by the length of the cylinder. The cylinder configuration is chosen so that a four-lobed symmetrical hoop mode is obtained. The magnetic pickup coil is mounted orthogonal to the drive coil in the spool body, and the spatial relationship of the two helps in setting the cylinder into this oscillatory mode. The repeatability of this type of transducer is reported as within 0.0001%. The temperature error is on the order of 0.01%/°C; it can be reduced by a factor of about five by incorporating a proportionally controlled heater within the transducer. Other options include linearization circuitry to improve the end-point linearity from the inherent ±5% to ±0.04%, and to ±0.008% when additional circuitry provides a binary (linear) output, according to the manufacturer.

A few other types of vibrating-element pressure transducers were also developed. One uses the force obtained from a deflecting sensing element to actuate a lever linkage which causes the oscillating frequency of a quartz crystal to vary in proportion to pressure. Another design, developed for very low-pressure measurements, uses capacitor plates as well as electrostatic forcing in a feedback network to maintain a diaphragm in oscillation while increasing gas density within the sensing chamber tends to dampen such oscillations.

### 15.4.10 *Pressure Switches*

Pressure switches are probably the most abundant type of all pressure sensors. Millions of them are used in residential and automotive applications, besides their widespread industrial use. Most of these devices share the following design features: a pressure-sensing element, often counterbalanced by one or two adjustable-tension springs; a mechanical linkage actuated by the sensing element; and a push-button type switch (typically with snap action). The sensing element can be a bellows, but is most frequently a diaphragm made of a metallic or nonmetallic (e.g., silicone rubber, Buna N) material. The counterbalancing spring is used to set the switching point. Many units include an externally accessible adjustment to set the spring tension and switching point. A variety of lever linkages are used, some with provisions for dead-band adjustment. The basic electromechanical switch is of the type that requires a very small amount of plunger motion to actuate it. Several types of switching configurations are available: normally open only, normally closed only, make/break separate pairs of contacts, or, most commonly, a single switch which provides a normally closed as well as a normally open contact. Many types of housings are used for pressure switches, ranging from simple plastic-molded enclosures with slip-on terminals to explosion-proof and corrosion-resistant metallic housings incorporating a terminal block and provided with threads for conduit couplings.

## 15.5   TRANSDUCER CHARACTERISTICS

Although various groups of characteristics can be of major importance in selecting a pressure transducer for a given application, the characteristics that most often govern such a selection are range, accuracy, output, frequency response, the nature of measured fluids, and environmental conditions.

Specification characteristics for pressure transducers start with the *range* and type of pressure reference (absolute, gage, differential, sealed-reference differential, and, if gage or differential, a unidirectional or bidirectional range). When the application requires it, the range of the transducer can be expressed in terms of water depth or altitude, provided that the transducer's calibration curve uses such units rather than units of pressure.

*Mechanical characteristics* to be considered include configuration, mountings, and all pertinent dimensions; flush mounting or location of pressure ports and their pertinent dimensions, including threads; location and type of electrical connections; measured fluids (and any limitations on their physical and chemical characteristics); sealing of case or housing (e.g., whether it is hermetically sealed, explosion-proof, and/or waterproof); isolation of sensing elements(s), if any, by isolation membrane and transfer fluid; mounting or pressure-coupling torque; mass; location of nameplate and nameplate contents; and type of damping (and of damping fluid) if used. If the transducer is expected to be cleaned by solvent, the cleaning fluid should be considered a "measured fluid." If a transducer must be hermetically sealed, details of the type of sealing should be stated; gaskets and O-rings may deteriorate, and brazed joints may have flux entrapped in the braze, with the entrapped material subject to outgassing and creation of small voids in the joint. Special sealing of the transducer and electrical cable is needed for water-depth transducers, of course.

*Electrical characteristics* include nominal and maximum excitation (voltage or current or both); output impedance and, for some types of transduction elements, also input impedance; insulation resistance or breakdown-voltage rating; electrical input/output isolation; wiper noise (in potentiometric transducers); output noise (when an amplifier is incorporated in the transducer); and any integral provisions for simulated calibration.

Additional specifications of design characteristics are necessary or desirable for certain types of transducers or certain applications. For differential-pressure transducers it is, of course, necessary to identify the two pressure ports either as "HIGH" and "LOW" or, more correctly, as "M" or "MEAS." and "R" or "REF.," since these transducers measure one pressure with reference to another pressure. (The "M" and "R" designations make sense in many languages besides English as well.) When a fairly high frequency response is required of a non-flush-mounted transducer, the total volume of the sensing cavity (*dead volume*), together with its change over the measuring range, can become important. Details of sensing and transduction elements are often shown and can provide information useful to the application. For some applications it is useful that the axis in which the transducer is most vibration sensitive be known. For high-pressure transducers it is useful to specify a case-burst-pressure rating in addition to that rating as it applies to the sensing element. Leakage of pressure into the case may build up a high enough pressure to rupture the case eventually, and the end plate carrying the electrical connector may then become a dangerous projectile.

*Performance characteristics* include those generally applicable to most transducers and explained in Chapter 2. Accuracy characteristics are often specified in terms of error band instead of in terms of such individual characteristics as linearity (and type of linearity), hysteresis, repeatability, zero balance, and zero shift. The warm-up period is sometimes specified, as is creep. For certain types of pressure transducers, their sensitivity is shown as a primary performance characteristic. *Reference-pressure* range and error (effects) must be specified for differential-pressure transducers, and over-range (*proof pressure*) must be specified for all types of pressure transducers. (Note that reference pressure is sometimes called "line pressure.")

*Dynamic performance characteristics* are usually shown in terms of a "flat" frequency response (with a tolerance applied to the flatness of the response curve), sometimes as a natural frequency (lowest-frequency resonant peak should be used) and damping ratio, and sometimes (for over-damped transducers) as a time constant and (when damping is close to critical damping so that a small amount of underdamping may occur) overshoot.

*Environmental performance characteristics* comprise those defined in general for transducers in Chapter 2. Limits on thermal effects are shown in terms of temperature error band or individual specifications for thermal zero and sensitivity shifts, over the operating temperature range (ambient temperature) or the measured-fluid temperature range, whichever is more severe (some specifications lack clarity in this respect). In some specifications error limits are shown as applicable over a "compensated temperature range," with no limits on performance shown between the limits of this range and a maximum (or "safe") temperature, or, similarly, a minimum temperature. Thermal effects on damping should be shown for artificially damped transducers. The frequency response at specified elevated or cold temperatures is sometimes shown for wide-response transducers. Temperature-gradient error may need to be considered (and pyroelectrically induced thermal errors in piezoelectric transducers). Errors due to acceleration, shock, and vibration (the latter can induce errors significant only at certain vibration frequencies) are shown where such environments are expected. Errors and defects can also be introduced in many types of transducers by high sound-pressure levels, high-intensity magnetic fields, nuclear radiation, and, in some designs, changes in ambient pressure.

*Reliability characteristics* include operating and cycling life, errors and defects seen after long periods of storage, and stability.

## 15.6   APPLICATION NOTES

Pressure transducers are used in a wide variety of applications for the measurement of pressures of a few kPa to over a hundred MPa. The applications are in stationary installations ranging from household appliances through petrochemical and other industrial plants to nuclear reactors, as well as in vehicular installations ranging from automobiles, trucks, and locomotives to ships, aircraft, and spacecraft. The transducers are used for information as well as control purposes. The applications are strongly related to availability and cost: only a relatively few transducer models are available for such low-cost, mass-production applications as are found in household appliances and automobiles; transducer designs in large numbers and at competitive cost are available for the intermediate applications (process measurement and control, other industrial installations, diesel-engine trucks and cabs, locomotives, and many types of aircraft); and, again only a limited number of transducer models at a fairly high cost are "space qualified" (approved for use in manned and unmanned spacecraft) or usable in nuclear reactors.

Pressure transducers are also used to measure flow (see Section 12.3.1) and liquid level (see Section 14.3.1), and can be designed and calibrated to indicate water depth as well as aircraft altitude and airspeed. When connected to a set of right-angle-bent tubes, with the tube tips at incremental distances from each other (*pressure rake*), a group of pressure transducers can provide a profile of pressure distribution in a stream (of air or liquid). Small, thin flush-mounted transducers are used to evaluate aerodynamic shapes during wind-tunnel testing. Short, high-pressure phenomena such as explosive pressures are sensed by piezoelectric transducers. Some of the following notes should prove useful in selecting a pressure transducer on the basis of the general transducer characteristics explained in Section 15.5.

*Range.* Virtually all types of pressure transducers can be used in the most popular ranges, between 100 to 7500 kPa. (Table 15–2 shows psi-to-kPa and -MPa conversions.) For very low pressures below about 10 kPa, the choice is essentially limited to capacitive transducers. Potentiometric, strain-gage, and reluctive transducers are most commonly available for the high-pressure ranges, between 7.5 and 75 MPa (a few models to 150 MPa). Piezoelectric transducers are also used in this range when high-frequency response is needed. Some strain-gage and piezoelectric transducer designs are usable to pressures as high as 350 MPa, but their *calibrated range* is usually less than that. For pressures in excess of 350 MPa, the choice is limited to manganin gages. Part of the range specification for a differential-pressure transducer is the reference-pressure ("line-pressure") range, which is often much higher than the measured differential pressure (e.g., " ± 100 kPa differential, reference pressure 500 to 10 000 kPa").

*Accuracy.* Errors are smallest for vibrating-element and servo pressure transducers. Most transducer types, for most ranges, can provide a terminal-based static error band between about 0.5 and 1.5% FSO. Some manufacturers specify a reasonable error band as "standard" and a smaller error band as "special" (at somewhat higher cost).

*Output.* Potentiometric transducers provide an inherently high output voltage (0 to 5 or 10 V, usually dc) since their full-scale output voltage is equal to their excitation voltage. The output voltage of semiconductor-strain-gage transducers is in the hundreds of mV, full scale. With the use of appropriate signal-conditioning circuitry, often integrally packaged with the transducer, virtually any type of output voltage or current, and even digital output in some designs, can be obtained from almost any type of pressure transducer. The output impedance should always be significantly lower than the load impedance, especially for potentiometric transducers whose output impedance varies over their range.

*Frequency Response.* The transducer flat-response frequency range is generally highest for piezoelectric types, somewhat lower for many semiconductor-strain-gage designs, significantly lower for most other types, very low for potentiometric transducers, and lowest for servo types. As a rule, sensing-element stiffness increases with range, and natural frequency (as well as frequency response) increases with stiffness. The volume of the plumbing from the point of measurement and the volume of the transducer's internal cavity (*dead volume*) both affect frequency response: frequency response is reduced when the fluid in the plumbing is gaseous rather than liquid, and also when the dead volume is relatively large. To get the highest frequency response from a given transducer, the transducer should be flush mounted (i.e., the sensing element should be located at the point of measurement and wetted by the measured fluid). High frequency response is not needed in a majority of pressure measurements. In fact, some measurements call specifically for a low frequency response; this can be achieved by artificial damping within the transducer, or by mounting a baffle-type or capillary "snubber" into the pressure port of the transducer (or into the plumbing leading to the pressure port, if plumbing is used). In some data systems it is adequate to filter out the unwanted high-frequency components electronically.

*Measured Fluids.* The pressure port of the transducer, any cavity behind the port, the sensing element, and any other internal portions not sealed off by the sensing element must all be compatible with the liquids and gases expected to come in contact with them. When such compatibility cannot be established, a (compatible) isolation membrane can be placed in front of the noncompatible portion and the gap between this portion and the membrane filled with a transfer fluid. By proper choice of materials and assembly methods, pressure transducers can be constructed for compatibility with virtually all types of measured fluids. Differential-pressure transducers tend to be more difficult to design when conductive, corrosive, or otherwise "difficult" liquids must be applied to both pressure ports; the design is simpler when the fluid applied to one port is air or a nonconductive, noncorrosive gas. The latter type of transducer is called a "dry-wet" type, whereas the former is of the "wet-wet" variety.

*Temperature.* Special considerations apply to pressure transducers that must either operate in a very hot or very cold environment or that have to measure a very hot or very cold fluid. Many error-causing thermal effects can be minimized by the proper choice of materials in the sensing portion of the transducer. When individual pieces must be assembled to create this portion, their temperature coefficients must be matched very closely and assembly methods are critical. In many designs this problem has been greatly reduced by simply machining the sensing portion, including the sensing element (particularly if it is a diaphragm), from a single piece of metal. Many transducer designs use a temperature sensor and associated circuitry to compensate for thermal effects; in such cases it is important to reduce the thermal path between the sensing element and the temperature sensor to a minimum. Any circuitry located within the transducer and used, for example, for signal conditioning can also be subject to undesirable thermal effects and usually needs some type of compensation. Some such circuitry simply cannot withstand the specified temperature extremes and must then be located in a more benign thermal environment. Transducers that are extremely temperature sensitive can be enclosed in a temperature-controlled oven or jacket. Water-cooling jackets can be used around transducers exposed to extremely high temperatures, with a constant water flow maintained.

*Vibration and Acceleration.* Acceleration forces, either steady state or vibratory, can act upon the transducer sensing element just like pressure forces. Additionally, vibration at certain frequencies can set up sympathetic vibrations (*resonances*) in internal transducer elements. As a result, not only can significant errors appear in the output signal, but there are known occurrences of catastrophic failures due to resonances within transducers. The effects can be minimized by using simple, stiff sensing elements that have a natural frequency higher than the highest expected significant vibration frequency. They can be further minimized by mounting the transducer so that acceleration forces will not act along the transducer's most acceleration-sensitive axis. However, some designs are inherently limited to operation in a low-vibration (and low-acceleration) environment.

# 16

# Sound-Measuring Microphones

## 16.1 BASIC DEFINITIONS

*Sound* is an oscillation in pressure, stress, particle displacement, particle velocity, or density that is propagated in an elastic or viscous medium or material, or it is the superposition of such propagated oscillations (*sound wave*); it is also the auditory sensation evoked by such oscillations (*sound sensation*).

*Sound energy* of a portion of a medium is the total energy in that portion minus the energy that would exist if no sound waves were present.

*Sound pressure* is the total instantaneous pressure, at a given point, in the presence of a sound wave, minus the static pressure at that point.

*Peak sound pressure* is the maximum absolute value of the instantaneous sound pressure within a specified time interval.

*Effective sound pressure* is the root-mean-square value of the instantaneous sound pressures, over a specified time interval, at a given point.

*Sound pressure level* is normally expressed, in decibels, as 20 times the logarithm to the base 10 of the ratio of the rms (effective) sound pressure $P$ to an rms reference pressure $P_{ref}$, or

$$L_p = \text{SPL} = 20 \log_{10} \frac{p(\text{rms})}{p_{ref}(\text{rms})}$$

where $L_p$ and SPL are both acceptable symbols for sound pressure level, expressed in dB. The reference pressure must be stated. It is usually taken as either $2 \times 10^{-4}$ μbar ($2 \times 10^{-5}$ Pa) or 1 μbar (1 dyne/cm$^2$, 0.1 Pa).

*Sound level* is a weighted sound-pressure level (in dB), at a point in a sound field, averaged over the audible frequency range, and as displayed on a *sound-level meter* that complies with a prescribed national and/or international standard. Examples of such standards are IEC R 179 (international); ANSI S1.4 (U.S.); and DIN 45 633 (Germany).

*Impulsive sound* is sound that consists of short bursts (rather than sustained tones).

*Sound intensity* is the average rate of sound energy transmitted in a specified direction through a unit area normal to this direction at a given point.

*Sound power (of a source)* is the total sound energy radiated by the source per unit of time.

*Sound absorption* is the process by which sound energy is diminished by being partially changed into some other form of energy, usually heat, while passing through a medium or striking a surface.

A *simple sound source* is a source that radiates sound uniformly in all directions under free-field conditions.

A *sound field* is a region containing sound waves.

A *free sound field* is a sound field in a homogeneous medium that is free of any acoustically reflecting boundaries.

*Noise* (in acoustics) is any unwanted sound.

*Noise dose* is the accumulated noise exposure a person is subjected to, with reference to a specified sound level [typically 90 dB(A)] and over a specified period of time (e.g., 8 h).

*Propagation velocity* is a vector quantity that describes the speed and direction with which a sound wave travels through a medium.

*Reverberation time* is the time required for the average sound-energy density, at a given frequency, to decrease to $10^{-6}$ ($-60$ dB) of its initial value when the source has been in a steady state and is stopped.

*Pressure frequency response* (of a sound-pressure measuring transducer) is the ratio, as a function of frequency, of the transducer output to sound pressure which is equal in phase and amplitude over the entire sensing-element surface of the transducer.

*Free-field frequency response* (of a sound-pressure measuring transducer) is the ratio, as a function of frequency, of the output of the transducer in a sound field to the free-field sound pressure that would exist at the transducer location were the transducer not present.

*Free-field normal incidence response* is the free-field frequency response when sound incidence at a specified sensing surface of the transducer is from a direction normal to that surface.

*Free-field grazing incidence response* is the free-field frequency response when sound incidence at a specified sensing surface of the transducer is from the direction parallel to that surface.

*Random incidence response* is the diffuse-field frequency response (of a sound-pressure measuring transducer) where sound incidence at a specified sensing surface of the transducer is from random directions.

*Diffuse sound* is sound, in a given region, which has uniform sound-energy density and is such that all directions of sound-energy flux are equally probable at all points in the region.

*Directivity* (of a sound-pressure measuring transducer) is the solid angle, or the angle in a specified plane, over which sound incident on the sensing element is measured (within specified tolerances) at a specified sound frequency or in a specified band of sound frequencies.

The *directivity factor* is the ratio of the square of the output (of a sound-pressure measuring transducer) produced in response to sound incident from a specified direction to the mean-square output that would be produced in a perfectly diffuse sound field, of the same frequency or band of frequencies and of the same mean-square sound pressure.

A *directional response pattern (directivity pattern)* of a sound-pressure measuring transducer is a description, usually in graphical form, of the transducer's response as a function of direction of incidence of sound waves in a specified plane and at a specified frequency or band of frequencies.

The *equivalent volume* of a sound-pressure measuring transducer is the transducer's acoustic input impedance expressed in terms of the acoustic impedance of an equivalent volume of a gas enclosed in a rigid cavity.

## 16.2 UNITS OF MEASUREMENT

*Sound pressure* can be expressed in *pascals (Pa)* or in equivalent non-SI units such as the *dyne/cm²* $(= 0.1\ Pa)$; however, it is generally expressed as *sound-pressure level,* in *decibels (dB),* as explained in the previous section. Sound-pressure level (*SPL* or $L_p$) is usually referenced to (referred to a reference pressure of) $2 \times 10^{-4}$ μbar $(= 0.0002\ \text{dyne/cm}^2 = 20\ \mu\text{Pa})$ unless a different reference pressure is specified.

Sound power is expressed in *watts (W)*.

Acoustic impedance, resistance, and reactance are expressed in N·s/m⁵; this unit is also referred to as the *mks acoustic ohm*.

## 16.3 SENSING METHOD

Sound is sensed by *pressure-sensing* elements; the *flat diaphragm,* or design variations thereof (see Section 5.3.2), is used in virtually all sound sensing devices. In some cases the transduction element itself provides the sensing function, and the diaphragm then acts as an isolation membrane, responding to variations in sound pressure. The sensing element is usually configured for gage-pressure sensing, that is, ambient pressure is admitted to the reference side of the diaphragm. Hence, sound pressure is measured with respect to ambient static pressure, while static pressures acting on the outer and inner diaphragm surfaces are equalized. The gage vent can also be made to act as a "low-pass filter" acoustic leak, which prevents the sound pressure variations from being seen by the reference side of the diaphragm. Some sensing elements exist in sealed-reference differential-pressure configurations, with their reference side sealed and sometimes partly evacuated.

## 16.4   DESIGN AND OPERATION

### 16.4.1   *Condenser Microphones*

Condenser microphones are capacitive sound-pressure transducers whose design is optimized for acoustic measurements. A typical design is shown in Figure 16–1. The metallic diaphragm acts as one electrode, and the backplate acts as the other electrode of the capacitive sensing/transduction element. The diaphragm is electrically and mechanically bonded to the housing, and the housing is also used as the ground terminal of the sensor. The backplate is retained in a silicone-treated quartz (or synthetic ruby) insulator which forms the rear wall of the sensing cavity and provides an acoustic reference surface. The perforations in the backplate equalize the pressure within this cavity. A capillary duct provides equalization between the sensing cavity and ambient pressure in a controlled manner. A silver wire is used for adjusting this "acoustic leak" so as to give a well-defined and low low-frequency limit. Electrical contact with the backplate is provided through the (gold) output terminal. An end cap with a symmetrical protecting grid is mounted, by threading, over the diaphragm.

The unit illustrated is intended for mounting to a housing which provides for either hand-held or clamped operation and also contains a preamplifier. When so installed, a contact on the preamplifier module comes in firm electrical contact with the output terminal of the "microphone cartridge." A cable (or rigid or flexible conduit) is furnished integrally with the preamplifier. The cable is then connected to an amplifier/display unit (which contains the required power supply) or to a power supply unit which also includes a line-drive amplifier (Figure 16–2). The basic condenser microphones require a stable dc polarization voltage, applied, through a high resistance, across the two capacitor electrodes in order to maintain a constant charge on the electrodes. This polarization voltage (usually 200 V, although lower voltages can also be used for certain microphone designs) is furnished by the power supply unit (or built-in power supply in other associated equipment) together with the power required by the preamplifier. The preamplifier also provides the low output impedance needed by the measuring system.

The need for a stabilized polarization voltage is obviated in condenser microphones incorporating a permanently polarized dielectric layer (*elec-*

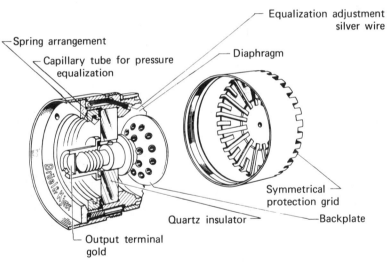

**FIGURE 16–1.**   Cutaway and sectional view of condenser microphone. (Courtesy of Brüel & Kjaer Instruments, Inc.)

*tret*) between the capacitor electrodes. Two microphone designs provided with such a charge-carrying layer are illustrated in Figure 16–3. There are various approaches to producing an electret. Barium titanate/lead titanate ceramic layers can be polarized by first raising their temperature above the Curie point and then cooling the layer while exposing it to a high-voltage

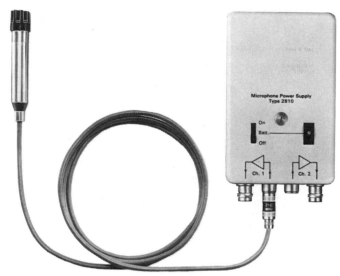

**FIGURE 16–2.** Condenser microphone system, consisting of preamplifier with microphone cartridge and integrally attached cable, and (two-channel) power supply unit. (Courtesy of Brüel & Kjaer Instruments, Inc.)

electric field. Another approach uses a metallized plastic foil processed through a high-voltage electric field at a high temperature. The use of a solid dielectric also permits closer spacing between the electrodes, which increases the capacitance of the sensor.

Some condenser microphones are specifically designed to operate with

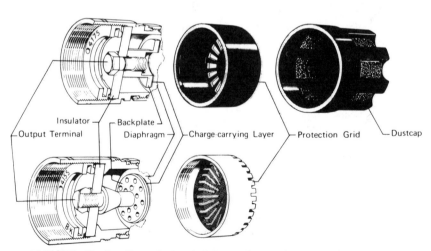

**FIGURE 16–3.** Prepolarized (electret) condenser microphones, with frequency response to 12 kHz (upper unit) and 20 kHz (lower unit). (Courtesy of Brüel & Kjaer Instruments, Inc.)

a high-frequency carrier system instead of a power supply. The carrier (typically 10 MHz) is modulated by the capacitance variations. This type of system extends the lower limit of the frequency response from the usual 2.5 to 7 Hz downward, to as low as 0.01 Hz.

The protecting-grid end caps on condenser microphones are usually designed for free-field measurements. For diffuse-field measurements this end cap can be replaced by one having a different grid configuration appropriate for such measurements ("random incidence corrector"). A dust cap is normally kept over the microphone when it is not in use. For use in certain environments, umbrella-like rain covers, spherical sponge windscreens, or other specialized protective devices can be placed over the microphone. Microphones with quartz coatings over their (nickel or nickel alloy) electrodes are available for use in humid, salty, or corrosive atmospheres. Silica gel dehumidifiers can be mounted between the microphone cartridge and the preamplifier to keep the sensing element dry. Other accessories include floor stands (tripods) and rotating booms.

### 16.4.2   Piezoelectric Microphones

The design and operation of piezoelectric microphones (often called *piezoelectric sound-pressure transducers*) is very similar to the design and operation of piezoelectric pressure transducers (see Section 15.4.8). Ceramic crystals are more frequently used than quartz crystals. The housing is usually cylindrical, allowing hand-held use as well as clamping in a fixed location. Designs with a partly threaded housing are also available, for flush-mounting into a threaded port in the wall of a chamber or duct. Many piezoelectric microphones have a built-in solid-state preamplifier which also acts as an impedance converter, providing a low output impedance (1000 ohms or lower) which obviates the many problems that would be encountered with the very high output impedance (typically hundreds of megohms) of the crystal itself. The microphone output is then usually high enough for display, recording, or analysis equipment, making an additional external amplifier optional. The diaphragm/crystal combination is typically designed and mounted so as to minimize the effects of vibration along the sensing axis; some designs even incorporate a vibration-compensating element.

### 16.4.3   Piezoelectric Hydrophones

A *hydrophone* is, generally, a microphone designed for use while immersed in liquid, especially in water. Most hydrophones are piezoelectric and are well sealed, since they must withstand underwater pressures at depths down to 1000 m or more. Their electrical connections are usually brought out through a waterproof cable, sealed at its transducer interface. As is the case for piezoelectric pressure transducers, hydrophones tend to have a high frequency response, up to about 16 kHz. Additionally, however, good fre-

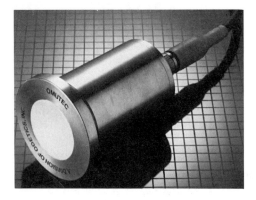

**FIGURE 16–4.** Piezoelectric hydrophone. (Courtesy of Omutec, a Division of Odetics, Inc.)

quency response at very low frequencies (to less than 0.1 Hz) is desirable for underwater work. The hydrophone shown in Figure 16–4 has a frequency response extending down to 0.03 Hz, making it usable for wave height monitoring as well as the other functions hydrophones are typically designed to perform (see Section 16.6.3). This transducer also incorporates a charge amplifier that produces high-level output signals while providing a low output impedance (50 ohms). The size of this hydrophone is 70 mm dia. × 120 mm long, and its mass is 2 kg.

Hydrophones are often designed and connected so that they can transmit as well as receive sound. Piezoelectric transducers are inherently capable of doing this, since the piezoelectric effect is reversible: a piezoelectric crystal, coupled to a diaphragm, will emit sound when excited by an ac or pulsed voltage. The process of pulsed sound emission by such a device has been referred to as "pinging."

## 16.5   MICROPHONE CHARACTERISTICS

Essential *mechanical design characteristics* of microphones include their configuration and dimensions, mounting means and dimensions, and the dimensions of any protecting-grid end cap. For those hydrophones whose sensing element is concealed by a waterproof sheath, the location of the center of the acoustic field should be defined on the outline drawing. Housing and sensing element materials and type of sealing should be described and any limitations on constituents and contaminants of the ambient atmosphere should be stated. A statement of equivalent volume (due to sensing element compliance) is often desirable also. Any enclosed or semienclosed volume associated with the sensing element, and its effect on frequency response, should be described in adequate detail.

The location and type of external electrical connections should be explained. If the microphone (or microphone/preamplifier assembly) is furnished with an integral cable, the type and characteristics of the cable and connector should be described.

*Electrical design characteristics* include polarization voltage requirements or carrier voltage amplitude and frequency requirements for condenser microphones, and power supply requirements for integrally packaged or integrally assembled preamplifiers. Other characteristics to be stated include output impedance, insulation resistance, output noise, grounding, and permissible range of load impedance of user-furnished associated equipment.

*Performance characteristics* are stated either for the basic sensor ("microphone cartridge") or for an integrally packaged (not capable of disassembly) microphone/preamplifier combination. The conditions under which the performance characteristics are applicable (e.g., excitation supply voltage and stability, and load impedance) must be explained.

*Range* (dynamic range) is usually specified in terms of sound-pressure level (SPL or $L_p$) together with a statement of the reference pressure (e.g., "Range: 48 to 170 dB re 20 $\mu$Pa"). In some cases range has also been expressed in units of pressure (Pa, $\mu$bar, psi, dynes/cm$^2$). *Overrange* capability or maximum sound pressure are often stated additionally.

*Linearity* (independent linearity unless otherwise specified) is often shown in conjunction with the range over which a specified linearity applies ("linear dynamic range"). It is expressed in percent or in dB. Since nonlinearity in an acoustic system produces harmonic distortion, the linear dynamic range is sometimes defined by sound-pressure levels at which a specified percent distortion occurs.

*Sensitivity* is usually shown in *mV/Pa* (pressure sensitivity). It is also often shown in terms of *sensitivity level*, expressed in decibels referred to a stated reference sensitivity, typically 1 V/$\mu$bar (e.g., "−94 dB re 1 V/$\mu$bar"). The sensitivity (level) in dB is equal to 20 $\log_{20} (V/p_s)$, where $V$ is the output in volts rms and $p_s$ is the effective sound pressure in $\mu$bar. Sensitivity can also be shown (for piezoelectric transducers without integral preamplifier) as *charge sensitivity level* (e.g., "____dB re 1 picocoulomb per microbar"). Sensitivity, regardless of the manner in which it is expressed, is always the *nominal* sensitivity. Its statement should be accompanied by a set of tolerances; however, microphones are usually accompanied by an individual calibration chart which includes a statement of their actual measured sensitivity.

*Frequency response* specifications should state whether pressure response, random incidence response, or free-field response is meant; when free-field response is indicated, normal incidence is implied unless grazing incidence is specified. Preferably, the frequency response should be referred to a specific amplitude and frequency. The flatness of frequency response is expressed in dB as a tolerance on the sensitivity, over a stated frequency range [e.g., "Free-field normal-incidence frequency response: within ±2 dB from 4.2 Hz to 40 kHz (as referred to the response at 100 Hz and at 160 dB SPL)''].

The pressure frequency response is generally equal to the free-field random- or grazing-incidence response at wavelengths which are long compared to the maximum dimension of the microphone. Frequency response can sometimes be calculated from the microphone's response to transient pressure, or from its geometry and mechanical properties when these are relatively simple; when determined in this manner, it must be stated as "calculated frequency response." If the microphone is intended for use in fluids other than air (or if it is a hydrophone), and its frequency response and other specification characteristics are applicable when used in such a fluid, the specification must be annotated appropriately.

*Directivity* can be specified as directivity factor or as solid angle symmetrical about (bisected by) the principal axis of the microphone. It is often shown, in manufacturers' literature, as one or more typical directional response patterns, based on test data and plotted on circular graph paper.

*Threshold* is sometimes specified. Resolution is not included in specifications since all commonly used measuring microphones have continuous resolution.

*Temperature sensitivity* is shown either as a tolerance, in $\pm$ dB, on sensitivity over a specified operating temperature range, or as thermal sensitivity shift, in dB/°C, for such a temperature range. *Temperature gradient error* is often shown, especially for piezoelectric microphones (which can exhibit pyroelectric-effect potentials); this is sometimes called "transient temperature response." Some microphone designs can be equipped with a water cooling jacket for high-temperature operation.

*Acceleration and vibration effects* have been covered, in specifications, in terms of the output, due to acceleration, equivalent to an output corresponding to a stated sound-pressure level. The acceleration error is then shown in dB per rms $g$ (or per m/s$^2$). Vibration error is also shown in this manner, with an additional statement of the vibration frequency range. Both of these specifications should indicate along which microphone axis the tolerances are applicable. (When not stated, acceleration acting along the longitudinal axis, normal to the sensing diaphragm, is implied, as the worst-case condition). Vibration and acceleration error can also be specified in terms of equivalent sensitivity level, expressed as equivalent SPL per $g$, and determined as 20 $\log_{10}$ of the ratio of the apparent rms sound pressure due to acceleration to the rms applied acceleration (vibration) amplitude.

*Ambient-pressure error* is specified as a tolerance on sensitivity, in $\pm$ dB, over a stated range of ambient pressures, or, as "altitude error," over a stated range of altitude above sea level. For hydrophones, this type of error can be shown for a stated range of pressures or depths below sea level.

*Other environmental effects* on performance may include those due to contaminating or corrosive atmospheres, to shock, or to exposure to nuclear radiation or electromagnetic fields.

## 16.6 APPLICATION NOTES

### 16.6.1 *Sound-Pressure Measurements*

A basic sound-pressure measuring system consists of a microphone with its preamplifier and an amplifier which has a flat frequency response matched as closely as possible to that of the microphone. For sound-pressure telemetry applications the compact, ruggedly packaged amplifier has a fixed gain setting such that full-scale output, as accepted by the telemetry set, is produced at the upper SPL range limit. Voltage amplifiers are used in all cases except when the output of a piezoelectric microphone without a built-in preamplifier must be amplified, in which case a charge amplifier is used instead. Since a wide band of frequencies must be examined, a wide-band telemetry channel is needed for the transmission of analog signals. In digital telemetry systems this requirement translates into a very high data rate for a single measurement. When a reduced data rate is required and a reasonable approximation of the measured signal is acceptable for reconstruction at the data reception, processing, and display facility, a *spectrum analyzer* can be used with the amplifier. Such an analyzer typically contains a number of narrow-band-pass filters to which the signal is applied. The output of each filter is an analog signal representative of the amount of acoustic energy in the respective narrow band of frequencies. The filter outputs are then read out sequentially.

Most sound-pressure measuring systems, however, are in the form of bench-test setups and are controllable by human operators. The amplifier is equipped with a calibrated attenuator so that selected portions of the overall measuring range can be displayed. The amplifier may also be equipped with one or more filters (or provisions for connecting an external filter or filter set). The output of the amplifier can either be applied directly, as ac signal, to display or analysis equipment, or it can be rectified into a dc signal. This signal can also be passed through a linear-to-logarithmic converter to provide a logarithmic display of sound-pressure level. The signal, in either of its forms, can also be recorded for non-real-time analysis. Analog as well as digital analyzers are available for obtaining a cathode-ray tube (CRT) or other graphic display of the frequency characteristics of the measured signal.

There are many applications in sound-pressure measurement where either the continuous or short-term sounds of a sound-producing source or the sound field existing at a particular location are measured and characterized as to amplitude vs. frequency distribution. Continuous sounds are emitted by propulsion systems (automotive, ship, rail, aircraft, and rocket engines) and by essentially all types of machinery. Measurements of *sound power* of machinery or other equipment are made by an array of microphones placed around the equipment or by a single microphone at the end of a rotating boom. Use of a sound source of known characteristics in conjunction with one or more microphones allows measurements of the *sound-insulating* qualities of walls, of *reverberation time* and *sound distribution* (e.g., in rooms, auditoriums, and concert halls), and of *sound absorption*.

Measurements of the *acoustic emissions* from materials undergoing stress, and the characterization of such emissions, have become a useful tool in stress and fatigue analysis. Such characterizations are related to the "signatures" of rotating parts and assemblies that are obtained by vibration analysis techniques from which such factors as bearing wear can be determined.

### 16.6.2  *Sound-Level Measurements*

Sound level is a weighted sound-pressure level; the signal from the microphone preamplifier is fed to an amplifier which incorporates a *weighting network* that emphasizes certain frequencies while deemphasizing others. Weighting networks carry a reference designation letter (A, B, C), and the characteristics of each network are defined in national and international standards. A sound-pressure level measuring system that incorporates one or more weighting networks is called a *sound-level meter*. A typical instrument is shown in Figure 16–5. It is a compact, hand-held, battery-operated device which contains a measuring condenser microphone and preamplifier in the necked-down stem at its top, two selectable weighting networks (A and C) as well as a selectable linear response (unweighted), and a calibrated attenuator (range selector) which shifts the range of the displayed reading by 10-dB increments. Since many noise level measurements are based on the "A" weighting network, some sound-level meters include only that network and then provide a display in "dB(A)" (sometimes simply shown as "dBA").

Sound-level measurements are specifically related to human sound sensation, and the characteristics of aural perception form the basis of the weighting networks. The frequency response of networks A, B, and C is nominally flat between about 1 and 10 kHz; however, the response of the A network drops off sharply below 1 kHz, the response of the B network drops off only moderately below 400 Hz, and the response of the C network is essentially flat between 100 Hz and 3 kHz and drops off only by 5 dB between 100 and 20 Hz and between 3 and 10 kHz.

Sound-level meters can include a number of additional features. *Impulse* sound-level meters are specifically designed for impulse sound (short

**FIGURE 16–5.** Sound-level meter. (Courtesy of Brüel & Kjaer Instruments, Inc.)

bursts) measurement (e.g., for sounds produced by a punch press or forge hammer). Other features include adjustable response time, octave (or octave-fraction) filter sets (or provisions for interconnection with external filter sets), peak value hold, overload indication, and determination and calculation of $L_{eq}$, the integrated ("equivalent continuous") sound level over adjustable periods of time. Extension cables can often be inserted between the microphone/preamplifier and the meter.

*Noise dose meters* measure accumulated noise exposure over a specified period of time (typically an 8-h working day). Such meters are usually compact enough to be worn in a worker's breast pocket. The meter is not resettable except by a supervisor, who reads and records the worker's noise dose reading at the end of the day. Some models provide a noise hazard warning to the worker when a specified sound level, such as 140 dB(A), is exceeded.

Sound-level measurements are the basis of noise control. With increased awareness of the hazards of "noise pollution," which range from hearing impairment and nervous disorders to interference with speech and the creation of sales resistance to a product, government agencies in many countries (e.g., OSHA in the United States) have been specifying limits on noise in work areas and have been enforcing such regulations. Additionally, manufacturers of a large variety of products have been encouraged to reduce the amount of noise made by their products. These range from industrial equipment to home appliances and cars. All this has increased the importance and use of sound-level meters substantially. Noise is measured, designs are modified, or insulation is added to reduce the noise to acceptable levels; the noise is then remeasured and often monitored continuously thereafter. As a result, work areas, highways, airports, and even homes have been getting quieter.

There are, of course, many applications of sound-level meters in areas other than noise control, e.g., to obtain proper acoustical designs of studios, classrooms, and other architectural work. They are also used by musical performers and performing groups and in many commercial establishments, such as department stores.

### 16.6.3  Underwater Sound Detection

The underwater version of the microphone, the *hydrophone,* is commonly used as an underwater listening device. Analysis of the frequencies, amplitudes, and wave shapes of its output signals can be interpreted to obtain information about the nature of the object from which the sound emanates. Two or more hydrophones can be used for *ranging,* that is, obtaining information about the location of the sound source.

Many hydrophone designs are capable of operating in the transmit mode (using the inverse piezoelectric effect) as well as the receive mode. They can then be used in such systems as the *echo-ranging sonar* (the acronym "sonar" is derived from "*so*und *na*vigation *a*nd *r*anging"). The usual operating mode of such a system involves the generation of a pulse of sound energy and its transmission by the hydrophone. The hydrophone is then immediately switched into its receive mode, and the travel time of the pulse (*ping*) to and from the target is measured by the equipment. The travel time, combined with knowledge of the directional characteristics of the hydrophone, provides range information about the target. Further analysis of the "echo" from the target can yield additional information about the target (e.g., its configuration and the material it is made of). Most hydrophones are piezoelectric, but some are magnetostrictive, in transduction principle.

Besides its many military applications, sonar is widely used for the detection of schools of fish, in salvage operations, for ocean-floor mapping, and in other oceanographic research.

### 16.6.4  Ultrasonics

Ultrasonic waves are electromagnetic waves that are made to propagate through a gaseous, liquid, or solid medium. The frequencies of ultrasonic energy extend from the upper limit of the audio-frequency band into the

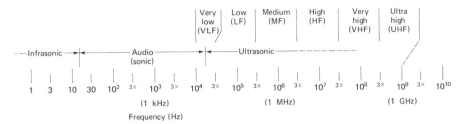

**FIGURE 16–6.** Frequency-band nomenclature below 3 GHz.

very-high frequency band used for telecommunications (see Figure 16–6). It is primarily the propagation velocity of ultrasonic waves and the effects on this velocity by the density, viscosity, and elasticity of the gaseous, liquid, or solid medium through which the sound waves propagate that establish the most useful characteristics of ultrasound for measurement purposes.

The theory and applications of ultrasound and ultrasonic equipment are well covered in available technical literature. Ultrasonic devices used for the measurement of specific types of quantities, such as flow and liquid level, are discussed in applicable sections of this book. The ultrasonic *transducers* (piezoelectric or magnetostrictive) are related to hydrophones (see Section 16.4.3) in that they can operate in either the receive or the transmit mode, or, alternatingly, in both modes. Ultrasonic measuring techniques are related to echo-ranging sonar techniques in that a pulse of ultrasonic energy is transmitted and the echo from the medium or target is analyzed.

The interaction of ultrasonic waves with substances is similar to the interaction of light with substances. Besides the reflectance mode that allows a comparison with sonar techniques, the transmission mode is used in some applications. In materials testing, for example, the reflectance mode is commonly used for flaw detection in solids and other materials; however, the transmission mode, in combination with ultrasonic emissions from a reference transducer, is used in *acoustic holography imaging* systems for nondestructive testing. Ultrasonic scanning, using reflectance, is used for subcutaneous mapping of biological organisms, sometimes in combination with computerized image processing techniques. Since the penetration depth of ultransonic energy, including into nonhomogeneous solids, is largely dependent on frequency, empirically defined strata of tissues and organs of the body can be scanned sequentially and three-dimensional images can be derived from the data.

# 17

# *Vacuum Sensors*

## 17.1 BASIC DEFINITIONS

A *vacuum* is, theoretically, the complete absence of any matter in a volume or a region of space; in practice, the term is applied to pressures (see Section 15.1) significantly below 1 standard atmosphere (about $10^5$ Pa) and typically to those below 7 kPa (about 1 psia). By convention (not by any formal standard), the vacuum range is divided into the following regions (note that, in vacuum terminology, a *low* pressure is a *high* vacuum):

> *Low vacuum*: 760 to 25 torr
>
> *Medium vacuum*: 25 to $10^{-3}$ torr
>
> *High vacuum*: $10^{-3}$ to $10^{-6}$ torr
>
> *Very high vacuum*: $10^{-6}$ to $10^{-9}$ torr
>
> *Ultra high vacuum (UHV)*: below $10^{-9}$ torr

## 17.2 UNITS OF MEASUREMENT

Vacuum has traditionally been expressed in *torr*; where 1 torr = 133.32 Pa. The torr equals one mm Hg at 0°C. Vacuum is sometimes expressed in "microns," a misleading term which, in the case of vacuum measurement, refers to micrometers of mercury at 0°C and has also been called "millitorr" ($10^{-3}$ torr). Occasionally, the mbar (see Section 15.2) is used as a unit for vacuum. With some rounding off, the following conversions can be used:

$$1 \, torr = 133 \, Pa = 1.33 \, mbar = 1 \, mm \, Hg = 1/760 \, standard \, atmosphere$$
$$1 \, Pa = 7.5 \times 10^{-3} \, torr$$

## 17.3  SENSING METHODS

### 17.3.1  *Pressure-Diaphragm Vacuum Sensing*

Sensors using the same basic operating principles (see Section 15.3) and some of the same design features as capacitive pressure transducers (see Section 15.4.1) have been designed for vacuum measurements down to about $10^{-4}$ torr. Sufficient deflection can be produced, even at such low pressures, in an appropriately designed and fabricated pressure-sensing diaphragm to vary the capacitance between it and either one or two electrodes sufficiently to produce a usable output signal. For measurements between about 1 torr and over 760 torr, pressure-sensing diaphragms and capsules are also used with other transduction elements (e.g., reluctive).

### 17.3.2  *Thermal-Conductivity Vacuum Sensing*

As the vacuum increases (i.e., pressure decreases) in a small chamber, the gas molecules in the chamber decrease in number. When a heat source is placed in the chamber, the flow of heat from source to wall surface decreases as the number of gas molecules decreases. At low pressures the thermal conductivity of a gas decreases linearly with pressure. If the heat source is a filament heated by a constant current through it, the reduction in heat transfer away from the filament, with decreasing pressure, will cause the temperature of the filament to rise. This effect is utilized in resistive as well as thermoelectric vacuum sensors.

In *resistive* sensors the increase in filament resistance due to heating is measured, typically by connecting the hot filament as one arm of a Wheatstone bridge (Figure 17–1). This type of sensor is commonly referred to as a *Pirani gage*. The filament is generally heated to about 200°C (at the lower end of the measuring range). A second sensor of the same configuration, but sealed so that it does not respond to vacuum changes, can be connected as a second, or reference, arm of the bridge circuit. A variation of this design employs a heated thermistor (*thermistor gage*) instead of a filament.

In *thermoelectric* sensors that utilize the thermal-conductivity principle the temperature (rise) of the filament is sensed by a thermocouple (see Section 19.3.1) welded to the center of the filament (Figure 17–2). This type of sensor is called a *thermocouple gage*. Some such designs use a thermopile instead of a single-junction thermocouple to increase the output signal.

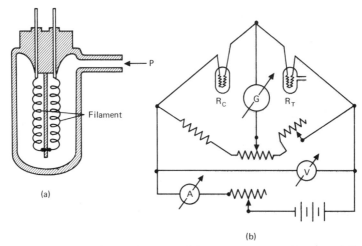

(a)

(b)

**FIGURE 17–1.** Thermal-conductivity vacuum sensing—resistive: (a) transducer; (b) typical bridge circuit.

### 17.3.3 Ionization Vacuum Sensing

In this general category of vacuum sensors, vacuum is measured as a function of gas density by measuring ion current. The ion current results from positive ions which are collected at a negatively charged electrode when the gas is ionized by a stream of electrons or other particles. The ion current is

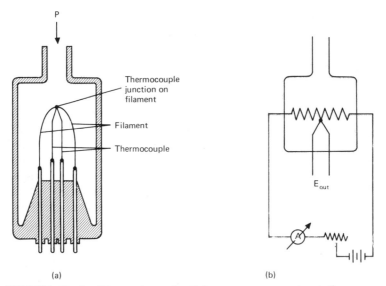

(a)                                        (b)

**FIGURE 17–2.** Thermal-conductivity vacuum sensing—thermoelectric: (a) transducer; (b) typical circuit.

proportional to the gas molecular density and, hence, inversely proportional to vacuum when the number of electrons and their average path length are constant and all ions are collected. Because such sensors respond to variations in gas density, which is different for different gases, the calibrations of the sensors will also be different for different gases.

*Ionization gages (ion gages)* represent the simplest form of *thermionic vacuum sensors.* An ionization gage (Figure 17–3(a)) resembles a triode-type vacuum tube that has an opening in its envelope. A filamentary cathode is surrounded by a helical grid around which a cylindrical anode is placed. Positive ions are collected at the anode, which is kept at a low negative voltage with respect to the filament. A basic measuring circuit is shown in Figure 17–3(c). Switching in the grid outgas supply allows heating the grid

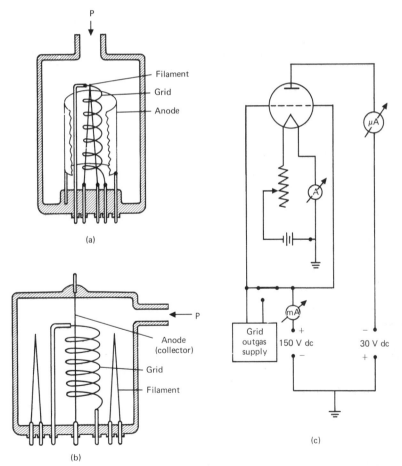

**FIGURE 17–3.** Ionization vacuum sensing—thermionic: (a) triode type; (b) Bayard-Alpert type; (c) typical circuit.

electrically to drive off gas molecules from the grid surface and thus to avoid any *outgassing* during the operation of the sensor. Outgassing products are detected just like the measured gas; hence they cause errors. Commonly used accessories, which take the place of some of the elements shown in the basic circuit, include an ion-current amplifier for the output current and a feedback-controlled power supply for the filament.

Various techniques and design modifications have been developed to extend the range of the basic ionization gage, which extends between about $10^{-8}$ and $10^{-3}$ torr. At pressures above 0.1 Pa (about $10^{-3}$ torr) the space-charge effect, as well as recombination of ions due to a reduction in their mean free path, causes extreme nonlinearity and effectively limits the usable range. In the *Shulz-Phelps gage* a second electron collector electrode is added to minimize space-charge effects and thus yield a usable range up to about 1 torr.

The lower end of the triode type's measuring range is limited to about $10^{-8}$ torr because of X-ray effects. The electron stream produces soft X-rays at the grid. When these X-rays strike the collector electrode, they drive electrons from it by secondary emission. The resulting ion current is of the same polarity as the normally produced ion current. At about $10^{-8}$ torr, then, the calibration curve of the sensor becomes asymptotic to this residual X-ray current. This phenomenon is substantially reduced in the *Bayard-Alpert gage* (Figure 17–3(b)). The arrangement of filament, grid, and anode is inverted. The anode is a thin wire whose surface area is very small, and only a small number of the X-rays produced at the grid strike the collector. The resultant much lower X-ray current extends the usable range down to $10^{-10}$ torr.

Because the Bayard-Alpert gage does not contain the large outer anode plate of the basic triode type, more electrons and ions strike the inside wall of the glass envelope, where they build up an electrostatic charge that can cause erratic operation of the sensor. In the *Nottingham gage* this problem is minimized by a metallic coating which is applied to the inside of the glass envelope and electrically connected into the circuit of the sensor. Also, the grid is provided with end shields. The Nottingham modifications to the Bayard-Alpert gage extend the usable range down to $10^{-11}$ torr.

Additional modifications to the Bayard-Alpert gage have resulted in further extensions of the low end of the range, mainly by effecting further reductions of the X-ray effects. Suppression of photoelectrons in the *Schueman modification* (W. C Schueman, 1962) extends the range close to $10^{-12}$ torr. The use of *modulation techniques* (P. A. Redhead, 1963, and J. P. Hobson, 1964) extends it respectively to $8 \times 10^{-13}$ torr (with use of an aluminosilicate glass envelope), and $7 \times 10^{-15}$ torr (with a portion of the measuring system immersed in liquid helium). The *buried collector gage* (F. P. Clay and L. T. Melfi, 1966, and L. T. Melfi, 1969) also extends the range to below $10^{-12}$ torr. Use of a low geometry and obtaining a high sensitivity by means of ion optics in the *extractor gage* (P. A. Redhead, 1966) extends the range to $5 \times 10^{-13}$ torr. A range extension to close to $10^{-13}$ torr was achieved in the bent-beam *Helmer gage* (J.C. Helmer and W. H. Hayward, 1966). Use of a *photomultiplier* (only the dynode-type electron multiplier portion is used) in place of a single-electrode anode allows the ion current to be amplified; hence, much lower electron currents can be used for the formation of positive ions, with a resulting decrease in the production of X-rays permitting additional low-range limit extensions. Employing a *channel electron multiplier* instead of a photomultiplier has been reported to extend the range to about $10^{-15}$ torr (D. Blechschmidt, 1973). Using the *deflected-beam* technique in conjunction with a channel electron multiplier has reportedly resulted in a range limit near $10^{-14}$ torr (D. Blechschmidt, 1975).

(In the preceding discussion of methods of extending the lower range limit of the Bayard-Alpert gage by modifications, the modifications extend over a considerable range of complexity. Some experimental results were much easier to achieve and proved more repeatable than others. The names and dates in parentheses are an attempt to indicate the original developer who published a paper on the modification and results, and the year the report was published in conference proceedings or professional journals. Range limits are shown in *torr (equivalent nitrogen)*, as is customary in ultra-high-vacuum work (i.e., the range limit that would apply had nitrogen been used as the calibration gas).)

The addition of a *magnetic field* to ionizing vacuum sensors increases the electron path length at very low pressures. This increases the ionization obtainable, since the probability of electron collisions with gas molecules is in turn increased. Two basic versions of such a *magnetron gage* have been developed, one using hot-cathode electron emission, the other cold-cathode electron emission.

In a *hot-cathode magnetron gage*, such as the *Lafferty gage* (Figure 17–4), the electric and magnetic fields are crossed. The electrons emitted from the filamentary cathode and accelerated radially toward the anode are also subjected to an axial magnetic field, which forces them into a helical path that is much longer than the basic linear path. The shield, collector, and plates are negative, and the anode is positive, with respect to the cathode. Ion current is measured between collector and cathode. A range of $10^{-4}$ to $10^{-5}$, extendable down to $10^{-17}$, torr by adding an electron multiplier, has been reported for this type of vacuum sensor.

The cold-cathode magnetron gage is illustrated by the design example shown in Figure 17–5. In this *Penning gage* (sometimes called *Philips gage* after its first manufacturer) the electrons are accelerated by a high-voltage field between the cathodes and the anode. When the cathode surfaces are bombarded by high-energy ions, these surfaces emit electrons which join

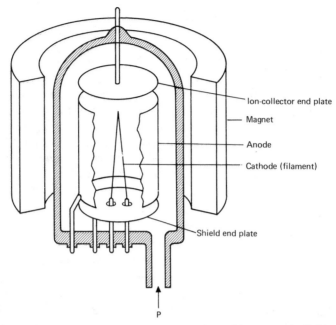

**FIGURE 17–4.** Ionization vacuum sensing with magnetic field—hot cathode.

the total electron stream and produce additional ions, which, in turn, produce additional electrons (*avalanche effect*). The transverse magnetic field forces the electrons to travel along a helical path. The total current, as seen by a microammeter (Figure 17–5(b)) between the cathodes and the anode, is the sum of the ion current and the electron current and, hence, is not linear with

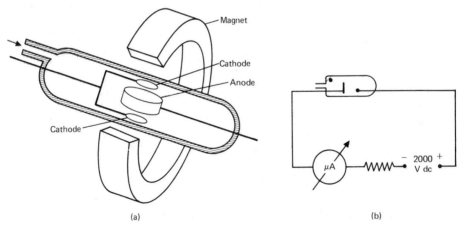

**FIGURE 17–5.** Ionization vacuum sensing with magnetic field—cold cathode: (a) transducer; (b) typical circuit.

pressure. The sensitivity of this type of sensor is about 0.5 A/torr. The usable range, $10^{-7}$ to $10^{-3}$ torr, can be extended downward to nearly $10^{-12}$ torr by the addition of a flash filament which triggers the discharge at very low pressures when the filament is briefly activated. Cold-cathode designs require a large-diameter pressure port and plumbing (*tubulation*) because of their relatively high ionic pumping speed. An advantage of such sensors is the absence of a hot filament which could burn out. (Some ionization vacuum sensors contain a standby filament that can be turned on and used when the other filament has burned out.)

In a design variation of the cold-cathode magnetron, the *Redhead gage* (named after its developer, as is frequent practice for vacuum sensors), the field emission and ion collection are separated by means of an auxiliary cathode (Figure 17–6). The normal magnetron type uses a magnetic field of about 0.1 T, and the ion current vs. pressure relationship is a straight line on a log-log plot. The inverted magnetron type uses a field of about 0.2 T, and its calibration is linear down to about $5 \times 10^{-10}$ torr and exponential below that level to the range limit (about $10^{-13}$ torr; the upper limit is about $10^{-4}$ torr). The auxiliary cathode acts as an electrostatic shield and prevents field emission from the edges of the opening in, or at, the collector electrode. The sensitivity of the Redhead gage is approximately 4.5 A/torr.

The use of particles other than electrons to ionize the measured gas is exemplified by the alpha-particle ionizing sensor illustrated in Figure 17–7. An appropriate radioisotope source provides a steady alpha-particle flux. The particles cause ionization of the gas molecules and their flow to the ion collector electrode. The principle employed is similar to that of the ionization chamber used for particle detection (see Chapter 19) except that in the vac-

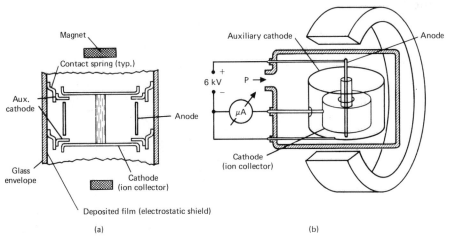

**FIGURE 17–6.** Use of auxiliary cathode in cold-cathode, magnetic field, and ionization vacuum sensors: (a) magnetron type (sectional view); (b) inverted magnetron type.

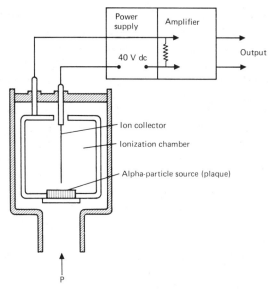

**FIGURE** 17–7. Ionization vacuum sensing using radioisotope source.

uum sensor the particle flux is constant and the pressure of the gas varies in the chamber. This type of sensor is normally used for a measuring range in the low and medium vacuum regions ($10^{-1}$ to 100 torr), but is capable of measuring between $10^{-5}$ and 1000 torr. The use of a beta source instead of an alpha source has also shown promising results and could extend the measuring range down to about $10^{-8}$ torr.

### 17.3.4   Gas-Analysis Vacuum Sensing

Ultra-high-vacuum measurements, below about $10^{-12}$ torr, can be performed by gas analysis equipment that allows the exact determination of the abundance of molecules of each gas that is a constituent of the mixture of gases within the volume in which vacuum is to be measured. The equipment most commonly used for this form of analysis is the *mass spectrometer*, notably that of the *quadrupole* type. Good quadrupole mass spectrometers are readily available commercially, perhaps more readily than one of the special types of ion gages that would be usable reliably for such vacuum ranges.

   The molecule of each gas constituent (e.g., $O_2$, $N_2$, Ar) is characterized by a mass number (mass unit), and the mass spectrum obtained from the equipment shows abundance vs. mass unit. The molecular density of each gas can be determined from this, and the equivalent partial pressure can be calculated. When all the partial pressures are known, the total pressure can be determined. (The molecular density of air is on the order of $10^3$ molecules/$cm^3$ at $10^{-14}$ torr, at a temperature of 300 K.)

## 17.4 DESIGN AND OPERATION

### 17.4.1 *Capacitance Manometers*

Capacitive pressure transducers (see Section 15.4.1) have been developed for vacuum measurements down to $10^{-5}$ torr, using special techniques and materials for construction of the sensing assembly, and using sophisticated electronics to detect the very small capacitance changes (on the order of $10^{-5}$ pF) provided by diaphragm deflections of a fraction of a nanometer. Figure 17–8(a) depicts two types of sensor construction, in absolute and differential pressure configurations. In the "double-sided" version the stretched (prestressed) welded diaphragm is positioned between two parallel electrodes, typically ceramic disks to which a metal layer has been applied. Leads from the electrodes are connected to the associated electronics. The excitation circuitry (see Figure 17–8(c)) drives the sensor, which acts as the variable element in a phase-sensitive *LC* bridge arrangement, at a carrier frequency of 10 kHz. The diaphragm displacement modulates the carrier, and a synchronous demodulator receives the amplified signal and provides dc output signals (which are bipolar for bidirectional differential-pressure sensors) which are then amplified to a 10-V full-scale output. Filters keep any electrically conductive or contaminating particles out of the sensing chamber. In the absolute-pressure configuration, the reference cavity is evacuated and then sealed. The vacuum is usually maintained by incorporating a *getter*, a metallic deposit (e.g., titanium) which absorbs gas molecules.

In the "single-sided" version (Figure 17–8(a)) the electrode assembly is removed from the measured-pressure cavity; a small baffle assembly replaces the filter and acts to keep high-speed particles from impinging on the diaphragm. A center electrode and an off-axis electrode are both placed in the reference side. One of the advantages of this type of design is that the measured gas, whose dielectric constant may vary, does not enter the gap between diaphragm and electrode. The two electrodes are used in a "curvature-sensing" mode: the diaphragm deflects much more at its center than at the off-axis location of the second electrode; hence, the differences in capacitance sensed by the two electrodes will be more pronounced with increasing diaphragm curvature due to its deflection. Developmental efforts have also resulted in a single-electrode, single-sided design which operates in capacitance/frequency-change circuitry and appears to have the capability of vacuum measurements down to $10^{-7}$ torr.

Figure 17–8(b) shows a typical capacitive vacuum transducer in its absolute-pressure configuration. (The same design also exists in a differential-pressure configuration.) Its measuring range extends from $10^{-5}$ torr to 15 000 torr. The device employs the single-sided design explained. The sensing assembly is temperature controlled at 50°C by a proportionally controlled heater, to minimize temperature errors. The transducer contains a

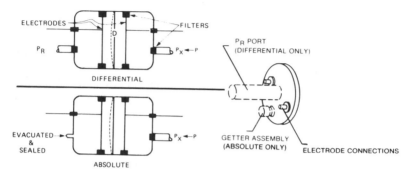

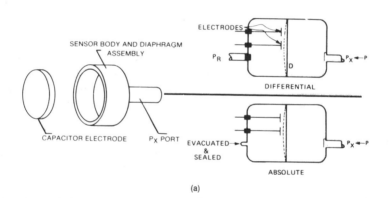

(a)

(b)

**FIGURE 17–8.** Capacitance manometer: (a) construction alternatives; (b) typical transducer and electronics unit; (c) block diagram. (Courtesy of MKS Instruments, Inc.)

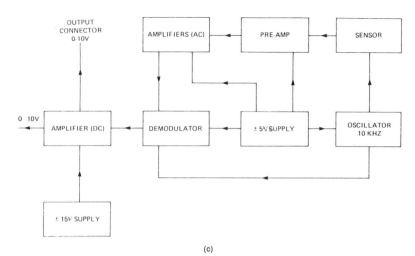

(c)

**FIGURE 17–8.** (*continued*)

solid-state follower and preamplifier and operates in conjunction with the separate electronics/display unit shown in the illustration.

The differential-pressure designs lend themselves to use for precision molecular flow measurement when used in conjunction with a flow element across which a pressure drop proportional to flow is developed. Capacitive vacuum sensing heads are available in nonbakeable as well as bakeable versions. Many applications require *bake-out* of elements that come in contact with the measured fluid, so that gas molecules are driven off surfaces prior to operation of the measuring system and will not then *outgas* (and introduce errors by increasing the molecular density of the fluid) during operation. Bake-out is performed at high temperatures.

### 17.4.2 Pirani Gages

Pirani gages, a version of thermoconductive (heat-loss) vacuum sensors, use a heated resistive element whose resistance varies with varying heat loss due to varying vacuum. The element is made of a pure-metal wire characterized by a high-temperature coefficient of resistance; nickel or tungsten is frequently used. One design uses a gold-plated tungsten wire heated to 125°C. Filament temperatures can run up to 400°C in some designs, but tend to be closer to 200°C. The element is typically connected as one arm of a Wheatstone bridge, with a second temperature-sensitive resistor that is not exposed to the vacuum used as a second (compensating) arm of the bridge. The sensor also contains the two fixed-resistance arms of the bridge. It is connected to an electronics/display unit, which also provides excitation and signal conditioning by a cable. Bakeable sensing heads are available, and thermistors have been used as the heated resistive element in a few designs. Most Pirani gages have a usable measuring range from $10^{-3}$ torr (1 millitorr) to about

10 torr; however, at least one modified design is capable of an extended upper range to 1000 torr.

### 17.4.3 Thermocouple Gages

Earlier thermocouple gage designs used the thermocouple to detect the changes in heat loss of a heated filament; thus, they employed the basic sensing principle explained in Section 17.3.2. Newer designs, however, do away with the filament and heat the thermocouple itself. Also, the single-junction thermocouple is often replaced by a thermopile of two or more junctions to increase the output signal. In the design shown schematically in Figure 17–9(a), two thermocouples are connected in series (but with polarities bucking) across the secondary of a transformer which supplies ac heating power to them. Since thermocouples are self-generating, they provide a dc output voltage that is proportional to temperature. The temperature increases with increasing vacuum due to decreasing heat loss from the thermocouples. An unheated thermocouple, connected between the center of the two-junction thermopile and the transformer's center tap, provides compensation for changes in ambient temperature.

In the full-bridge thermopile circuit (Figure 17–9(b)) two dissimilar noble-metal wires, welded in the center, form two thermocouples. A constant alternating current from a dual-secondary transformer heats each wire independently to the same temperature rise above ambient temperature. The dc thermoelectric emf caused by this temperature rise is dependent on the applied heating power as well as the heat loss through the surrounding gas (measured fluid). Since the reference junctions are all at the same (ambient) temperature and the hot thermopile junction is the hottest spot on any of

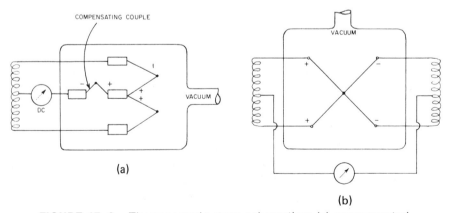

**FIGURE 17–9.** Thermocouple gage schematics: (a) compensated half-bridge thermopile; (b) full-bridge thermopile. (Courtesy of Teledyne Hastings-Raydist.)

the four wires, the bridge is symmetrical and is inherently temperature compensated.

Figure 17–10 shows a configuration typical for most thermocouple gages. The pins in the octal header provide all external electrical connections. The housing is usually metal, but can be a high-temperature glass. Measuring ranges are usually in the region $10^{-4}$ to $10^{-2}$ torr, but some extend close to $10^{-5}$ torr; others are available for medium vacuum ranges, $10^{-3}$ to 1 torr or 0.1 to 100 torr.

### 17.4.4  Hot-Cathode Ion Gages

Most triode-type ion gages are Bayard-Alpert gages. The anode is a central wire surrounded by the coiled grid, and the filamentary cathode is located outside the grid. Since a common failure mode of Bayard-Alpert gages is filament burnout, many designs incorporate a second filament as backup. This filament can be switched in when the first filament has burned out. A typical design is shown in Figure 17–11. The small-diameter grid (made of platinum-iridium alloy), the closed grid structure, and the thin (0.1-mm diameter) collector (anode) are design features that aid in minimizing the X-ray limiting current equivalent to and indicative of $2 \times 10^{-11}$ torr). A backup filament is included, the material of which is typically thoria-coated iridium. The gage *tubulation* (equivalent to the pressure port of a pressure transducer) is 38 mm in diameter for the design illustrated and can be either glass or Kovar alloy; the latter can be sealed by soldering or welding. Alternatively, the tubulation can be furnished with a flange.

In some applications of ion gages it is desirable that all tubulation be eliminated. This requirement can be met by a *nude ion gage* (Figure 17–12). The nude gage illustrated is similar in its design features to the gage shown in Figure 17–11, except that it has no glass envelope. The internal

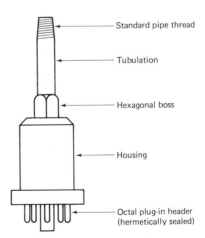

Standard pipe thread

Tubulation

Hexagonal boss

Housing

Octal plug-in header
(hermetically sealed)

**FIGURE 17–10.** Typical thermocouple gage configuration.

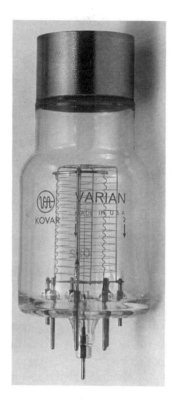

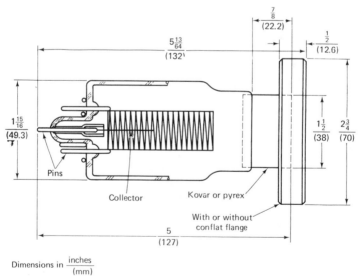

Dimensions in $\frac{inches}{(mm)}$

**FIGURE 17–11.** Bayard-Alpert ion gage. (Courtesy of Varian Associates, Vacuum Products Division.)

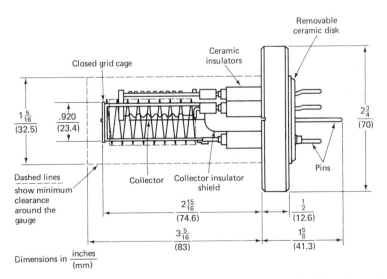

Removable
ceramic disk

Ceramic
insulators

Closed grid cage

$1\frac{5}{16}$
(32.5)

.920
(23.4)

$2\frac{3}{4}$
(70)

Dashed lines
show minimum
clearance
around the
gauge

Collector

Collector insulator
shield

Pins

$2\frac{15}{16}$
(74.6)

$\frac{1}{2}$
(12.6)

$3\frac{5}{16}$
(83)

$1\frac{5}{8}$
(41.3)

Dimensions in $\frac{\text{inches}}{\text{(mm)}}$

**FIGURE 17–12.** Nude ion gage (Bayard-Alpert type). (Courtesy of
Varian Associates, Vacuum Products Division.)

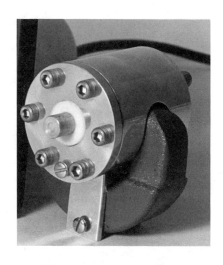

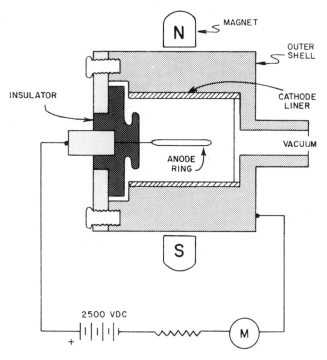

**FIGURE 17–13.** Cold-cathode ion gage. (Courtesy of Teledyne Hastings-Raydist.)

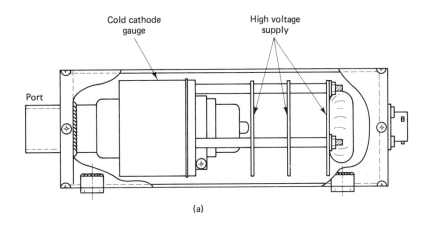

(a)

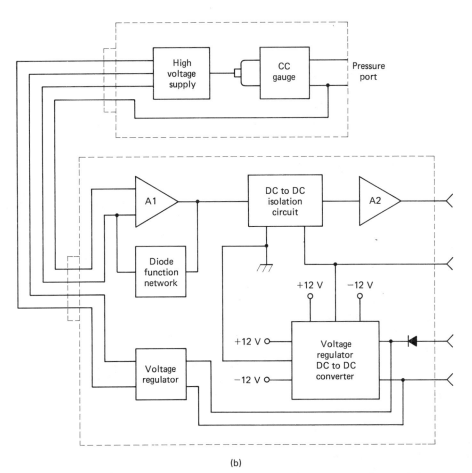

(b)

**FIGURE 17–14.** Cold-cathode ion gage: (a) sensor package; (b) sensor system block diagram. (Courtesy of SenTran Co.)

elements are supported by ceramic insulators, so that the gage can be baked out at 450°C. Since the gage has no envelope, there is easy access to the filament when it needs to be replaced.

The range of the Bayard-Alpert gages typified by the preceding designs is from $10^{-3}$ to $2 \times 10^{-11}$ torr (the X-ray limit). In many other designs the range is shifted upward in pressure; such ranges as $10^{-3}$ to $10^{-9}$ torr or 0.6 to $10^{-6}$ torr are often seen in catalogs. Ion gages that incorporate one or more of the modifications that extend their range down to $10^{-13}$ torr or lower (see Section 17.3.3) are rarely commercially available. In most cases they have been custom built in the research facility that established a need for them.

### 17.4.5   Cold-Cathode Ion Gages

Cold-cathode ionization gages that use a magnetic field for their operation have been developed in many experimental and some practical designs. Figure 17–13 shows a commercially available design and its electrical operating principle. A high dc voltage discharge in a magnetic field ionizes the particles of the measured gas, producing an ionization current that is inversely proportional to the density, and hence pressure, of the gas. The high-voltage potential is applied between an elliptical-ring-shaped anode and the cathode liner of the outer shell. The magnetic field is supplied by a strong permanent magnet. The measuring range of this design is from $10^{-2}$ to $10^{-6}$ torr, with some visibility down to $10^{-7}$ torr. The associated signal conditioning and display equipment provides an expanded-scale indication ($10^{-6}$ to $5 \times 10^{-5}$ torr), as well as a full-range indication. Figure 17–14 shows a cold-cathode ion gage packaged with its high-voltage supply; the associated block diagram represents the sensor package as well as the associated electronics unit.

## 17.5   TRANSDUCER CHARACTERISTICS

The most essential characteristics of vacuum sensors are measuring range, sensitivity, and provisions for degassing and bakeout. Typical measuring ranges of vacuum sensors (vacuum gages) are given in the list that follows this paragraph. The ranges shown are intended to apply to generally available "off-the-shelf" gages built by more than one manufacturer. It is recognized that range extensions, usually to lower pressure limits, can be obtained on special order (and at extra cost and delivery time) from several manufacturers and that, in some cases, there is one manufacturer of a given type of gage that can offer such a feature "off the shelf." Note that developments leading toward extending the lower range limit have been directed somewhat more at hot-cathode ion gages and capacitance manometers than at other types. Some gage types, in conjunction with their electronics/readout unit, lend themselves to having the entire measuring range shifted upward.

*Reluctive diaphragm gages*: from over $10^3$ to 1 torr

*Capacitance manometers*: from over $10^3$ to $10^{-4}$ torr

*Pirani gages*: from 10 to $10^{-3}$ torr

*Thermocouple gages*: from 1 to $10^{-4}$ torr

*Cold-cathode ion gages (discharge type)*: from $10^{-2}$ to $10^{-6}$ torr

*Hot-cathode ion gages*: from $10^{-3}$ to $5 \times 10^{-11}$ torr

*Quadrupole mass spectrometers*: from $10^{-6}$ to $10^{-14}$ torr.

Ranges for specialty ion gages, e.g., hot-cathode and cold-cathode magnetron gages, are given in Section 17.3.3.

Vacuum sensors are usually procured together with their electronics/display (readout) unit. The readout device shows vacuum units in terms of "equivalent nitrogen," since calibrations are normally performed with this gas. These calibrations apply, however, to air as well as nitrogen. When the measured gas is something other than air or nitrogen, an appropriate correction must be applied to the readings obtained from most types of gages; a notable exception is the single-sided type of capacitance manometer in which the transduction element does not come in contact with the measured gas. Applying such corrections to readings is quite simple, since most manufacturers will furnish conversion charts for other commonly measured gases, such as helium, krypton, argon, neon, oxygen, Freon, carbon dioxide, and methane.

The sensitivity of a vacuum gage whose output is an ionization current is customarily shown in amperes per torr (A/torr) or in amperes per pascal (A/Pa). When the output is a different quantity, such as a voltage, the sensitivity is shown in units of that quantity per torr. The torr is still the most commonly used unit of vacuum; however, the pascal is used increasingly, and some scales have markings in both units. Submultiples of the torr or the mm Hg are still seen in some scale markings.

Sensors for the high- to ultra-high-vacuum ranges usually incorporate provisions for degassing in situ, by applying a potential across one element or between two elements. Specifications for such gages also show the temperature at which bakeout can be performed; the bakeout temperature for components of UHV systems is usually 450°C.

## 17.6 APPLICATION NOTES

Applications of vacuum sensors have been increasing in several fields as technology in these fields expands. In the low- to high-vacuum regions sensors have been in use for quite some time in food processing, chemical research and processing, and backfilling of gas tubes. Measurements of the *residual gas* in vacuum chambers and vacuum systems, i.e., the gas remaining in the system after pump-down, can be anywhere from high to ultra-high vacuum in range. Gas analysis equipment is also used for such measurements (*residual gas analysis*).

Ultra-high-vacuum measurements are needed primarily in research, notably in particle accelerators and high-energy physics. In some semiconductor research and even manufacturing processes an ultra-high vacuum is preferred, but these processes usually accept the very high vacuum in which much of the routine semiconductor and high-purity materials processing is customarily performed. Thin-film metal deposition and optical coating fall into the latter category. Leak detection devices extend into the high and very high vacuum ranges. Applications in a variety of ranges are found in aerospace research and testing, e.g., shock tubes, high-altitude measurements, aerodynamics, and space vehicle and component testing.

The higher the vacuum, the more important the reliability and stability of a vacuum sensor become. This is primarily due to the increase in complexity, cost, and time involved for pump-down of the vacuum system, which would be necessitated by a failure and removal of a component such as the vacuum sensor.

# 18

# Viscometers and Densitometers

## 18.1 BASIC DEFINITIONS

*Viscosity* is a fluid's resistance to the tendency to flow.

*Kinematic viscosity* is the ratio of the viscosity of a fluid to its density.

*Relative viscosity* is the ratio of the viscosity of a liquid to the viscosity of a liquid considered as standard; in the case of a solution, it is the ratio of the viscosity of the solution to the viscosity of the pure solvent in it.

The *Prandtl number* of a fluid is the ratio of its kinematic viscosity to its thermal conductivity.

*Density* ("absolute density") is the ratio of the mass of a homogeneous substance or body to its volume (mass per unit volume).

*Relative density* is the ratio of the mass of a substance to the mass of a reference substance of the same volume.

*Specific gravity* is the ratio of the density of a substance at a given temperature to the density of a substance considered as standard; pure distilled water at 4°C or at 15.5°C (60.0°F) has been used as such a standard in reference to liquids and solids. Specific gravity varies with temperature.

## 18.2 UNITS OF MEASUREMENT

*Viscosity* is expressed in units of force per unit area per unit time, e.g., $N{\cdot}s/m^2$.

To convert from *centipoise* to $N{\cdot}s/m^2$, multiply by $10^{-3}$.

To convert from *slug/foot second* to N·s/m², multiply by 47.88.

*Kinematic viscosity* is expressed in units of area per unit time, e.g., *m²/s*.

To convert from *centistoke* to m²/s, multiply by $10^{-6}$.

To convert from *foot²/second* to m²/s, multiply by $9.3 \times 10^{-2}$.

*Density* is expressed in units of mass per unit of volume, e.g., *kg/m³*.

*Specific gravity* is typically expressed in the form "$t_f/t_s$ *specific gravity*," where $t_f$ is the temperature of the fluid under consideration and $t_s$ is the temperature of the fluid considered as standard (e.g., "the 20/20°C specific gravity"); it is a dimensionless number since it is a density ratio.

To convert from *gram/centimeter³* to kg/m³, multiply by 1000.

To convert from *pound mass/inch³* to kg/m³, multiply by $2.77 \times 10^4$.

To convert from *slug/foot³* to kg/m³, multiply by 515.4.

## 18.3   SENSING METHODS AND TRANSDUCER DESIGNS

### 18.3.1   Viscosity Sensing

A limited number of sensing devices have been developed for the direct measurement of viscosity. The generic sensor types are described here briefly. Some sensor types measure *absolute viscosity*, whose SI unit is the $N·s/m^2$ and whose U.S. Customary unit is the *poise* ($= 0.1$ N·s/m²), whereas other types measure *kinematic viscosity* (the ratio of absolute viscosity to density), whose SI unit is the *m²/s* and whose U.S. Customary unit is the *stoke* ($= 1.0$ cm²/s $= 1 \times 10^{-4}$ m²/s). The ranges of viscosity sensors (*viscometers*) are typically shown in *centipoise* or *centistokes*.

*Torque sensing* is employed in *rotating element* devices, such as rotating-cone, gyrating-cone, rotating-disk, and rotating-spindle viscometers. The element is made to rotate in the measured fluid, and a torque, or a reaction torque, developed due to viscosity (viscous drag) is detected by a torque or reaction-torque transducer. One rotating-spindle design uses a radial spring to couple the spindle mechanically to the motor shaft; the angular displacement of the spring due to torque is then transduced. Different measuring ranges are obtainable for this device by using spindles of different sizes.

*Differential-pressure flow-rate sensing* is used in *capillary flow* devices in which the measured fluid is forced to flow through a restriction, typically a capillary tube, and the pressure differential across the restriction, at a constant flow rate and under laminar flow conditions, is measured by a differential-pressure transducer whose output is then proportional to kinematic viscosity. When the pressure drop is due to a fixed force (e.g., a positive-displacement pump or fixed high-pressure source) that is independent of density, the transducer output is proportional to absolute density.

*Time measurement* is employed in *falling-piston* and *falling-ball* devices. A piston or ball of known density is first lifted and then permitted to drop (due to gravity) through a sample of the measured fluid. (The sample can be contained in the process line.) The time it takes the piston or ball to drop to the bottom of the measuring chamber, with its final position sensed, for example, by a magnetic switch, is measured and is indicative of viscosity.

*Oscillation damping* is sensed in *vibrating-element* viscosity transducers. The amplitude of a vibrating reed, paddle, or sphere is reduced with increasing viscosity of the measured fluid. The element is driven electromechanically, and a transduction element detects the amplitude changes. Such devices respond to changes in viscosity as well as density; however, effects of density variations can be eliminated by maintaining the measured liquid as well as the transducer at a closely controlled constant temperature.

*Linear-displacement sensing* is used in *float-type* viscometers. This device consists of a variable-area flowmeter with a viscosity-sensitive float. With the flow rate held constant (by a separate flow control system), the linear displacement of the float is indicative of kinematic viscosity.

### 18.3.2  Density Sensing

Several different sensing methods have been successfully employed in density transducers (*densitometers*, not to be confused with optical densitometers, which respond to attenuation of light transmission through materials such as exposed photographic film).

*Capacitive* sensors make use of the fact that the dielectric constant of a liquid is proportional to the density of the liquid. If the temperature of the liquid is known (because it is measured simultaneously), density readings can be obtained from the output of a fixed-plate, variable-dielectric type of capacitive sensing device. Two concentric electrodes, immersed in the measured fluid, are typically employed as capacitor electrodes.

*Photoelectric* sensors in the form of *refractometers* (see Section 22.6) are used for density determinations; the refractive index correlates with mass density.

*Sonic* sensing of density is based on the changes in propagation velocity of sound through a liquid; the velocity changes are proportional to density (at constant pressure and temperature).

*Nucleonic* density measuring systems rely on the increasing attenuation with increasing density of gamma rays passing through a liquid. One such system feeds the output of the radiation detector to a microprocessor which performs the desired signal conditioning; simultaneous inputs from a temperature transducer are used for automatic compensation of density readings for changes in temperature.

*Vibrating-element* densitometers use the changes in natural frequency, due to changes in density, of a mechanical element that is maintained in vibration electronically and is either located within the fluid or has the fluid

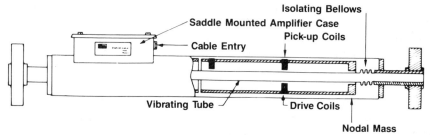

**FIGURE 18–1.**   Vibrating-tube densitometer. (Courtesy of Solartron Transducers.)

flowing through it. The mechanical element can be a straight tube (as in the densitometer shown in Figure 18–1), or it can be a U-shaped tube, twin tubes arranged in the shape of a tuning fork, or an end-fixed beam or vane positioned so that it is parallel to the direction of flow. Various methods are used to drive the sensing element into oscillation, to maintain this oscillation, and to detect the frequency of the oscillation. The output frequency generally decreases with increasing density of the measured fluid.

## 18.4   TRANSDUCER CHARACTERISTICS

Essential performance characteristics of viscometers are range, accuracy, resolution, and warm-up time. The physical and chemical characteristics of the measured fluid must be well understood to interpret the data properly. Response time varies for different designs and types.

Essential characteristics of density sensors include the type of measured fluids the sensors can be used for and the physical and chemical characteristics of these fluids, including viscosity and temperature range. Operation with cryogenic fluids has been demonstrated for several designs, and their use for density determinations of, for example, liquified natural gas (LNG) is quite common. Other measured-fluid characteristics that must be considered are pressure or range of pressures and corrosiveness.

The measuring ranges of density sensors extend from below 0 to 50 kg/m$^3$ to 0 to about 400 kg/m$^3$ for gases and to over 0 to 2000 kg/m$^3$ for liquids. Accuracy characteristics may be different for different measured fluids, but error bands between ±0.2 and 0.5% FSO are typical for most fluids. Where the output is in terms of frequency, the output frequencies are usually on the order of 1 to a few kHz.

## 18.5   APPLICATION NOTES

Viscometers and densitometers are used in many industrial applications, on samples in research and quality control laboratories and, increasingly, for closed-loop (on-stream) control.

Viscometry is of particular importance in food processing; in the dairy industry, for example, viscometers are used to measure the curd product in making cheese. Other examples of applications are found in plastics processing equipment, in controlling the combustion of heavy fuel oils, in Diesel-fuel preheaters (used to optimize combustion), in monitoring coal/water slurries, in monitoring and control of lubricants and paints, and in chemical processing (control of polymerization reactions is just one example). Other applications are in the biomedical field, such as in viscosity determinations of whole blood and blood plasma, and in the production of liquid and compressed gases.

Densitometers also have many industrial uses. Examples of density monitoring and control are found in mining (ore slurries, flotation, heavy media separation), in the pulp and paper industry (liquor concentrates, paper coating), in food processing (bulk fill, starch solutions, evaporator cooking), in many areas of the petrochemical industry (distillation, foam control, polymerization, multiphase blending), and even in pollution control (sulfur dioxide scrubbers, wastewater).

# 19

# *Thermometers*

## 19.1  DEFINITIONS AND RELATED LAWS

Thermometers, radiation pyrometers (Chapter 20) and heat flux sensors (Chapter 21) are all interrelated because they are used to measure heat-related (*thermal*) quantities. For this reason, the definitions and related laws for all thermal quantities are contained in this section.

### *19.1.1  Basic Definitions*

The *temperature* of a body is its thermal state considered with reference to its power of communicating heat to other bodies. It is a measure of the kinetic energy of the molecules of a substance due to heat agitation; or, looked at another way, it is the potential of heat flow.

*Absolute temperature* is temperature measured on the thermodynamic scale—temperature measured from *absolute zero* (0 K, $-273.15°C$).

The *boiling point* is the temperature of equilibrium between the liquid and vapor phases of a substance (at 1 standard atmosphere unless otherwise specified).

The *freezing point* is the temperature of equilibrium between the solid and liquid phases of a substance (at 1 standard atmosphere unless otherwise specified).

The *triple point* is the temperature of equilibrium between the solid, liquid, and vapor phases of a substance.

The *ice point* (273.15 K, 0°C, 32.0°F) is the temperature at which ice is in equilibrium with air-saturated water at a pressure of 1 standard atmosphere.

The *sublimation point* is the temperature at which a substance changes from its solid phase directly to its vapor phase without passing through a liquid phase.

*Heat* is energy, specifically energy in transfer, due to temperature differences between a system and its surroundings or between two systems, substances, or bodies. It has also been defined as the energy contained in a sample of matter comprising potential energy resulting from interatomic forces as well as kinetic energy associated with random motion of molecules in the sample.

*Heat transfer* is the transfer of heat energy by one or more of the following methods:

1. *Conduction:* by diffusion through solid material or through stagnant fluids (liquids or gases).

2. *Convection:* by the movement of a fluid (between two points).

3. *Radiation:* by electromagnetic waves.

*Heat flux* is the amount of heat transferred across a surface of unit area per unit time.

*Heat capacity* is the quantity of heat required to raise the temperature of a system, substance, or body by one degree of temperature (in a specified manner).

*Specific heat* is the ratio of the heat capacity of a body to the mass (or to the volume, as specified) of the body. An earlier definition made specific heat the (dimensionless) ratio of the quantity of heat required to raise the temperature of a material (without change of phase) by one degree to the quantity of heat required to raise the temperature of water, having the same mass as the material, by one degree (in temperature), under specified conditions.

*Thermal conductivity* is the ratio of the time rate of heat flow per unit area to the negative gradient of the temperature per unit thickness in the heat-flow direction (see Fourier's law, Section 19.1.2.2).

*Thermal diffusivity* is the ratio of the thermal conductivity to the product of specific heat and density; the magnitude of thermal diffusivity determines how fast temperature differences that exist in a body or substance will equalize.

*Thermal resistance* is a measure of a body's ability to prevent heat from flowing through it; it is equal to the difference in temperature between opposite faces of the body divided by the heat-flow rate.

*Thermal equilibrium* is a condition of a system and its surroundings (or of two or more systems, substances, or bodies) such that no temperature differences exist between them (i.e., no more heat transfer occurs between them).

A *black body* (ideal black body) is an ideal body that would absorb all and reflect none of the radiation incident on it (its *absorptivity* would be 100% and its *reflectivity* would be 0%); it is also the ideal radiator.

*Emissivity* is the ratio of the radiation emitted by a given surface to the radiation emitted by the surface of a black body heated to the same temperature as the surface, under the same conditions and within the same solid angle.

The *emissive power* of a black body is the total hemispherical radiation from the body per unit area of radiation surface (see the Stefan–Boltzmann law, Section 19.1.2.3).

### 19.1.2  Related Laws

**19.1.2.1   Gas Temperature.**   *Boyle's Law* says that in a given quantity of gas the product of pressure and volume remains constant as long as the temperature is held constant. According to *Charles's Law*, in a given quantity of gas the ratio of the absolute temperature to the volume remains constant if the pressure is held constant, and the ratio of the absolute temperature to the pressure remains constant if the volume is held constant. Also, by the *Ideal-gas Law*,

$$\frac{pv}{T} = R = \frac{p}{\rho T}$$

where $p$ = pressure
  $v$ = specific volume
  $T$ = absolute temperature
  $\rho$ = density
  $R$ = gas constant

**19.1.2.2   Heat Conduction.**   *Fourier's Law* is given by

$$\frac{dQ}{dt} = -kA\frac{dT}{dx}$$

where $Q$ = quantity of heat transferred across a body
  $t$ = time
  $k$ = thermal conductivity of the body
  $A$ = cross-sectional area of the body
  $T$ = temperature
  $x$ = distance, in the direction of heat flow (taken as normal to $A$)

In this equation, $dQ/dt$ is the time rate of heat flow and $dT/dx$ is the *temperature gradient*.

### 19.1.2.3 Thermal Radiation. According to the *Stefan–Boltzmann Law*,

$$Q_T = \sigma A T^4$$

where $Q_T$ = total heat radiated from the surface of an ideal black body (total hemispherical emission in all wavelengths)
$\sigma$ = Stefan–Boltzmann constant
$A$ = area of emitting surface
$T$ = absolute temperature of emitting surface

| Units of $Q_T$ | Units of A | Units of T | Value of $\sigma$ |
|---|---|---|---|
| W | m² | K | $5.67 \times 10^{-8}$ |
| W | cm² | K | $5.67 \times 10^{-12}$ |
| Btu/h | ft² | °R | $1.72 \times 10^{-7}$ |

*Wien's Displacement Law* says that

$$T\lambda_{max} = 2.8978 \times 10^{-3} \text{ (Wien's constant)}$$

where $T$ = absolute temperature, K
$\lambda_{max}$ = wavelength of maximum radiance from black body, m

*Note:* When the wavelength is expressed in μm rather than in m, Wien's constant becomes 2,898 μm · K.
By *Planck's Radiation Formula*,

$$W_\lambda = \frac{C_1 \Delta \lambda \epsilon_\lambda}{\lambda^5 \left(e^{C_2/\lambda T} - 1\right)}$$

where $W_\lambda$ = radiation intensity per unit area of source, at wavelength $\lambda$, over a spectral range $\Delta\lambda$, W/m²
$\lambda$ = wavelength, m
$\epsilon_\lambda$ = emissivity of source at wavelength $\lambda$ (emissivity = 1.0 for a black body)
$e$ = Naperian base ($\approx 2.71828$)
$C_1$ = first radiation constant, $3.7413 \times 10^{-16}$ W · m²
$C_2$ = second radiation constant, $1.4388 \times 10^{-2}$ m · K
$T$ = absolute temperature, K

(*Note:* This formula can also be shown for radiation intensity per $\Delta\lambda$, expressed in $W/cm^2/\mu m$, with the wavelength expressed in $\mu m$ (and $\Delta\lambda$ deleted from the numerator); when so expressed, $C_1 = 37\ 413 \times 10^{-12}$ and $C_2 = 14\ 388$.)

## 19.2   UNITS OF MEASUREMENT AND TEMPERATURE SCALES

Temperature scales were originally established on an arbitrary basis; much later, international agreement was reached on an appropriately defined temperature scale. The *Fahrenheit* scale is based on the mercury-in-glass thermometer, with the ice point defined as 32°F and the *steam point* (boiling point of water) defined as 212°F; the ice point–steam point difference is 180°F. The *Rankine* scale is an absolute temperature scale, with the same difference (180°R) between ice point and steam point as the Fahrenheit scale, but with the *absolute zero* of temperature defining 0°R; hence, the ice point is 491.7°R and the steam point is 671.7°R. Since both of these temperature scales have the same difference in degrees between ice point and steam point, 1°F = 1°R. The *Réaumur* scale was used in some European countries, but is now essentially obsolete; on this scale the ice point was 0° and the steam point was 80°. The *Celsius* scale (a now obsolete name for what was the *centigrade* scale) is also based on the mercury-in-glass thermometer, but with the ice point defined at 0°C and the steam point at 100°C. This scale is widely used, and the °C is considered a unit of the "metric system."

The unit of temperature in the International System of Units (SI) is the *kelvin* (*K*). It should be noted that the word "degree," or its symbol, is not used in conjunction with this unit. The kelvin is the unit of *thermodynamic* temperature and is defined as the fraction 1/273.16 of the thermodynamic temperature of the triple point of water. The triple point of water is 0.01 K above the ice point; hence, on the kelvin scale, the ice point is 273.15 K and the steam point is 373.15 K. The difference between these two temperatures is 100.00 K, and 1 K = 1°C, exactly. Absolute zero is 0 K.

An International Temperature Scale was adopted in 1948. Between the oxygen point (boiling point of oxygen, −183°C) and the antimony point (freezing point of antimony, +630°C), it was based on the resistance of a standard platinum resistance thermometer; temperatures within this range were determined by calculations based on measured resistance values at 0°C and at the temperature to be determined. The relationship used was the *Callendar-Van Dusen equation* (see Section 19.3.2). Between 630°C and the gold point (freezing point of gold, 1064°C), the 1948 Temperature Scale was based on the emf vs. temperature relationship of a "standard" platinum (actually, 90% platinum, 10% rhodium) thermocouple. Above the gold point, the Temperature Scale was based on the optical pyrometer.

In 1968 The International Committee on Weights and Measures (CIPM) adopted a revised temperature scale, the International Practical Temperature

Scale of 1968 (*IPTS-68*). The main reasons for devising the new scale were to extend a unified scale down to about 10 K (instead of 90 K, the oxygen point) and to reflect more accurately measured temperatures between the oxygen point and the gold point.

The IPTS-68 is based on the assigned values (*defining fixed points*) of the temperatures of a number of equilibrium states (freezing points, boiling points, triple points) and on standard instruments calibrated at those temperatures. These points are established by realizing specified equilibrium states between phases of pure substances. All except one of these points (the 17.042-K point, the boiling point of equilibrium hydrogen at $\frac{25}{76}$ standard atmosphere) are included in Table A–1 in the Appendix. Interpolation between the fixed points is provided by formulas establishing the relationship between standard-instrument indications and temperature values. Formulas and reference functions are available in the applicable documents published by the BIPM, 92 Sèvres, France, and reproduced by governmental standards agencies. The standard instrument used from 13.81 K (triple point of equilibrium hydrogen) to 630.74°C (freezing point of antimony) is the platinum resistance thermometer, with a winding of pure, strain-free, annealed platinum. For the range between 630.74 and 1064.43°C (freezing point of gold), the standard instrument is the platinum–10% rhodium/platinum thermocouple. Above 1064.43°C the IPTS-68 is defined by Planck's radiation formula (see Section 19.1.2.3), with this temperature (which equals 1337.48 K) as reference temperature. Secondary reference points, to supplement the defining fixed points, are available, and many of the freezing and melting points shown in Table A–1, additional to those designated as defining fixed points, are usable as such additional reference points.

Research is continuing on reducing uncertainties in the temperature scale, especially in the very low and the very high temperature regions, with the ultimate aim of generating a formal revision to the IPTS-68.

The following relationships are used for temperature scale conversions:

$$°C = (°F - 32) \times \tfrac{5}{9} \quad \text{and} \quad °F = \tfrac{9}{5} \times °C + 32$$

$$K = °C + 273.15$$

$$°R = °F + 459.67 \text{ (which can usually be rounded off to 459.7)}$$

Table A–1 is a convenient scale conversion table which includes defining fixed points for the temperature scale as well as many melting and freezing points of general interest.

## 19.3 SENSING METHODS

*Temperature sensors (temperature transducers)*, or (electric) *thermometers*, are characterized by having their sensing element also perform the transduction. Such sensing/transduction elements are referred to here simply as "sensing elements." All sensing methods described here pertain to *contacting* thermometers. Noncontacting temperature-sensing methods are covered in Section 20.3.

### 19.3.1   *Thermoelectric Temperature Sensing*

The sensing and transduction of temperature by thermoelectric means is based on the *Seebeck effect* (Thomas J. Seebeck, 1770–1831, German physicist): when two dissimilar conductors *A* and *B* (see Figure 19–1) comprise a circuit (by being joined at both of their ends), a current will flow in that circuit as long as the two junctions are at different temperatures, one junction at a temperature $T$, the other at a higher temperature $T + \Delta T$. The current will flow from *A* to *B* at the colder junction when conductor *A* is positive with respect to *B*. Two other effects are related to the Seebeck effect. (1) When a current flows across a junction of two dissimilar conductors, heat is absorbed or liberated at the junction, depending upon the direction of current flow (*Peltier effect*). (2) When a current flows through a conductor along which a temperature gradient exists, heat is absorbed or liberated in the wire (*Thomson effect*).

The *thermocouple circuit* is based on the Seebeck effect. Two dissimilar conductors are joined at the point where temperature is to be measured (*sensing junction*) and are terminated at a point where both terminals are at the same temperature, which must be known (*reference junction*). Although the temperature $t_{ref}$ of the reference junction can be any temperature, it is usually kept at the ice point (0°C). Thermocouple calibration curves are normally based on this as the reference-junction temperature, and the conductors are designated as positive (*P*) and negative (*N*) when the measured temperature $t$ is higher than the reference temperature (see Figure 19–2). The connecting leads between the reference junction and the load (across which a voltage as a function of temperature is obtained), represented as $R_L$, can be copper wire.

When the load (display device or data system) is located some distance away from the sensing junction, it is often economical to use *thermocouple extension wire* between the measuring thermocouple and the load. This type of wire is made of materials that match the emf-vs.-temperature characteristics of the thermocouple very closely and, in effect, transfer the reference junction from the thermocouple side to the load side of the extension wires (see Figure 19–3).

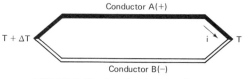

**FIGURE 19–1.**   Seebeck effect.

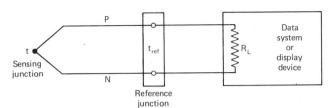

**FIGURE 19-2.** Basic thermocouple circuit.

The magnitude of the *thermoelectric potential (thermal emf)* produced by a thermocouple depends on both the wire materials and the temperature difference between the two junctions. Table 19-1 shows the thermal emf of a number of different wire materials that would be produced if the material would be formed into a thermocouple with platinum as the other conductor and with a reference-junction temperature of 0°C. It can be seen that the polarity of some materials is negative for temperatures above 0°C, whereas it is positive for other materials. Since the physical and chemical characteristics as well as the cost of some materials are more suitable for their use in thermocouples than others, only a limited number of materials are used for thermocouples and most of those materials are alloys that were developed specifically for such use. Materials which, in combination with platinum, exhibit an increasingly positive potential are used as the positive (*P*) leg of thermocouples, whereas those exhibiting an increasingly negative potential are used as the negative (*N*) leg. The thermal emf produced by a thermocouple, then, is the algebraic difference of the two potentials that would be produced if each material would be combined with platinum; for example, the emf produced by a Type K (Chromel-Constantan) thermocouple at 500°C is $+16.21 - (-4.43) = 16.21 + 4.43 = 20.64$ mV.

The most commonly used and available thermocouples have been standardized. In the United States, American National Standard (ANSI Standard) MC96.1 (Revised 1982) covers coding (by type designation as well as by color) of thermocouple wire and extension wire (single wire and insulated duplex wire), terminology, wire sizes (wire gages), and limits of error; temperature-vs.-emf tables for thermocouples are included, and appendices cover fabrication, checking procedures, selection, and installation. Several other countries have issued similar standards for thermocouples. Interestingly, the color coding for specific thermocouple wire and duplex insulated wire is quite different in the ANSI as against the BS (British) and as against the DIN (German) standards.

**FIGURE 19-3.** Thermocouple circuit with extension wire.

**Table 19–1.   THERMAL EMF (IN mV) OF CONDUCTORS RELATIVE TO PLATINUM AT $t_{ref} = 0°$ C**

| Conductor Material | Junction Temperature, $t$ (°C) | | | | | |
|---|---|---|---|---|---|---|
| | $-200$ | $0$ | $+200$ | $+500$ | $+1,000$ | $+1,400$ |
| *Pure metals* | | | | | | |
| Aluminum (Al) | +0.45 | 0 | +1.06 | +3.93 | | |
| Antimony (Sb) | | 0 | +10.14 | +25.10 | | |
| Bismuth (Bi) | +12.93 | 0 | −13.57 | | | |
| Cadmium (Cd) | −0.04 | 0 | +2.35 | | | |
| Cobalt (Co) | | 0 | −3.08 | −9.35 | −14.21 | |
| Copper (Cu) | −0.19 | 0 | +1.83 | +6.41 | +18.16 | |
| Germanium (Ge) | −46.0 | 0 | +72.4 | +63.5 | | |
| Gold (Au) | −0.21 | 0 | +1.84 | +6.29 | +17.05 | |
| Iridium (Ir) | −0.25 | 0 | +1.49 | +4.78 | +12.57 | +20.47 |
| Iron (Fe) | −2.92 | 0 | +3.54 | +6.79 | +14.28 | |
| Molybdenum (Mo) | | 0 | +3.19 | +10.20 | +27.74 | |
| Nickel (Ni) | +2.28 | 0 | −3.10 | −6.16 | −12.11 | |
| Palladium (Pd) | +0.81 | 0 | −1.23 | −3.84 | −11.61 | −20.40 |
| Rhodium (Rh) | −0.20 | 0 | +1.61 | +5.28 | +14.02 | +22.99 |
| Silicon (Si) | +63.13 | 0 | −80.57 | | | |
| Silver (Ag) | −0.21 | 0 | +1.77 | +6.36 | | |
| Tantalum (Ta) | +0.21 | 0 | +0.93 | +4.30 | +15.15 | |
| Tungsten (W) | +0.43 | 0 | +2.62 | +9.30 | +27.73 | |
| *Standard thermocouple alloys*[a] | | | | | | |
| Chromel (KP, EP) | −3.36 | 0 | +5.96 | +16.21 | +32.47 | +44.04 |
| Alumel (KN) | +2.39 | 0 | −2.17 | −4.43 | −8.78 | −11.77 |
| Constantan (JN, EN, TN) | +5.35 | 0 | −7.45 | −20.79 | −43.85 | |
| Pt—13% Rh (RP) | | 0 | +1.47 | +4.47 | +10.50 | +16.04 |
| Pt—10% Rh (SP) | | 0 | +1.44 | +4.23 | +9.58 | +14.37 |

[a] See the upper portion of the table for iron (JP) and copper (TP); platinum (RN, SN) is the reference conductor in this table.

Reference tables which show the electromotive force (emf) generated by the "standard" thermocouples over their applicable range of temperatures have been published (in the United States) by the National Bureau of Standards at various times. The tables, shown for temperatures in °F as well as °C, were updated after the revised International Practical Temperature Scale IPTS-68 was adopted and are currently contained in National Bureau of Standards (NBS) Monograph 125 (available from the U.S. Government Printing Office, Washington, DC). Favorable results have also been reported for Nicrosil/Nisil thermocouples, which are similar in output to Type K but seem to have better stability.

The output (emf)-vs.-temperature curves for various commonly used thermocouples are shown in Figure 19–4. Three popular alloys used for thermocouples are Chromel (90% Ni, 10% Cr), Constantan (55% Cu, 45% Ni), and Alumel (95% Ni, 2% Al, 2% Mn, 1% Si). The names of the alloys

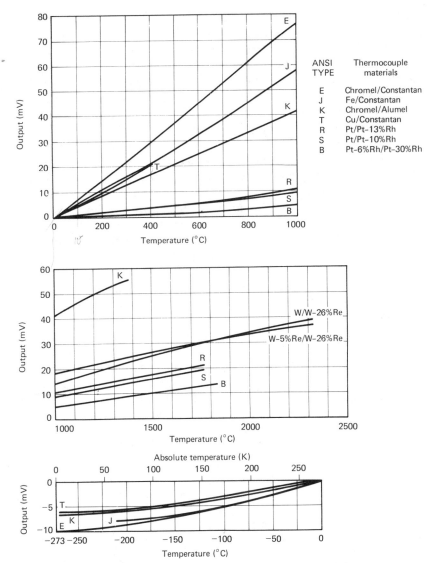

**FIGURE 19–4.** Output voltage (thermoelectric emf) of thermocouples ($t_{ref}$ = 0°C).

are trade names assigned to them by their original manufacturers; additional trade names now exist for them as well. The tungsten-rhenium combinations are now in common use for high-temperature measurements. The original pure-metal thermocouple (W/Re) is rarely used since higher output at high temperatures can be obtained by the alloy combinations, notably W/W— 26%, W–5% Re/W—26% Re, and W–3% Re/W–25% Re. Considerable research has been done on other high-temperature thermocouples, but most of the combinations developed and tested are not generally available commercially and stability characteristics are, in most cases, not yet well enough established. Developments have included combinations of iridium with iridium-rhodium alloys (e.g., Ir/Ir—40% Rh), which are usable to about 2100°C; molybdenum-rhenium alloys in combination with alloys of the same materials but in different proportions (e.g., Mo—10% Re/Mo—50% Re) and in combination with tungsten–rhenium alloys (e.g., Mo—3% Re/W—25% Re), which seem to be usable to about (and possibly somewhat above) 2000°C; and nonmetallic thermocouples. The latter are combinations of carbides, carbon, graphite, and oxides, including graphite/titanium carbide, columbium carbide/zirconium carbide, and carbon/titanium carbide; usable maximum temperatures for such materials range from 1800 to over 3000°C, depending largely on the environment in which they are applied. Some thermocouples were also developed for the purpose of obtaining higher sensitivities in the low cryogenic range; these low-temperature thermocouples include Chromel/gold—0.07 at. % iron and Cu/Au—2.1% Co.

When a higher output is required from a point of measurement, a number of thermocouples of the same combination of materials can be connected in series to form a *thermopile* (see Figure 19–5). The output of a thermopile is equal to the emf produced by a thermocouple of the material combination chosen, multiplied by the number of thermocouples in the thermopile. A common reference junction is used for all sensing junctions in the thermopile.

When averaging of temperatures at a number of measurement points is required, thermocouples can be connected in parallel, and, provided that they are made from identical wire materials, sizes, and lengths, the emf generated will be indicative of the mean of the temperatures at the sensing junctions (see Figure 19–6). When such identity cannot be assured, "swamping resistors" can be added, with one such resistor (about 200 to 500 Ω) connected in series with each wire.

The integrity of the thermocouple material in a thermocouple circuit,

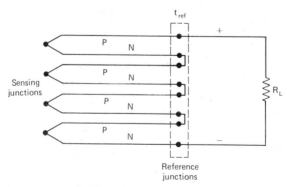

**FIGURE 19–5.** Thermopile circuit.

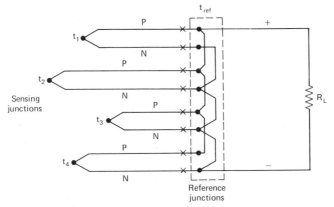

**FIGURE 19–6.** Parallel connection of thermocouples for temperature averaging. When used, swamping resistors are inserted at all points marked X.

between the sensing and reference junctions, must be maintained; that is, the material of each leg must either be the same for the entire run of the circuit, or it must be a material having the same (or very close to the same) temperature-vs.-emf characteristics. When electrical connections are made between sections of a thermocouple leg, they must bring the same materials into good electrical contact with each other. Lead–tin solder and silver solder have thermal emfs different from those of most thermocouple materials, and so do materials typically used for contacts in electrical connectors, such as beryllium–copper, brass, or phosphor bronze; all these materials exhibit a thermal emf (relative to platinum, at a $t_{ref}$ of 0°C) on the order of $+0.6$ mV.

### 19.3.2 Resistive Temperature Sensing Using Conductors

Conductors generally exhibit an increase in resistance with temperature. The change in resistance is given by the conductor's *temperature coefficient of resistance* $\alpha$, and the *base resistance* of a conductive temperature sensing element (usually its resistance at the ice point, sometimes at another specified temperature) is given by the cross-sectional area (the wire gage in case of a wire conductor) and the *resistivity* $\rho$ of the material. Table 19–2 shows values of $\alpha$ and $\rho$ for some typical conductors, including those which have been used for temperature sensing as well as some that exhibit particularly small alphas. For temperature measurement purposes, it is desirable to have a conductor which has (1) a reasonably high $\alpha$, so that a substantial resistance change is obtained; (2) an $\alpha$ that remains fairly constant over a wide temperature range, so that the resistance-vs.-temperature characteristic is close

**Table 19-2.   RESISTANCE PROPERTIES OF PURE, ANNEALED METALS AND ALLOYS[a]**

| Material | Temperature Coefficient of Resistance $\alpha$ between 0 and 100°C ($\Omega/\Omega/°C$) | Resistivity at 20°C ($\mu\Omega\cdot cm$ or $10^8\ \Omega\cdot m$) |
|---|---|---|
| Aluminum | 0.0042 | 2.69 |
| Constantan (55% Cu, 45% Ni) | ±0.00002 | 49.0 |
| Copper | 0.0043 | 1.673 |
| Gold | 0.0039 | 2.3 |
| Indium | 0.0047 | 9.0 |
| Iron | 0.00651 | 9.71 |
| Manganin (86% Cu, 12% Mn, 2% Ni) | −0.00002 | 43.0 |
| Nickel | 0.00681 | 6.844 |
| Nichrome (60% Ni, 16% Cr, 24% Fe) | 0.0002 | 109.0 |
| Palladium | 0.00377 | 10.8 |
| Platinum | 0.00392 | 10.6 |
| Rhodium | 0.00457 | 4.7 |
| Silver | 0.0041 | 1.63 |
| Tungsten | 0.0046 | 5.5 |

[a] Selected from various recent sources.

to linear; and (3) a reasonably high resistivity, so as to minimize the amount of material required for use in a practical temperature sensor.

An approximate relationship for the resistance-vs.-temperature characteristics of conductors between 0 and 100°C is given by the equation

$$R_T = R_0(1 + \alpha t)$$

where $R_t$ = resistance at temperature $t$, °C
$R_0$ = resistance at 0°C
$\alpha$ = temperature coefficient of resistance

Originally, copper wire was used in such resistance elements; its main disadvantage was the low resistivity of copper and the resulting impractically long wire needed for winding a practical element. Later, platinum came into use increasingly; it is now used to the virtual exclusion of all other conductors except for some specialized applications. Elements made of nickel wire and some nickel–alloy wire were used for measurements over a relatively narrow temperature range; they were increasingly replaced by platinum elements, but are still used occasionally because of their lower cost and acceptably high base resistance values. Attempts at using tungsten wire have been essentially discontinued. Certain alloys have been found usable for measurements in the low cryogenic range, notably a rhodium—0.5 at. % iron alloy wire which exhibits a higher sensitivity than platinum below 20 K. For all wire-wound elements, it is important that they be annealed after they are wound; annealing is essential to obtaining a stable *R*-vs.-*T* (resistance-vs.-temperature) relationship.

Platinum is used for wire elements and in a few cases also for film elements. Wire elements are used predominantly, mainly because manufacturing techniques resulting in a predictable and repeatable *R*-vs.-*T* relationship are better established. Elements wound of pure platinum wire in such manner that they are strain-free, and which are annealed after winding, are used in most platinum resistance thermometers. The same basic type of element, but with additional precautions taken, is used in reference-standard-grade thermometers used to define the interpolation between fixed points of the IPTS-68 scale, between the triple point of hydrogen and the freezing point of antimony (see Table A–1). Research and development are continuing to make such standard thermometers usable up to significantly higher temperatures. Some platinum-wire-element temperature sensors have been used (with somewhat less than reference-grade accuracy and not for extended periods of time) up to temperatures around 1100°C.

The *R*-vs.-*T* relationship of a platinum-wire element between − 183 and + 630°C is given by the *Callendar–Van Dusen equation*,

$$R_t = R_0 + R_0\alpha[t - \delta(0.01t - 1)(0.01t) - \beta(0.01t - 1)(0.01t)^3]$$

where $R_t$ = resistance at temperature $t$, °C

$R_0$ = resistance at 0 °C

$\alpha, \delta, \beta$ are constants; $\alpha$ is usually determined by measuring the element's resistance at + 100°C and $\beta$ from a resistance measurement below 0°C, usually at the oxygen point (− 182.96°C); $\delta$ is determined by a resistance measurement well above 100°C, such as the boiling point of sulfur (+ 444.7°C); typical values are:

$\alpha = 0.00392$

$\beta = 0$ (if $t$ is positive) and 0.11 (if $t$ is negative)

$\delta = 1.49$

The temperature values of the relationship established by this equation can be corrected for IPTS-68 by adding to them a small temperature value $\Delta t$ which is generally less than 0.05°C and can be calculated, if necessary, as

$$\Delta t = 0.045(0.01t - 1)(0.01t)(t/419.58 - 1)(t/630.74 - 1)$$

The constant $\alpha$ is given by the purity of the platinum used in the element. It increases with purity, and values of up to 0.003927 have been obtained.

A value of 0.003925 is acceptable for reference-standard-grade elements, and this value is used for reference tables of $R$ vs. $T$ in the United States. For industrial resistance thermometers, a value of 0.00390 is usually acceptable. For the so-called European curve (shown, e.g., in DIN standards), a value of 0.003850 is used. Resistance values obtained by the U.S. and European curves are shown in the typical values of Table 19–3.

Platinum-film elements, in various shapes, have also been manufactured; the film is usually deposited on a glass or ceramic substrate. Similar techniques have been applied to integral circuits containing a metal-film temperature sensor. Yet another form of temperature sensing element is a foil grid, photoetched from thin foil stock; the sensing material is typically not platinum, but an alloy such as Manganin/nickel.

### 19.3.3 Resistive Temperature Sensing Using Semiconductors

The most widely used semiconductor temperature sensing elements are *thermistors*. Their measuring range is typically between $-50$ and $+300°C$, although some designs have been developed for temperatures in the low cryogenic range. The semiconductive materials, usually sintered mixtures of sulfides, selenides, or oxides of nickel, manganese, cobalt, copper, iron, and uranium, are formed into small beads, disks, or rods which are then encapsulated (e.g., in glass). Thermistors are characterized by high resistivities and high negative temperature coefficients of resistance. Their $R$-vs.-$T$ characteristics are nonlinear (see Figure 19–7). Their characteristics are stated as "zero-power" characteristics, including the base resistance, which is normally shown as $R_{25°C}$ rather than at $0°C$, and the $R$-vs.-$T$ relationship. The term "zero power" refers to a power dissipation low enough that self-heating

**Table 19–3. RESISTANCES OF A PLATINUM RESISTANCE THERMOMETER HAVING $R_{0°C} = 100.00\ \Omega$ BETWEEN $-200$ AND $+800$ °C (OHMS)**

| Temperature (°C) | "U.S. Curve" ($\alpha = 0.003925$) | "European Curve" ($\alpha = 0.003850$) |
|---|---|---|
| $-200$ | 17.02 | 18.61 |
| $-100$ | 59.49 | 60.26 |
| 0 | 100.00 | 100.00 |
| $+100$ | 139.25 | 138.50 |
| $+200$ | 177.33 | 175.84 |
| $+300$ | 214.24 | 212.03 |
| $+400$ | 249.97 | 247.08 |
| $+500$ | 284.53 | 280.99 |
| $+600$ | 317.93 | 313.73 |
| $+700$ | 350.15 | 345.26 |
| $+800$ | 381.19 | 375.56 |

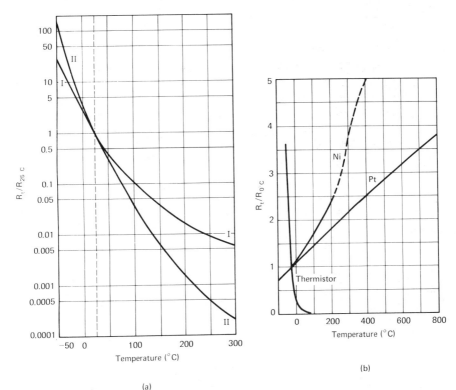

(a)

(b)

**FIGURE 19–7.** *R*-vs.-*T* characteristics of thermistors, nickel and platinum: (a) typical thermistor curves; (b) comparison with conductors.

is negligible; the excitation current for temperature-measurement thermistors is typically below 100 μA.

The basic resistance-vs.-temperature characteristic of a thermistor, at "zero power," is expressed by

$$\frac{R_T}{R_{ref}} = e^{\beta(1/T - 1/T_{ref})}$$

where $R_T$ = zero-power resistance at measured absolute temperature $T$

$R_{ref}$ = zero-power resistance at known absolute temperature $T_{ref}$, usually at 298.15 K (25°C)

$\beta$ = material constant, K

$e$ = 2.718, base of natural logarithm

Typical values of $\beta$ are between 3000 and 4500 K, as determined from resistance measurements at the ice point and at a temperature somewhat above room temperature, usually 50°C. The values of zero-power resistance (sometimes called "cold resistance") at 25°C can be between 500 Ω and over 10 MΩ.

*Germanium* temperature-sensing elements were developed primarily for the very low cryogenic region, between about 1 and 35 K. The element is single-crystal doped germanium in which various controlled impurities are used as the dopants. Their *R*-vs.-*T* characteristics exhibit a nonlinearity similar to that of thermistors. Temperature sensors using germanium elements have been favored when measurements with close repeatability are required from below the helium boiling point to around the hydrogen boiling point; repeatabilities within 0.001 K have been reported for the former and within 0.05 K for the latter.

*Gallium arsenide (GaAs) diode* elements are also used primarily for cryogenic temperature measurements, although their usable range is generally between 1 and 300 K. The forward voltage of a GaAs diode, at a constant current (of about 100 μA), increases fairly linearly (compared to curves for thermistors or germanium elements) with decreasing temperatures. The excitation current must be controlled within small tolerances and the associated load resistance must be quite high, since the output impedance of a GaAs diode is relatively high (typically around 15 kΩ at 20 K).

*Carbon resistors* of commercially available types have been used for cryogenic temperature measurement, but are now largely replaced by other devices. At temperatures below 60 K their *R*-vs.-*T* characteristics are somewhat similar to those of thermistors and germanium elements. Although carbon resistors exhibit good sensitivity at very low temperatures, they tend to be deficient in long-term stability and can be affected by variations in ambient pressure, such as encountered in pressurized vessels or ducts.

*Silicon* crystals, doped typically with phosphorous impurities, were developed for use in the medium-temperature range as well as for the low cryogenic range. Electrical conduction takes place in the diffused surface layer. The *R*-vs.-*T* characteristic is unique in that it exhibits a positive and fairly linear slope above −50°C, whereas below this temperature the slope becomes sharply negative. Different doping techniques can alter these characteristics. The use of silicon elements for liquid-hydrogen temperature measurements has been reported.

### 19.3.4  Oscillating-Crystal Temperature Sensing

*Quartz crystals*, used as the frequency-controlling element in oscillator circuits, can be used for temperature sensing. The sensitivity of such crystals to temperature changes has long been accepted in the field of communications engineering. This phenomenon is utilized for temperature measurements in the −50 to +250°C range, using quartz crystals specially designed and cut from synthetic single-crystal quartz at an orientation that optimizes the linearity of the frequency-vs.-temperature relationship. When connected into an oscillator circuit and excited at its third overtone resonance (typically around 30 MHz), such a crystal can provide a sensitivity of about 1 kHz/ °C. The oscillator output is usually mixed (heterodyned) with the output of a reference oscillator so that a beat frequency is obtained. This difference frequency can then be displayed on a frequency counter.

### 19.3.5 Capacitive Temperature Sensing

*Capacitive* temperature sensors depend on temperature-induced changes in the characteristics of the dielectric for their operation. Such sensors have been used primarily for measurements in the cryogenic and low cryonic regions. The material of the dielectric is selected for optimized temperature dependence in the intended measuring range. The effect is multiplied in multilayer capacitors. Developments have included small thin-film capacitive sensors with fast response times (below 1 ms). The relative absence of errors due to magnetic fields is a reported advantage of capacitance thermometers.

### 19.3.6 Experimental Temperature-Sensing Methods

*Thermal-noise* temperature sensors (*noise thermometers*) have been developed for experimental use. The operating principle here is the temperature dependence of (thermal) noise generated in a resistor. The resistor is connected to a high-gain amplifier which is so designed that it will amplify the thermal noise in the resistor without contributing any significant amount of noise of its own. When used for low-temperature measurements between about 10 and 90 K, some means of obtaining a reference temperature is required to minimize calibration uncertainties such as those due to amplifier gain and bandwidth uncertainties. However, at temperatures sufficiently low to make superconducting devices operable (below 10 K), absolute temperature measurements have been obtained using a Josephson junction as preamplifier, typically with this junction connected into an oscillator circuit so that a frequency output proportional to temperature is obtained with variations in thermal noise in a shunt resistor. Absolute temperature measurements well below 1 K have been obtained with Josephson-junction noise thermometers.

*Acoustical* temperature sensing has been employed in experimental measurements of not only low cryogenic temperatures (below 20 K), but also high and very high temperatures up to about 17 000 K (e.g., for plasma temperature measurements). Various techniques have been used to obtain temperature measurements on the basis of temperature effects on the propagation velocity of sound through a hot (or cold) solid, liquid, or gas. This velocity tends to decrease with increasing temperature for solids and liquids, and to increase with increasing temperature for gases. Either changes in frequency or changes in pulse transit time are used for temperature determinations.

Nuclear magnetic resonance (*NMR*) and *nuclear quadrupole resonance* (*NQR*) techniques have also been applied experimentally to temperature sensing.

## 19.4   DESIGN AND OPERATION

### 19.4.1   Thermocouple Thermometers

The simplest type of thermocouple is made from two insulated wires of the "positive" and "negative" materials by stripping the insulation from the wires where they are to be joined and then forming a good electrical junction between the two bare wire tips. More commonly, *wire thermocouples* are made from two-conductor (*duplex*) insulated, shielded or unshielded thermocouple cable, with solid or stranded conductors (depending on the mechanical flexibility required), or from metal-sheathed, swaged, ceramic-insulated thermocouple cable which exists in a large variety of diameters and with various conductor, sheath, and insulator materials. For a given type of junction, response time decreases with wire size (*wire gage*) used.

*Junctions* (see Figure 19–8) can be made by a number of different techniques of wire joining. The junction is usually welded, but is sometimes silver soldered or brazed. *Butt-welded junctions* are made by pressing the two sanded-flat wire ends together (using spring loading) and joining the ends by resistance welding. *Beaded junctions* are usually formed by heating the joined wire ends to their melting points and then rotating the weld until a bead is formed at the tip. For *lap-welded junctions*, the wire ends are formed so that they are in contact with each other over a length of about 3 mm and then resistance welded. *Cross-wire junctions* are made by bending

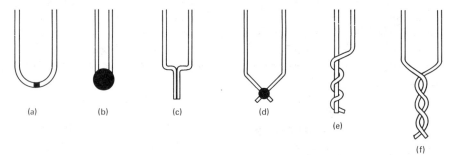

**FIGURE 19–8.** Thermocouple wire junctions: (a) butt welded; (b) beaded; (c) lap welded; (d) cross wire; (e) coiled wire; (f) twisted wire.

the two wire ends at an angle of less than 90°, laying one wire over the other, welding them together at this crossing, and then cutting excess wire material off the ends. In *coiled-wire junctions*, one wire is kept straight and the other is wound tightly over the straight wire end. This type, as well as *twisted-wire junctions*, is usually gas welded or arc welded, but can also be brazed or silver soldered.

Thermocouple junctions can be grounded or ungrounded, exposed or enclosed (see Figure 19–9). The shortest time constants (fastest response time) are obtained with butt-welded *exposed* junctions (for a given wire diameter). In many applications, however, the junction must be protected from the measured fluid. *Enclosed* ("protected") junctions are then used, and they can be grounded or ungrounded. *Ungrounded* enclosed junctions (again for a given wire diameter) have the longest time constants. *Grounded* enclosed junctions can be used when such grounding is compatible with the associated circuitry, when noise pickup in the wiring from the junction is not a significant problem, and when conduction error (error due to heat conducted away from the sensing tip) can be minimized or is acceptable. Grounded junctions can be made by joining and welding the wire ends together and to the inside of the sheath (probe) tip, or by welding the two wire ends separately to the probe tip, but in close proximity to each other. Wire thermocouple junctions become grounded junctions when they are welded to a measured surface which is conducting and which is at system ground potential.

*Thermocouple elements* are designed primarily for use in thermocouple thermometer assemblies. Typical elements are shown in Figure 19–10. Bare elements are those which are not equipped with insulators; however, the individual wires may be insulated. Insulated elements are used most frequently, with ceramic insulators most common. Such insulators have the form of short segments and can be intended for a single conductor (single-hole insulators) or two conductors (double-hole or double-bore insulators).

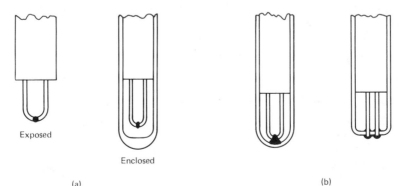

Exposed

Enclosed

(a)                                 (b)

**FIGURE 19–9.** Examples of ungrounded and grounded thermocouple junctions: (a) ungrounded; (b) grounded.

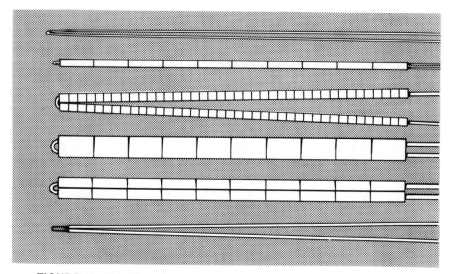

**FIGURE 19–10.** Bare and insulated thermocouple elements. (Courtesy of Omega Engineering, Inc., an Omega Group Co.)

They can be round or oval in cross section, and they can be square edged or fish-spine shaped.

Thermocouple thermometer assemblies, as typically used in industrial (as opposed to aerospace, research, or specialized) applications, are composed of a number of parts that are usually assembled nonpermanently, that is, by threading rather than by methods such as welding. A typical assembly (see Figure 19–11) consists of a *connection head*, an optional extension (connection head extension) with any required extension hardware (a union is shown in the illustration as an example), and a protection tube, which may be open or closed at its end. Other mountings and pressure-seal fittings are also available. In many applications, especially those in which the measured fluid is pressurized, a *thermowell* (see Figure 19–12) is used. This is a hollow fitting, closed at its tip, with provisions for its pressure-tight installation to a vessel, and with additional provisions (such as an internal thread) to receive a thermocouple (or other type of) thermometer. Use of a thermowell enables removal of a thermometer without breaking a pressure seal. Thermometers used in thermowells normally provide some form of mechanical protection, such as a sheath, for the thermocouple element. A terminal block is almost invariably included in the connection head to facilitate making (and changing) external electrical connections to the thermometer assembly. Some electrical circuitry (e.g., signal-conditioning circuitry) may also be included in the connection head.

Thermocouple probes used in aerospace and some other (e.g., nuclear) applications are typically assembled permanently, with as much as possible

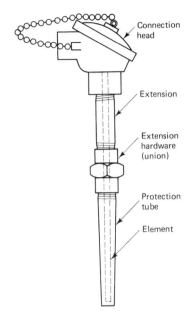

**FIGURE 19–11.** Thermocouple thermometer assembly.

of the probe integrally machined and any additional pieces welded. External electrical connections are made to a cable that is not removable from the probe assembly. Mounting threads are precision machined to provide a pressure seal (when this is a requirement) in conjunction with a compression seal (e.g., a metal O-ring). Alternatively, such probes can be designed for mounting by welding.

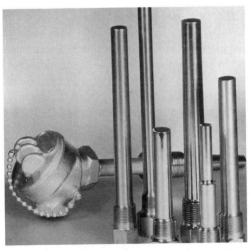

**FIGURE 19–12.** Thermowells (with thermometer assembly in background). (Courtesy of Omega Engineering, Inc., an Omega Group Co.)

Among the many other varieties of thermocouple probes are those with very thin, diagonally cut tips of the same size and shape as a hypodermic needle; implantable flexible probes; externally threaded probes that can be mounted directly into plastic extruder heads or die adaptors of injection molding machines; spring-loaded probes; and probes for surface temperature measurements intended either for fixed installations or for hand-held operation.

Various thermocouple assemblies other than probe types also exist. *Clamp-on* thermocouples are configured as pipe clamps with an integral sensing junction and connecting cable. Similarly, *washer-type* thermocouples (see Figure 19–13), intended for mounting under a bolt head or a nut, contain a sensing junction at the washer and are equipped with an integral cable.

*Foil thermocouples* (Figure 19–14) are designed for surface temperature measurements where very thin, flat sensing junctions are required. Each half of the symmetrical foil pattern is made of a different thermocouple material, and the junction is usually of the butt type. Among the design versions are a free-foil style with removable carrier base and a matrix type with the foil embedded in very thin plastic material. Either type can be bonded to flat or curved, conducting or nonconducting, surfaces. The foil construction method has also been applied to miniature thermopiles for surface temperature and heat-flow measurements. Related to foil gages are micro-miniature thin-film (e.g., evaporated-film) thermocouples and thermopiles used in specialized applications (e.g., as sensing elements in infrared radiometers).

### 19.4.2 Platinum Resistance Thermometers

Platinum-wire resistance thermometers predominate among all the types of electrical thermometers using a metallic resistance element. The other types, which include nickel-wire, nickel-alloy-wire and platinum-film designs, are now so rare (compared to platinum-wire types) that they will not be discussed here. Platinum (-wire) resistance thermometers ("PRTs") exist in two configurations: immersion probes and surface-mounted sensors.

*Immersion probes* (probe-type resistive temperature sensors) are usually mounted by their threaded portion near the housing. The threaded section is also machined or equipped with fittings, for obtaining a pressure seal. The seal is then completed by use of a gasket—a rubber or metal O-ring— or by compression of tubing or a mating fitting. A typical probe configuration is shown in Figure 19–15. The *sensing element* (resistance winding) is located near the tip of the probe *sheath* (or shield). The threaded mount is just below the housing, or *head*. The housing is usually filled with packing such as insulating fibers. An electrical connector (receptacle), or an end plate provided either with solder terminals or an integrally connected cable (or in-

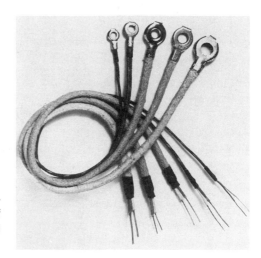

**FIGURE 19–13.** Washer thermocouple assemblies. (Courtesy of Omega Engineering, Inc., an Omega Group Co.)

dividual leads), caps the housing. The head may also contain one or more trimming or adjusting resistors or other circuit elements. A *trimming resistor*, when used, provides the function of adjusting the total sensor resistance to a specified value. An internal seal, located near the root of the mounting threads, enables the probe to be mounted into pressurized ducts or vessels (particularly when it has an exposed sensing element).

The probe illustrated contains a ceramic-*coated* (ceramic-encapsulated) sensing element, and the sheath is perforated in the sensing-element region to admit the measured gas or liquid so that it contacts the element thermally. The ceramic coating provides electrical insulation, so that this type of element can be used with conductive fluids. The construction of such an element is shown in Figure 19–16. A thin-walled platinum tube, coated with ceramic, is used as the mandrel for the platinum winding, which is then coated with the same type of ceramic. The thermal coefficient of expansion of the tube, ceramic, and wire are closely matched to assure strain-free performance of the element under wide temperature variations. The same illustration shows a miniature version of the coated element; the connecting leads of this design are carried through the platinum tube, and they are insulated by ceramic within the tube. When such an element is intended to be exposed to fluids with significant conductivity, an additional coating of organic composition or of Teflon can be applied over the element.

Two other types of element construction are also shown in the figure. In the *exposed* or open-wire element, the platinum wire is wound loosely over a supporting cage of thin platinum rods which are coated with ceramic at the points of wire contact to provide fixing of the element as well as electrical insulation between the wire and the cage. This type of element has the shortest time constants, but cannot be used with very rapidly moving or conductive fluids. Time constants of coated elements are about 5 to 10

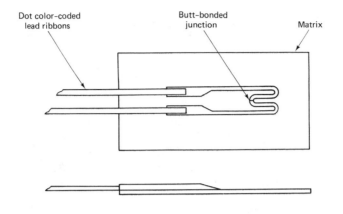

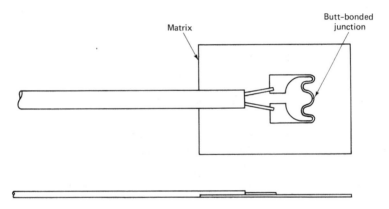

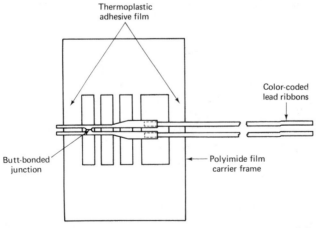

**FIGURE 19–14.**   Foil thermocouples. (Courtesy of RdF Corp.)

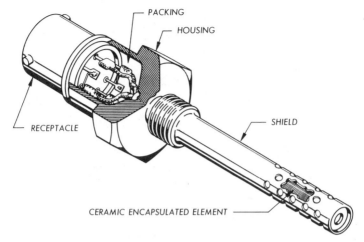

**FIGURE 19–15.** Platinum-wire immersion-probe temperature sensor. (Courtesy of Rosemount Inc.)

times as long as those of the exposed elements. For applications requiring full enclosure of the element, a "well-type" element can be used. Such elements can be used with highly corrosive and very rapidly moving fluids. However, the time constants of *enclosed* elements are on the order of 50 times as long for the type of construction shown as compared to those of the exposed element. In many applications such relatively long time constants are acceptable, and the "rugged-well-type" element does offer the necessary strain-free characteristics by employing a number of coiled windings fixed in the mandrel by ceramic only over a fraction of each wire turn. In all such elements using ceramic for fixing the winding, the wire becomes annealed when the ceramic is brought to its firing temperature.

Special and more complex methods of construction are required for platinum-wire resistance thermometers capable of being certified as temperature standards, that is, acceptable as interpolation standards to convert temperature, as defined by the IPTS-68, to resistance over the range 13.81 to 903.9 K. Such thermometers are also used as reference standards in calibration laboratories. One type of construction of such a temperature standard is shown in Figure 19–17. The platinum wire is run through holes in ceramic tubing without being fixed at any point, to assure the complete absence of even the smallest amount of strain. High-purity alumina are used as insulators.

Platinum-wire resistance thermometers intended for industrial applications (notably those in process measurement and control) are configured similarly to the thermocouple assembly shown in Figure 19–11. Enclosed elements are most frequently used, and thermometer assemblies with waterproof connection heads are frequently designed for installation into a ther-

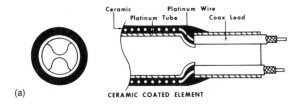

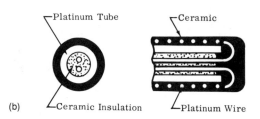

(a)  CERAMIC COATED ELEMENT

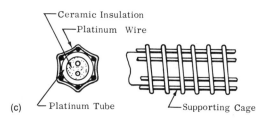

(b)

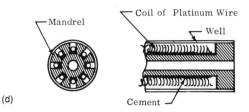

(c)

(d)

**FIGURE 19–16.** Typical forms of platinum-wire element construction: (a) ceramic-coated element; (b) miniature ceramic-coated element; (c) open-wire element; (d) rugged-well-type element. (Courtesy of Rosemount Inc.)

mowell (see Figure 19–12). The connection head may contain circuitry to convert the resistance changes into current changes (e.g., the range 4 to 20 mA commonly used in process measurement systems).

*Surface-mounted sensors* exist in many sizes and configurations. They are mounted by being cemented, welded, or clamped to a surface. Some designs are constructed in a manner similar to that of metal-wire or metal-foil strain gages (see Sections 9.4.1. and 9.4.2) and require mounting techniques applicable to these (see Section 9.6.1). The sensor shown in Figure

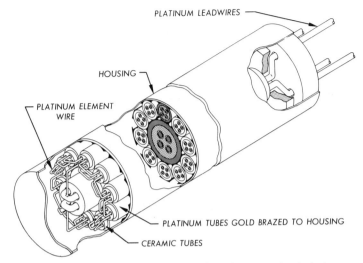

PLATINUM LEADWIRES

HOUSING

PLATINUM ELEMENT WIRE

PLATINUM TUBES GOLD BRAZED TO HOUSING

CERAMIC TUBES

**FIGURE 19–17.** Construction of a calibration-standard platinum resistance thermometer. (Courtesy of Rosemount Inc.)

19–18, for example, consists of a wire grid in an insulating carrier material. Bondable (cementable) sensors of this type are typically less than 1 mm thick and use carrier materials typical of strain gages. A high-temperature design uses an alumina-insulated element with a weldable stainless-steel support plate. A related clamp-on design uses an element overmolded with silicon rubber. When mounting a wire-grid sensor, care must be taken not to introduce strain errors.

Many surface temperature sensor designs use coiled platinum-wire elements. Such elements tend to be essentially free of strain effects. A (pat-

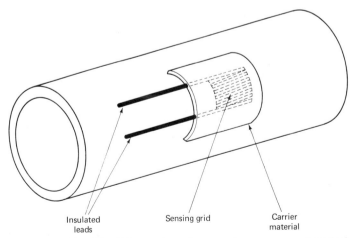

Insulated leads

Sensing grid

Carrier material

**FIGURE 19–18.** Flexible surface-temperature sensor bonded to pipe. (Courtesy of RdF Corp.)

ented) coiled-wire element in which the coil is fixed to the bottom plate of a sensor over only a small fraction of a turn is used in the sensors shown in Figure 19–19. An air gap between the element and the top plate insulates the element from ambient temperatures. Depending partly on the configuration, the sensors are mounted by cementing or by clamping (in the case of the rectangular sensors). Washer-type sensors are intended to be fastened under a screw head or nut. The thinnest of the sensors illustrated (0.75 mm thick), which is either cemented or spot welded to the measured surface, is shown in more detail in Figure 19–20. The connecting leads are Teflon-insulated copper or (bare) platinum leads.

In choosing an installation method for a surface temperature sensor, it is important to match the sensor configuration (including the nature of the connecting leads) to the installation so that heat conduction from the measured surface is maximized and so that the sensor sees a minimum of heating or cooling due to convection or radiation to or from other sources. In cementing a sensor to a surface, the amount of adhesive must be kept to a minimum. Insulation can be applied over the installed sensor to insulate it from ambient conditions. In some installations it is necessary to make the sensor an integral part of the measured surface, which can be provided with a cutout or spot-faced portion into which the sensor is cemented, potted, or welded. Care must also be taken to prevent conduction errors that may be introduced by the connecting leads and temperature gradients existing along these leads.

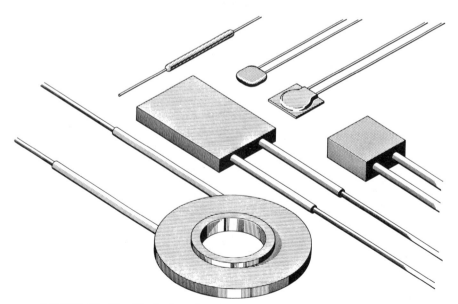

**FIGURE 19–19.** Typical surface temperature sensor configurations. (Courtesy of Rosemount Inc.)

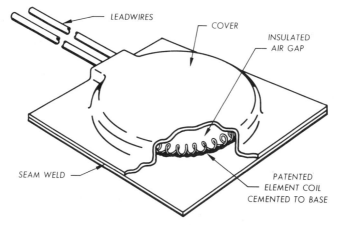

**FIGURE 19–20.** Construction of surface temperature sensor using minimally anchored platinum-wire coil element. (Courtesy of Rosemount Inc.)

### 19.4.3 Thermistor Thermometers

Thermistors are characterized by their small size, high negative temperature coefficient, fast time constant (obtainable with small bead types), and wide range of available base resistances (usually given at 25°C rather than at the ice point) ranging from hundreds of ohms to about 1 MΩ. When used for temperature measurement, the current flowing through thermistors must be kept very low (typically less than 100 μA) to assure near-zero power dissipation and near-zero self-heating.

Thermistors are available in many configurations and sizes and are either used in their basic form or assembled into probes. Examples of typical configurations are shown in Figure 19–21. Thermistors are generally very small in size. The small beads are about 0.5 mm in diameter. The dimension "D" of the epoxy-coated chips ranges from 2.4 to 3 mm. (Uncoated chips are about two-thirds that size and less than one mm thick.) The diameter of the miniprobe (hermetically sealed in glass) is 1.5 mm. An encapsulated thermistor sensor for use in medical electronics (not shown) has a diameter of 0.5 mm. The linear thermistor network consists of one twin thermistor and two precision resistors. It is designed to produce a resistance change or voltage output that varies linearly with temperature over selected temperature ranges when used with a regulated voltage source. Among additional thermistor configurations are small surface-mounted, end-banded (for contacts) chips, designed to be mounted onto printed-circuit boards using robotic techniques; glass-encapsulated chips with axial leads that are suited for automated assembly; and thin discs with leads. Several thermistor probe configurations are also shown in Figure 19–21; these include probes for liquid

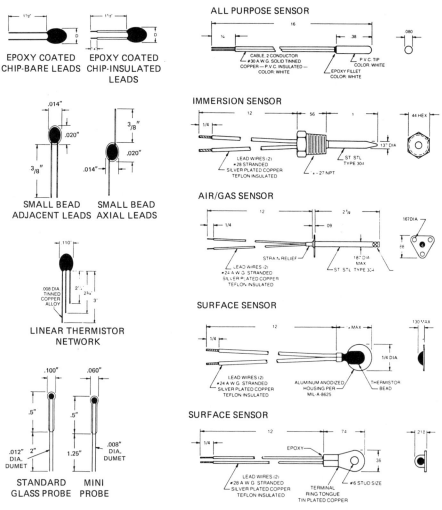

**FIGURE 19–21.** Examples of thermistor and thermistor-probe configurations (dimensions in inches). (Courtesy of Fenwal Electronics/ APD.)

or gaseous fluids as well as surface sensors that are mounted by cement or fasteners.

The fabrication of bead thermistors involves mixing metal oxide powders and a binder material in prescribed proportions, shaping, sintering, and encapsulation (bare beads are rarely used). Lead attachment is typically performed during the shaping process. Other thermistor types, such as disks, are pressed into their shape, sintered, metallized, trimmed to adjust their resistance value, provided with leads, and then encapsulated (e.g., in epoxy).

Beads can then be glass coated, made part of a small glass probe, or fixed within a glass tube or envelope (glass-coated beads need not have been encapsulated). The sintering temperature is quite high, on the order of 1300°C.

### 19.4.4 *Germanium Resistance Thermometers*

Thin slices of specially grown germanium crystals, typically doped with such materials as arsenic, gallium, and (in very slight amounts) antimony, are used in germanium resistance thermometers. The crystal in a typical thermometer of this type (see Figure 19–22) is provided with two pairs of gold wires 25 μm in diameter, using techniques to assure good ohmic contacts with the crystal. The crystal is mounted strain free to a header, a glass/platinum hermetic seal through which four 0.13-mm-diameter platinum wires pass. The four-wire connection is necessary since, to provide precise measurements, one of each pair of contacts is used as a current contact (to be connected to a current source) whereas the other contact in each pair is used as a voltage contact (from which the output signal is taken). The thin gold wires are attached to the platinum leads. A gold-plated copper case is placed over the assembly and then is evacuated, back-filled with helium, and sealed to the header. Later, color-coded, Teflon-insulated, stranded-copper leads are attached to the outside ends of the platinum wires, and the region in which the connections are made is encapsulated with epoxy, filling this region up to the outside edge of the hermetic seal. To stabilize the thermometer's characteristics, each unit is temperature cycled repeatedly

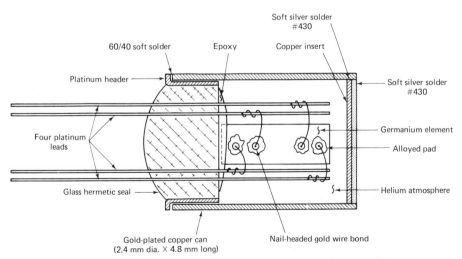

**FIGURE 19–22.** Internal construction of germanium resistance thermometer. (Courtesy of Scientific Instruments, Inc.)

from room temperature to 4.2 K. Specification limits call for a repeatability within ±0.0005 K at 4.2 K as the acceptance criterion.

The completed thermometer can be used in its basic configuration, or it can also be equipped with a bolt-down flange surface, or installed in the head of a small screw for mounting in a tapped hole, or mounted in the tip of the stainless-steel sheath of an immersion probe. The measuring range of germanium resistance thermometers is usually the full cryogenic range, 1.5 to 100 K; however, some models are available for portions of this range, including for very small portions near absolute zero (0.5 to 3.2 K) and for the $^3$He range (0.3 to 3.2 K). Over the range 2 to 20 K such thermometers are used as primary and reference standards. The *R*-vs.-*T* relationship is nonlinear and exhibits slope changes over the range 1.5 to 100 K. It is important to use the individual calibration curve for each thermometer; repeatability, as referenced to this curve, can then be expected to be very close.

### 19.4.5   Diode Thermometers

Silicon and gallium-arsenide diodes have been used primarily for low-temperature thermometry.

*Gallium-arsenide-diode thermometers* offer the advantage of being usable over a temperature range from 1 to 400 K. They exist in configurations based on standard transistor cans as well as in other configurations, such as small probes and threaded inserts. The slope of their *R*-vs.-*T* curve and their sensitivity are quite different below about 70 K (and especially below about 10 K) than for the higher portion of their overall measuring range. A well-regulated constant-current source is required for their excitation, with fixed polarity, and a sensitive voltmeter with a high input impedance (or a preamplifier with equivalent characteristics) is needed for their output. Their relatively small size (typically less than 5 mm) and mass (0.5 g or less, except for threaded-insert types) makes fast time constants not difficult to achieve.

*Silicon-diode thermometers* have characteristics somewhat similar to those of the GaAs diodes in that their output (voltage)-vs.-temperature curve also exhibits a radical slope change for very low temperatures (below about 30 K). The average sensitivity at low temperatures is approximately 20 times as great as at temperatures about 40 K. A small-probe configuration is usual for these devices.

### 19.4.6   Capacitance Thermometers

Capacitance thermometers have a usable range from near 0.01 K to about 60 K; they are also usable for temperatures between 70 and 400 K, but have different characteristics over this higher range and are rarely used there. The multilayer glass/ceramic-metallized electrode devices offer the advantage that they are essentially unaffected by even very strong magnetic fields. Small (about 10 mm long) configurations, flat as well as cylindrical, have been produced.

### 19.4.7 Acoustical and Noise Thermometers

*Acoustical thermometers*, which are usable to extremely high temperatures, and *noise thermometers*, which have been applied in the very low cryogenic region, belong in the categories of developmental and research devices. Both types have been applied experimentally to temperature measurements near absolute zero. For experiments involving the measurement of plasma temperatures in the tens of thousands of degrees, acoustical thermometry offers a viable means of temperature determination.

### 19.4.8 Temperature Switches

Temperature switches, which can be considered discrete-output thermometers, are frequently used in control applications, and rarely used in measurement systems. They can be grouped in two categories: direct-actuating types and analog thermometers with gating circuitry. Another type is the sealed metal bulb, liquid or gas filled, which provides a pressure change as temperature sensed by the bulb varies. Although such devices can be used in conjunction with a pressure switch, they are usually designed for directly operating a control element or providing a pneumatic output (e.g., 3 to 15 psi) rather than an electrical output.

The most commonly used directly actuating temperature switches employ a bimetallic element (two mechanically joined strips or disks having different thermal coefficients of expansion) which effects a switch closure when it deflects to a preset position due to heating or cooling. Thermostats used in home heating and air-conditioning systems typically use such elements; however, bimetallic temperature switches are employed in numerous other applications as well.

More flexibility, improved accuracy, and (when required) faster response time are provided by analog thermometric sensors whose output is converted into one or more discrete levels by appropriate signal-conditioning circuitry. This circuitry can be designed for on/off control outputs or for two-position control (e.g., high and low limit), and set points cannot only be changed by a simple manual adjustment, but can also be programmable. Additionally, the circuitry can be designed for proportional control in one, two, or three modes. The temperature-sensing devices used with such switch-output circuitry can be thermocouples, resistive sensors (conductive or thermistors), or radiation pyrometers.

Essential characteristics of temperature switches include span (the range over which the switching point or points can be set), set-point repeatability ("accuracy"), threshold (often incorrectly called "sensitivity"), hysteresis (dead band), response time (or time constant), and any delay effected by circuitry to prevent rapid cycling around the set point.

Temperature switches find their applications whenever temperature has to be controlled within more or less narrow limits; they are also used to provide alarm indications (e.g., overheating alarms). Associated circuitry and devices can provide control, alarm indication (visible or audible or both), and numerical display.

## 19.5  TRANSDUCER CHARACTERISTICS

Design and performance characteristics that should be considered for the specification or selection of thermometers (temperature sensors, temperature transducers) vary in both number and complexity for different designs. A limited number of characteristics apply to "basic" sensors, that is, sensing elements which either are installed or used as such, or are assembled into a sensor package by the user. A larger number of characteristics apply to such relatively simple "packaged" sensors as surface-temperature transducers. A far larger number of characteristics need to be considered for devices such as completely packaged immersion probes. The most prevalent "basic" temperature sensors are thermocouple wire and bead thermistors.

*Wire thermocouples*—with sensing junction, connecting hardware, and associated circuits completed by the user—are specified by their wire material, wire gauge or diameter, number of conductors in a cable, insulation and coverings, and "limits of error" (accuracy, usually just stated as either "standard" or "special"). For most commonly used thermocouple wire and extension wire types there are standard designations for these characteristics, established in readily available national, technical society, or governmental standards. Such standards are frequently reprinted in manufacturers' bulletins and should be adhered to and used to the greatest extent possible. Manufacturers' standards exist for numerous types of metal-sheathed, ceramic-insulated thermocouple cable, and cable characteristics can be selected from such tabulated data.

*Bead thermistors* can be specified by their bead dimensions and covering, the materials, dimensions, insulation (if any) and configuration of the leads, base resistance ("cold resistance," normally the zero-power resistance at 25°C), usable or intended temperature range, and either the material constant $\beta$ or, more simply, a *characteristic curve* (sometimes given as a table) of $R$ vs. $T$ and the expectable deviation of actual values from this curve. Specifications for the time constant and dissipation constant, together with applicable conditions, are usually also included.

The consideration of characteristics of immersion probes and surface-temperature sensors, which are widely used in a very large variety of applications, requires considerable attention to details. The design, performance, and environmental characteristics which require such attention are explained next. Note that some of the performance characteristics are unique to temperature transducers.

### 19.5.1 Mechanical Design Characteristics

Mechanical design characteristics for immersion probes and surface-temperature sensors include several important features. Characteristics for hand-held and ambient-air probes are somewhat similar to those of immersion probes, and surface-temperature sensors include not only bondable (cementable) designs, but also such configurations as threaded-insert types and washer types. Configurations and dimensions have to be chosen with extreme care. Typical examples are shown in Figure 19–23 for integrally assembled immersion probes and for bondable, weldable, or potted-in-place surface-temperature sensors. Mounting provisions and methods, together with their effect on configuration, should be determined; they are critical factors for having the sensor respond only to the temperature to be measured (to the best extent possible).

The stem length of an immersion probe intended for installation in a pipe or duct should, as a "rule of thumb," be so chosen as to place the center of a resistive sensing element, or the junction of a thermocouple, at a radial position located $0.72r$ (for turbulent flow) or $0.58r$ (for laminar flow) from the centerline of the pipe, where $r$ is the radius of the pipe (or duct).

For "industrial"-type immersion probes, which are usually capable of disassembly and may be assembled either by the user or the manufacturer, the individual piece parts that make up the assembly are selected, usually from a catalog. The parts include the connection head assembly, any required extension hardware (which, in effect, extends the head assembly from the mounting boss), the type and length of thermocouple element, any required protection tube or sheath, and the associated thermowell if used.

Configurational considerations include the protection of the sensing element, which may be exposed (bare) or enclosed, or, for certain types of platinum-wire sensors, coated (but without a metal enclosure). For some applications a "stagnation fitting" may be required around the sensing element; this is a cage designed to minimize impact pressure, due to flow velocity, acting on the element. Another provision that may have to be specified or considered is the nature of any spring-loading arrangement.

The physical and chemical properties of all measured fluids that will come in contact with the sensor (or a thermowell) have to be known, and sensor compatibility with the fluids must be established (see Appendix Table A–2). The maximum flow velocity of measured liquids and gases should be

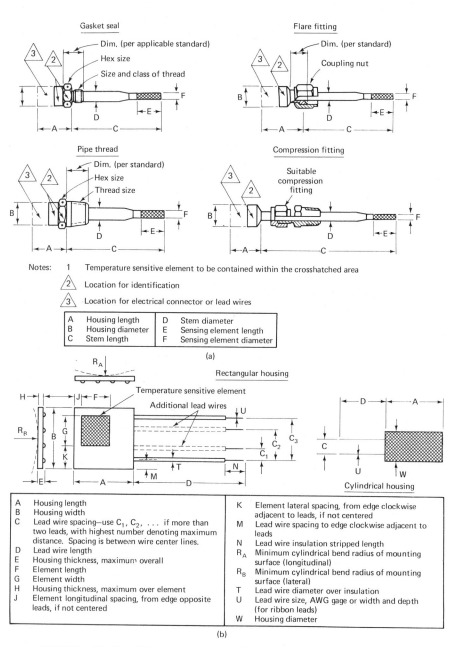

**FIGURE 19-23.** Dimensional specifications for temperature sensors: (a) typical immersion-probe configurations; (b) typical surface-temperature transducer configurations.

stated for immersion probes, and the operating pressure range, proof pressure, and burst pressure rating of the probe must be specified. If necessary, appropriate leak-test requirements can be imposed.

Mechanical characteristics to be considered for surface-temperature sensors should include the carrier or matrix material, lead pullout strength, and any limitations on mounting procedures, including maximum torque for washer types and maximum clamping force for clamp-on types.

### 19.5.2 Electrical Design Characteristics

Electrical design characteristics pertain to the sensing element as well as to any additional electrical components that may be packaged integrally with the sensor (e.g., contained within the connection head). Some examples of internal schematics of thermoelectric and resistive temperature transducers are shown in Figure 19–24. Thermocouple junctions must be specified as either grounded or ungrounded. Insulation resistance or breakdown voltage rating or both should be specified for all sensors having an ungrounded element; this is of particular importance for surface-temperature sensors.

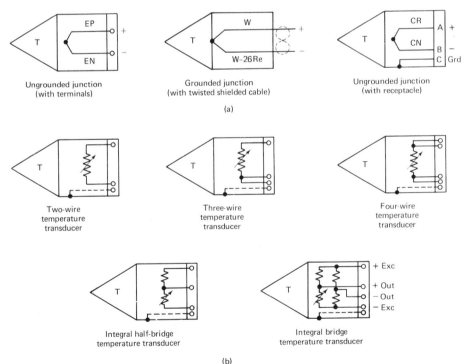

**FIGURE 19–24.** Typical internal schematic diagrams of temperature transducers: (a) thermoelectric transducers; (b) resistive transducers.

Additional characteristics for sensors with passive elements (all except thermocouples) include nominal and maximum excitation current, or voltage and power; type and power requirements, if any, of integrally packaged signal-conditioning circuitry; and sometimes the nominal resistance across the transducer's terminals at room temperature (for continuity checks). When ac excitation or pulsed dc excitation is to be applied to a sensor with a wire-wound element, the inductive reactance of the element at the specified form of excitation should be considered because it may cause errors or pulse-shape distortion in the sensor's output.

### 19.5.3   Performance Characteristics

Performance characteristics comprise primarily range, output, dynamic response, and accuracy characteristics. All these characteristics can be influenced significantly by the intended use. *Range* is usually shown as the lower and upper limits of temperature over which accuracy and other performance characteristics and their tolerances apply. In most cases this is the overall "usable" or "recommended" range of a sensor. In some cases, however, the user may have standardized on certain output spans that match the signal-conditioning equipment and data system. This is sometimes advisable when a significant number of temperature measurements must be made in a given installation. If the user has, for example, selected an amplifier that requires 50 mV for a full-scale deflection of his or her telemetry or display equipment, the range of a thermocouple that is used will be dictated by the temperature at which 50 mV is produced by the thermocouple. The range could be shifted up or down by biasing the input to the amplifier or, for a thermocouple, by appropriate selection of the reference-junction temperature. Similarly, in the case of resistive sensors, the user may have selected a resistance bridge standardized so that a resistance change of 100 $\Omega$ in the leg represented by the sensor (which may be in series with a trimming resistor) provides the required full-scale deflection. The range of the resistive sensor is then given by the limits between which the 100-$\Omega$ resistance change occurs.

*Output* is governed by similar considerations. For most applications the output of the sensor is given, for thermocouples, by applicable emf-vs.-temperature tables and, for resistive sensors, by a table or curve (*theoretical curve* or *reference curve*) of the nominal $R$-vs.-$T$ relationship (or that obtained by an individual calibration); equivalent reference curves (tabulated or in graphical form) define the output of frequency-output, voltage-output, or current-output sensors. However, the use of standardized signal-conditioning equipment may force the output to be shown by specific end points or by a specific change over a stated temperature range.

*Maximum and minimum temperature* are often shown as overrange conditions beyond specified range limits. When so specified, the accuracy characteristics do not apply over these extended portions of the range; however, the sensor is expected not to be damaged and, usually, not to lose its normal performance by exposure to such temperatures over stated periods of time.

*Repeatability* is the most essential accuracy characteristic of temperature sensors. It can be stated in terms of the temperature indicated by the output of the sensor (e.g., "within 0.05°C") or in terms of output (e.g., "within 0.5 Ω" or "within 0.03 mΩ"), and sometimes in terms of percent of (output) reading. *Stability* is sometimes specified additionally, as repeatability (with larger tolerances) applicable either over a long period of time or over a large number of temperature cycles. *Linearity* is rarely specified for temperature sensors, since all types are inherently nonlinear to some extent; it is sometimes shown for sensors that are permanently connected into some sort of linearization network. *Hysteresis* is normally negligible in temperature sensors.

*Error* (or an *error band*) refers to the maximum deviation, over the specified temperature range, of output values from a reference curve (which is usually shown in tabular form). For thermocouples the same is accomplished by "limits of error"; the standardized emf-vs.-temperature tables (or equivalent tables furnished by a manufacturer for nonstandardized thermocouples) serve as the reference curve. When the same error tolerances are specified for a number of sensors to all of which a given reference curve applies (as is the case for thermocouples), these tolerances are also indicative of *calibration interchangeability*. When error, limits of error, or an error band is specified, the tolerances apply to deviations due to causes attributable to the sensor in its intended use, regardless of the source of the error. A number of specific sources of error in temperature sensors are explained next.

*Thermoelectric potentials* can be generated within a resistive temperature sensor at internal connections between wires of different metals such that a temperature gradient exists along one or more of the wires. This error is best avoided by using internal connecting leads of the same metal as used for the sensing element if it is conductive, and assuring isothermal connection points when more than one lead material is used in a sensor containing a semiconductive element.

*Self-heating* ($I^2R$ *heating*) can cause errors in resistive sensors, particularly in semiconductor sensors. Self-heating is dependent on the power dissipation of the element as established by the element's resistance, the current through the element, and heat transfer to the measured fluid or surface. Self-heating will be more severe, for example, when the same sensor, using the same excitation current, is used in still air than when it is immersed in a rapidly flowing liquid. One way of determining self-heating error is to take a measurement with the sensor using the nominal excitation current and then, without changing any other measurement condition, repeating the measurement using a fraction (e.g., 20%) of the nominal current. The difference in output readings, if any, is the self-heating error. The conditions under which the measurement is made must, of course, correspond closely to the intended use of the sensor.

*Conduction error* is caused by heat conduction between the sensing element and the mounting of a temperature sensor. This occurs primarily in immersion probes and can be quite severe when the element is enclosed and the enclosing sheath is short. Conduction error can also occur in surface sensors, due to heat conduction along the connecting leads to the nearest terminal; it can be minimized by making the leads as thin as possible and mounting the nearest electrical terminals close to the point of measurement. The conduction error of an immersion probe can be roughly determined by a test during which the probe is first immersed into an agitated temperature bath (a liquid whose temperature is known and controlled within close limits) only a little above the sensing-element region while the head of the sensor is artificially cooled or heated to a temperature that differs from the bath temperature by the same amount that it would differ in the probe's intended application. Next, the sensor is immersed in the bath up to nearly the outermost edge of the head (so as to avoid immersing the electrical terminals or connector pins into the bath). The two readings are then compared, and their difference is due to conduction error.

*Mounting error* ("strain error") can occur in surface-temperature sensors after they are bonded to the measured surface, due to strain in the measured region. It can be determined by measuring the output of the sensor, at the same temperature, before and after mounting. Mounting error can also be introduced by using an improper mounting procedure.

*Time constant* is the most essential dynamic characteristic of a temperature sensor. Its specification is meaningful only when the value is accompanied by a statement of the two temperatures that constitute the step change as well as by statements of the type of fluid used at each of the two temperatures and the flow rate of both fluids. An example of a correct specification would read "Time constant: 50 ms, max., from still air at $25 \pm 2°C$ to distilled water at $80 \pm 2°C$ moving at 1 m/s." When only the second fluid and its flow rate are specified, still air at $25 \pm 5°C$ can usually be assumed as the first fluid. *Response time* is sometimes specified as 98% or 99% response time under the same test conditions applicable to determining the time constant.

*Recovery error* is sometimes stated for temperature sensors intended to be used at measured-fluid flow velocities in excess of Mach 0.2, especially when the flow is transverse to a probe-type sensor. Recovery error then is the error in total temperature (the temperature indicated by the output of the sensor) caused by the assumption of a unity *recovery factor*. The recovery factor is the proportion of kinetic energy converted into heat, expressed as the ratio of the difference between the recovery temperature and the absolute temperature to the difference between the total temperature and the absolute temperature. Recovery error can be determined by appropriate calculations.

### 19.5.4 Environmental Characteristics

Environmental characteristics of thermometers are often limited to industrial or governmental codes for hazardous, explosive, corrosive, etc., environments. In the case of immersion probes, such specifications apply primarily to their seal and the connection head, unless the sensing element is exposed to the same environment. Depending on the application, however, specifications governing other environmental effects can become very important. Temperature transducers installed in or on rotating or reciprocating machinery, vehicles of many types, and in most aerospace applications can be severely affected by shock and vibration. (Relatively long immersion probes are particularly vulnerable in such environments.) Vibration not only can cause output errors, but it can also result in a sensing element's opening inadvertently (and the associated measuring circuitry then responding to the open circuit as an indication of upper-range-limit temperature in the case of resistance thermometers). Indeed, it may even result in fracture, which can have extremely dangerous consequences if, for example, an immersion probe is installed at the inlet of a pump or turbine. Use of temperature sensors in a nuclear environment requires special attention to the materials used in them: many "off-the-shelf" thermometers cannot operate in such an environment for a significant length of time. The encapsulating material of surface-temperature sensors is subject to degradation when exposed to sunlight (solar ultraviolet radiation) for prolonged periods of time, and ambient pressure becomes important for temperature sensors used in underwater applications.

### 19.6 APPLICATION NOTES

#### 19.6.1 Selection Criteria

Selection of a temperature transducer can involve a relatively larger matrix of considerations than is necessary for selection of other generic types of sensors. Primarily, it is necessary to select a sensor design whose sensing element will attain the temperature of the measured fluid or surface within the time available for making the measurement. The output of a temperature sensor at any time is merely a measure of the temperature of its sensing element, and it is often difficult to assure that this temperature is indeed the same as the temperature that is required to be measured.

For most applications the following selection criteria are of primary importance: the nature and characteristics of the measured fluid or solid; the measuring range (and possible overrange limits); the time constant; and the type of the associated signal conditioning, data system, and display equipment available or intended to be used.

*Measured fluids*, whose temperature is sensed with immersion probes, can be liquids or gases; corrosive or noncorrosive; oxidizing or reducing; stagnant or moving at low, medium, high, or very high velocities; and freely flowing or contained in a pipe, duct, tank, vessel, or cavity.

The temperature of a contained fluid can usually be measured with an immersion-probe-type sensor. A thermowell can be permanently installed in an opening in the wall of the container when the operating life of the installation is much greater than the expected useful life of the sensor, or when it is expected that the sensor must be removed fairly frequently for calibration or maintenance, and when it is impractical or undesirable to have an "open hole in the container" for probe replacements. Thermowells lengthen the time constant of a measurement substantially; hence, when temperature fluctuations are slow, and when conduction error can be minimized, thermowells would not compromise measurement accuracy; when such fluctuations are rapid, the loss in measurement accuracy due to a relative inability of the sensing element to follow such fluctuations, and due to a probable increase in conduction error, must be traded off against the maintainability considerations described previously.

All materials of an immersion probe must be compatible with the fluids that can come in contact with them. When an enclosed-element probe is used, it is usually only the sheath material and the material of a portion of the mounting that need to be considered. When an exposed-element probe is used, the design must be analyzed carefully to determine all the various sensor materials (sensing element, connecting leads, supports, cage, partial sheath) that may come in contact with the measured fluid; compatibility of each of the materials must then be assured. When a thermowell is used, only the thermowell material needs to be considered. For considerations of material compatibility and structural/mechanical integrity of a probe, the term "measured fluid" must be extended to encompass all fluids that may come in contact with the probe in its installation; besides the measured fluid, these may include test fluids and purge fluids. Table A–2 in the Appendix can be used to determine recommended probe sheath and thermowell materials for immersion in various fluids.

Resistive elements are normally enclosed or at least coated when immersed in liquids and most gases. Exposed resistive elements cannot be immersed in conductive or contaminating fluids. They can, however, be used in dry air and in a limited number of other gases and some liquids when their fast time constant is required. For such applications in stagnant or slowly moving fluids, some sort of cage usually surrounds the element for mechanical protection. Somewhat more attention needs to be paid to the design of such a cage when it is also intended to shield the element from radiated heat, or when stagnation temperatures must be considered. For exposed-junction thermocouples, general material-compatibility considerations point to the use of Type J for reducing fluids; Types K, E, R, and S for oxidizing fluids; and Type T for oxidizing as well as reducing fluids or atmospheres. Exposed tungsten-rhenium junctions are limited to use in vacuum or clean inert gases; such junctions also tend to be brittle and require careful handling. Thermistors are usually coated and can be used in most fluids; however, material-compatibility considerations still apply to other elements that are part of a thermistor probe. The elements of specialized types of cryogenic sensors (e.g., Ge, GaAs) are usually enclosed, primarily for structural/mechanical protection, but also to prevent contamination. The choice of ungrounded vs. grounded thermocouple junctions is primarily given by a trade-off between the faster time constant of a grounded junction against the potential increase in electrical noise in the associated measurement circuitry; many data systems require single-point grounding, which dictates the use of ungrounded junctions.

A number of other considerations can affect immersion-probe design. The probe must be rated for the specified operating, proof, and burst pressure, and it must withstand the maximum specified transverse-flow velocity. Also, it should have an immersion length appropriate to obtaining a measurement of the mean temperature in a pipe or duct; at the same time, it should be long enough to minimize conduction error, and these two requirements may be in conflict and require a trade-off.

When an opening in a pipe, duct, or vessel is impractical or prohibited, so that an immersion probe or thermowell cannot be used, a temperature measurement whose accuracy is often adequate can still be obtained by clamping or bonding a surface-temperature sensor to the outside of the wall. To assure a reasonable amount of measurement accuracy, the sensor must be thermally insulated from the ambient atmosphere and from sources of heat radiation, and the heat conduction from the wall to which it is mounted to its sensing element must be optimized by sensor design and installation.

*Measured solid materials* can be metallic or nonmetallic, of different thickness and cross-sectional configuration, and exposed to various ambient environmental conditions. Any type of surface-temperature sensor will modify the characteristics of the surface or subsurface of the material and configuration to some extent. The sensor and its installation method must therefore be chosen so that such disturbances are minimized. An ideal sensor would be made entirely of the same material as the solid and would become an integral part of it so that the characteristics of the configuration at and around the point of measurement are not altered in any manner. A good practical sensor, together with its installation, introduces as little foreign material into the solid as possible and affects the configuration as little as possible. This rule also applies to associated electrical connections, housing, and mounting materials.

Most temperature measurements of solids are made at the surface, by cementing or welding a thin resistive or thermoelectric sensor to the surface. Any envelope or housing of such a sensor should match the surface material as closely as possible. A variety of such sensors are available with a thickness of 1 mm or less, including resistive (wire, film, and thermistor) as well as thermoelectric (foil, thin-wire, and thin-film) types. Metal-sheathed ceramic-insulated thermocouple cable (two-conductor) has been made with overall diameters down to 0.3 mm. Very small weldable clips are available for such cables; their connection to larger diameter extension cable should be made sufficiently far away not to affect the surface configuration near the point of measurement.

Subsurface temperature measurements are made with small embeddable sensors or threaded inserts. They are typically installed so that their outer edge, in the as-installed configuration, is flush with the outside surface of the measured material. The material of the threaded insert should be the same as, or at least very similar to, the measured material. When washer-type sensors are used, care must be taken to assure that the temperature attained by the washer will be as nearly as possible the temperature to be measured.

*Measuring range* strongly influences the selection of a temperature sensor. Platinum-wire sensors can be used when the range is any reasonably sized portion of their overall useful range. Relatively narrow ranges often require the base resistance of such a sensor to be fairly high so that a sufficiently large resistance change can be obtained. The large resistance changes provided by thermistors make these sensors very suitable for narrow measuring ranges (e.g., those of medical thermometers). Thermocouples tend to be more useful when a fairly large portion of their usable range is to be measured. The ice point should preferably be included in the range since thermocouple emf tables are referenced to that temperature (and reference junctions are normally kept at that temperature or adjusted to simulate it). The reason for using thermocouples for larger temperature ranges is given by their output and associated requirements imposed on signal-conditioning equipment. For example, even the most sensitive material combination, Type E, provides an output of only 3 mV at 50°C (referred to 0°C), and data systems usually require larger input signals than that. This means that the signal has to be amplified, and amplifier cost, complexity, instability, and noise are inversely proportional to input signal level as a general rule. For the same thermocouple type, the output signal would be 53 mV for the range 0 to 700°C, and it would be 76.4 mV for the range 0 to 1000°C, and such larger signal levels are easier to work with. Sensor linearity over a given range may be an additional factor in sensor selection. Specialized cryogenic-temperature sensors such as germanium resistance thermometers are particularly useful when narrow portions of temperature ranges in the low cryogenic region must be measured.

*Response time*, usually expressed in terms of *time constant*, is another essential criterion for sensor selection. When, for example, the temperature of a liquid in a storage tank is being monitored, the time constant of the temperature sensor is of no great importance. However, when temperature must be measured in a shock tube during an experiment, the time constant becomes the governing criterion for sensor selection. The time constant of a temperature-sensing element is proportional to the ratio of its heat capacity to the heat transfer between the measured material and the element. Relatively short time constants can be obtained using bead thermistors and exposed-junction fine-wire thermocouples, whereas ungrounded, enclosed-junction thermocouple immersion probes, particularly when inserted into a thermowell, will have relatively long time constants.

Time constants are usually stated for a condition where the step change in temperature terminates in a well-controlled, steady-state temperature of a moving liquid, primarily because properly controlled tests for such a condition are convenient. Such statements of time constant are useful for comparing different sensor designs since they can be tested under the same conditions. However, the knowledge of a sensor's time constant, established on such a basis, provides little useful information about temperature measurements obtained from the sensor when it is exposed to rapidly fluctuating temperatures such as would be experienced in turbulent gas flows. *Dynamic temperature measurement* is really a specialized field of study involving considerable expertise and complex calculations. Reasonable approximations of the dynamic behavior of a sensor can often be obtained by tests in which the conditions that will prevail in the sensor's intended use are simulated as closely as possible.

*Associated circuits*, primarily signal-conditioning circuits, will affect sensor selection, mainly when more than one temperature measurement is handled by a common data system and particularly when the design of the signal-conditioning circuitry was frozen prior to the time the sensor is selected. When reference junctions and thermocouple-signal amplifiers are readily available, a preference for thermoelectric sensors is quite understandable and insistence on using a resistive sensor may well meet with objections. For installations in which all signal levels are standardized (e.g., 4 to 20 mA, or 0 to 5V dc), either resistive or thermoelectric sensors can be used since both types can be equipped with the required integrally packaged signal-conditioning circuitry (at additional cost). Typical circuits associated with temperature sensors are described in the next section.

### 19.6.2 *Measurement Circuits*

*Thermocouples* require a reference junction in their circuit which must assure that both terminals (the terminals for the positive and the negative wire) are at the identical temperature and that this temperature is known; further, for most applications this temperature is chosen as exactly zero degrees

Celsius, primarily because thermocouple reference tables (*T*-vs.-emf tables) are based on this as the reference-junction temperature. The classic ice-point reference junction (Figure 19–25) is one that is immersed in an *ice bath*, a mixture of ice and water contained in a well-insulated container (e.g., dewar flask). Since the duration of the effectiveness of such ice baths is limited, alternatives to this technique have been developed. One method employs a refrigerator, such as one based on the (thermoelectric) Peltier effect, with good closed-loop control so that an ice/water equilibrium is maintained in the ice bath. Closely controlled ovens have also been used for maintaining a known reference-junction temperature. One reference-junction block may be used for two or more reference junctions. When the reference-junction temperature cannot be maintained at exactly the same value, a temperature sensor can be embedded in the reference-junction block so that the temperature can be monitored and appropriate corrections to the output readings made.

A resistive temperature sensor embedded in, or in very close thermal contact with, a reference-junction block is used in one arm of a bridge circuit for *reference-junction compensation*. In such *reference-junction compensators* a specified temperature is simulated rather than maintained by direct means. The compensation circuit (Figure 19–26) is initially adjusted so that the specified reference temperature (e.g., 0°C) is simulated. When the temperature of the reference junction changes from these initial conditions, usually because of changes in the ambient temperature, the resulting resistance change of the embedded temperature sensor is used to create an error signal which causes a compensating voltage to be inserted in series with the thermocouple circuit. The polarity and magnitude of the compensating voltage is such that the output voltage $E_{out}$ equals that voltage that would have been obtained had the reference-junction temperature remained at its specified value. In the typical circuit illustrated, resistance $R_N$ is related to the nonlinearity of the temperature sensor and resistance $R_B$ balances its leg of the thermocouple circuit for the equivalent resistance of the compensation network (including the power supply) in the other leg. The power supply can

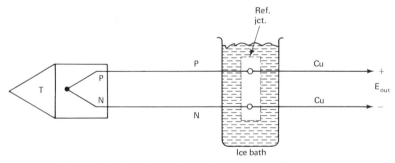

**FIGURE 19–25.**  Ice-point reference-junction control.

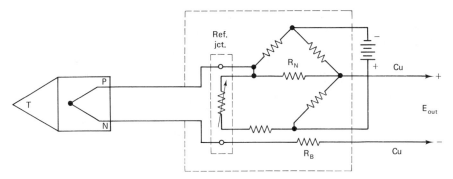

**FIGURE 19–26.**  Reference-junction compensation.

be external, or it can be a small battery (typically a mercury battery) integrally packaged with the compensator ("self-powered" compensator), in which provisions for verifying that the battery voltage is correct are usually included.

*Resistive sensors* usually require circuitry that converts their resistance changes into voltage changes. The voltages obtained can then be fed to a display device or into a telemetry system or other data system. The sensor can also be connected into a resistance bridge (Wheatstone bridge or more precise versions of it) in which variable resistors in the adjacent bridge arm are adjusted manually until bridge balance is indicated by a null indicator. The resistance settings of the variable resistors are then read. This technique, however, is usually employed only for bench tests and calibrations of such sensors.

The simplest resistance-to-voltage conversion circuit is the *voltage divider* (Figure 19–27(a)). The fixed resistor $R$ is usually chosen larger in value than the base resistance of the temperature sensor. The output voltage $E_{out}$ varies as the ratio between the variable (sensor) resistance and the fixed resistance changes. This circuit is often used with thermistors, but has also been used with conductive wire- and film-element sensors, with best results obtained when the excitation source is a constant-current supply. Such a supply is also used in the *voltage-drop* circuit (Figure 19–27(b)) in conjunction with a load (e.g., voltmeter) having a high input impedance. Lead-resistance effects on the output voltage are minimized in this circuit, which is often used with germanium resistance thermometers but can be used with any resistive sensor. Note that the current must always be kept within appropriate limits. A simple means of monitoring the current is by inserting a resistor between one supply terminal and the sensor terminal normally connected to it and then monitoring the voltage drop across this fixed resistor.

The *unbalanced-bridge* circuit shown in Figure 19–27(c) is typical for circuits used in many telemetry systems. The trimming resistor $R_T$ connected in series with the sensor $R_X$ can be adjusted to balance the bridge at a sensor resistance corresponding to the lower end point of the temperature range to

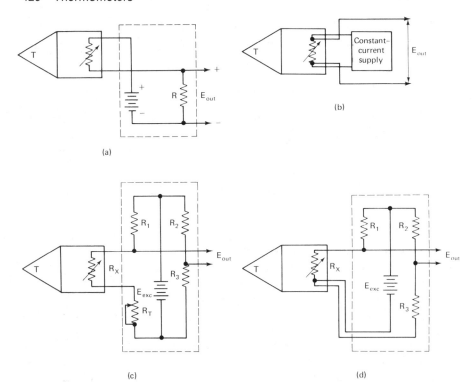

**FIGURE 19–27.** Typical resistive-sensor circuits providing a voltage output: (a) voltage divider; (b) voltage drop; (c) basic unbalanced bridge with trimming resistor; (d) unbalanced bridge for three-wire transducer.

be measured. This circuit, as well as the one shown in Figure 19–27(d), is typically used with platinum-wire sensors. The output voltage $E_{out}$ of the bridge can be calculated for any combination of resistance values and any excitation voltage $E_{exc}$ by

$$E_{out} = E_{exc}\left(\frac{R_1}{R_1 + R_X + R_T} - \frac{R_2}{R_2 + R_3}\right)$$

*Three-wire* sensors can be connected into an unbalanced bridge circuit (Figure 19–27(d)) so that changes in long sensor connecting leads are compensated for. All three sensor connecting leads have the same length (and resistance). Two of these leads are connected to the same sensor terminal, but one of them is placed in the excitation branch and the other in the opposite leg of the bridge. The lead from the other sensor terminal is, of course, in the same leg as the sensor. Hence, lead-resistance changes in the sensor leg of the bridge are compensated for by equal changes in the $R_3$ leg when the resistance $R_3$ is approximately equal to the resistance of the sensor.

More complex bridge circuits have been designed to compensate further for lead-resistance changes, to provide adjustments for zero and slope of the output, to linearize the output when sensor characteristics are nonlinear, and for differential-temperature measurements.

For diode-type sensors a circuit similar to the one shown in Figure 19–27(b) can be used, with the variable resistor replaced by a (GaAs) diode. Circuits additional to those shown have been devised to amplify output voltages to specified levels or to convert them into specified current changes (e.g., the very popular 4 to 20 mA) when data-system requirements demand this.

### 19.6.3 Typical Applications

*Thermocouples* of all types are the dominant temperature sensors in most industrial applications. They can be used over wide temperature ranges, tend to be low in cost, and lend themselves to "do-it yourself" fabrication. Their accuracy is usually sufficient for many types of temperature measurements, and their output-vs.-temperature curves, though inherently nonlinear, are well understood. When only a temperature difference must be measured, a *differential thermocouple* can be installed in which a sensing junction is placed at one point of measurement and the reference junction is placed at the other such point. Standard thermocouples (those designated by a letter such as E, J, or K) can be used for medium to high temperature ranges.

For very high temperatures the use of tungsten/rhenium combination thermocouples is indicated; these can provide accuracies on the order of ±1% of reading up to about 2300°C and are usually made from metal-sheathed, ceramic-insulated thermocouple cable. For very high temperature applications the sheath is made of tantalum or molybdenum, and beryllia (BeO) or thoria (ThO$_2$) is used as an insulating material. It is important that the cut end of the insulating material is well sealed, because the material tends to absorb moisture. Although some good sealers exist, the (grounded or ungrounded) junction is usually enclosed to assure a good seal. For special applications, such as in nuclear reactors, the high-temperature thermocouples are made leak-tight and are then back-filled with an inert gas such as argon.

*Resistance thermometers* (usually with a platinum-wire sensing element) are the preferred temperature sensors when ranges are relatively narrow and when accuracies over such ranges must be better than obtainable from thermocouples, for example, when a reliable indication of a 0.5°C temperature change is important. They can also be made very durable for use in severe environments. Platinum resistance thermometers with exposed elements are the prime candidate for measurements in gaseous fluids, including gases with high flow velocities, and in gaseous as well as liquid fluids when a short time constant is needed; they are very well suited to dynamic temperature measurement. Many vehicular telemetry systems (including aircraft, rockets, and spacecraft) incorporate signal-conditioning circuitry usable only for resistance thermometers; therefore, all temperature sensors in such systems, probes as well as surface types, are resistance thermometers using the same general type of sensing element (usually platinum wire).

*Thermistors* are normally usable between $-100$ and $+300°C$. They are particularly useful when fine resolution and repeatability are needed over narrow temperature ranges. A good example of this is medical temperature measurement, and thermistors are used exclusively in medical "electronic thermometers." These are commonly used in homes, many doctor's offices, clinics, and hospitals. The small size of bead thermistors also makes them usable for certain invasive procedures in the operating room, where they are introduced into the body by means of catheters. Thermistors are also often used for temperature monitoring and compensation in electronic circuitry. In special mounts, thermistor beads, supported only by their leads, are used for narrow-range, fast-response temperature measurements in air and gases. When embedded or encapsulated, thermistors make useful surface sensors and immersion probes, particularly when narrow temperature ranges have to be monitored. When good long-term stability is required, thermistors should be "aged" by applying rated excitation to them and then temperature-cycling them by immersing them alternatingly in two temperature baths, one at low temperature and the other at high temperature (within their intended measuring range). Thermistors have applications other than as thermometers; their use in, for example, liquid-level and dew-point sensors and in bolometers is described elsewhere in this book.

# 20

# *Radiation Pyrometers*

## 20.1 BASIC DEFINITIONS

Refer to Section 19.1.1 for applicable definitions, including those for *emissivity* and *black body*.

## 20.2 UNITS OF MEASUREMENT

Radiation pyrometers are based on sensing radiation intensity, expressed in $W/m^2$ or, more commonly, in $W/cm^2$. The output readings of radiation pyrometers, however, are displayed in the same units as temperature (see Section 19.2), viz., the *degree Celsius* (°C), the *degree Fahrenheit* (°F), and (not too commonly) the *kelvin* (K). Wavelength is usually expressed in *micrometers* (μm; often called "microns").

## 20.3 SENSING METHODS

Radiation pyrometry, as explained and described in this chapter, is the remote (noncontacting) measurement of temperatures of objects on the basis of thermal radiation emanating from them. Pyrometric sensing methods are based on the radiation laws shown in Section 19.1.2.3. The five curves shown in Figure 20–1 were plotted on the basis of Planck's radiation formula. Each curve represents the *radiation intensity,* or *radiant excitation,* per $\Delta\lambda$, as a

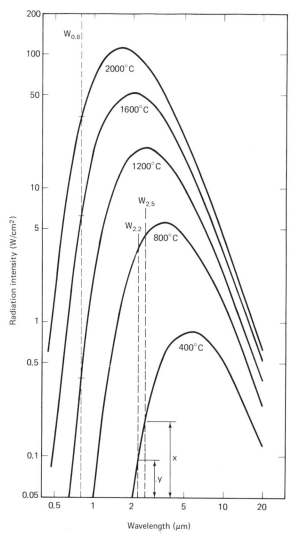

**FIGURE 20–1.** Radiation intensity as a function of wavelength for different source temperatures (assuming that $\epsilon = 1.0$).

function of wavelength (i.e., the *spectral radiance*) of a black body heated to a given temperature. This family of curves also demonstrates *Wien's displacement law:* since the product of peak wavelength and absolute temperature is a constant, the peak wavelength decreases as the temperature increases; this can be observed from the fact that the peaks of the curve shift toward the left as the temperature increases. Further, the curves also demonstrate the *Stefan-Boltzmann law:* the area under each curve (not shown completely on this graph) equals the total radiation emitted from a black body, which is proportional to the fourth power of its absolute temperature. Integration of the areas under the 1600°C curve and the 2000°C curve would give more accurate results; however, a coarse approximation can be obtained by calculating the ratio of the fourth power of the two temperatures ($27 \times 10^{12}/12.25 \times 10^{12} = 2.2$) and noting the radiation intensity at the peak of the two curves (111 and 51, respectively, for a ratio of 2.18).

The pyrometric sensing principles are also illustrated in Figure 20–1. The output of a *wide-band* ("total radiation") *pyrometer* is proportional to most of (only theoretically "all" of) the energy under a curve. The output of a *narrow-band pyrometer* ("brightness pyrometer," "monochromatic radiation pyrometer") is proportional to the energy at a specific wavelength (in practice, in a very narrow band of wavelengths); the dashed line marked "$W_{0.8}$" indicates the radiation intensity, for each of three temperatures, that would be observed at 0.8 μm. The output of a *ratio pyrometer* ("two-color pyrometer," "color-ratio pyrometer") is proportional to the ratio of two energies emitted at two different wavelengths; in the example illustrated, energy $x$ would be observed at 2.5 μm and energy $y$ would be observed at 2.2 μm, and the pyrometer output would be indicative of the ratio $x/y$.

Note that the dashed "$W_{0.8}$" line intersects only the three upper curves. Because of the effect of Wien's displacement law in shifting the peaks of lower temperatures toward longer wavelengths, different wavelengths are used in narrow-band and ratio pyrometers for different temperature ranges.

The curves in the figure are based on the *emissivity* of the source being 1.0 (i.e., that of an ideal black body). The emissivities of real materials, however, are generally significantly less than unity; additionally, for a given material and surface finish, they vary with temperature and with wavelength. *Total emissivity* is emissivity expressed for the total radiation (at "all" wavelengths), *spectral emissivity* is emissivity at a stated wavelength (or as a function of wavelength), and *monochromatic emissivity* is the emissivity in a very narrow band of wavelengths. Carbon has one of the highest emissivities, around 0.81 (total) and 0.85 (at 0.65 μm). The total emissivities of metals are much lower: 0.08 for aluminum, 0.09 for many steels, 0.25 for cast iron, and only 0.02 for gold. The latter shows why gold is often used when a good thermal reflector is needed: its reflectivity is $1.0 - 0.02 = 0.98$.

Figure 20–2, which was purposely plotted somewhat differently than Figure 20–1 (using linear rather than logarithmic coordinates to obtain a different form of graphic display), shows how a curve of spectral radiance as a function of wavelength, for a given temperature, changes for different emissivities. To simplify the illustration, the curves assume that the emissivity is not wavelength dependent. It can be seen that the area under each of the curves gets significantly reduced with decreasing emissivity. Furthermore, the ascending slope of each curve is different. (The descending slopes are also different, but the difference is not as pronounced.) The curves

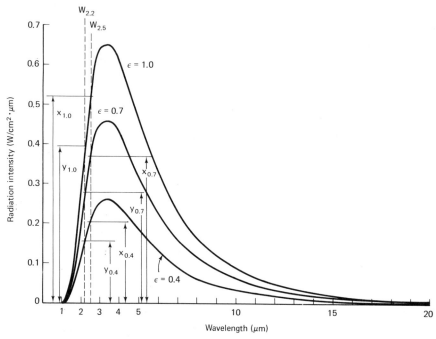

**FIGURE 20–2.**   Radiation intensity of a 600°C source as function of wavelength for different values of source emissivity (assuming that ε is not wavelength-dependent).

illustrate an advantage of the ratio pyrometer. Using the same two wavelengths as in Figure 20–1 (2.2 and 2.5 μm), and observing the energies at both wavelengths, $x$ and $y$, it can be seen (with appropriate scaling) that the ratio $x_{1.0}/y_{1.0}$ is the same as the ratios $x_{0.7}/y_{0.7}$ and $x_{0.4}/y_{0.4}$, namely about 1.33. Ideally, then, a ratio pyrometer can be used to obtain accurate temperature readings regardless of emissivity (i.e., it can be used with varying or unknown emissivities). In practice, some errors could still be introduced by not accounting for differences in monochromatic emissivity at the two wavelengths by such other factors as unequal absorptivity of smoke or fumes in the optical path between the source and the radiometer.

The "disappearing filament" pyrometer is one of the earliest versions of a narrow-band (brightness) pyrometer. This instrument cannot be considered an electronic sensing device, at least not in its original form, since it requires the eye of an observer. The observer sees the measured object as well as the tungsten ribbon filament of a "standard lamp." He varies the current through the filament until it matches the measured object in brightness. When the filament "disappears" (i.e., provides an exact brightness match), the current is read and the temperature of the measured object is determined from a calibration chart. Note that the IPTS-68, above the freezing point of gold (1064.43°C, 1337.58 K), is defined on the basis of thermal radiation, and instruments based on the "disappearing filament" principle are used as calibration devices. Automatic versions of this type of pyrometer alternately image the measured surface and standard-lamp filament on a photoelectric sensor (e.g., by use of a rotating sector disk or by an oscillating mirror). The output due to unequal brightness is used as an error signal in the filament control loop. The current through the filament is thus varied automatically until a null condition is reached with a brightness match. The temperature of the measured object (which may be a "standard black body") is then determined from the filament-current reading, using a calibration chart.

The detectors in pyrometers (i.e., the devices that convert the radiation incident on them into an electrical signal) are of the types described in Section 22.4. Wide-band pyrometers typically use bolometric or thermopile detectors, whereas narrow-band and ratio pyrometers usually employ photovoltaic or photoconductive detectors, especially those having sharply peaking spectral response, in conjunction with spectrum-limiting windows and filters. The former are classified as *thermal detectors* (or energy detectors), whereas the latter are classified as *quantum detectors* (or photon detectors).

## 20.4  DESIGN AND OPERATION

Radiation pyrometers are remote-temperature-sensing systems consisting of three subsystems: the optical subsystem, the detector subsystem, and the electronics subsystem. These subsystems, as well as typical pyrometer configurations, are described next. Note that additional information about detector characteristics and requirements, window characteristics, monochromators, and choppers is included in Chapter 22.

### 20.4.1  *Optics*

The primary function of pyrometer optics is to focus the incoming radiation onto the detector in the form required for proper operation of the pyrometer system. A secondary function of the optics, in most designs, is to make the target (portion of measured surface to be viewed by the pyrometer) visible to a human operator, so that the pyrometer can be pointed appropriately. This secondary function is sometimes performed by a separate sighting tube attached to the pyrometer housing; more frequently, it is performed by a beam divider in the optics which focuses a portion of the incoming energy onto a viewing port or eyepiece. The optics subsystem is the primary determinant of whether a pyrometer is of the wide-band, narrow-band, or ratio type.

*Wide-band pyrometer* optics are the simplest, since such devices as filters are generally excluded in order to sense as wide a band of wavelengths as possible (Figure 20–3(a)). Also, the detectors tend to be of the more slowly responding thermistor bolometer or thermopile types, and choppers are usually omitted. The wide-band requirements apply to any window in front of the entrance aperture as well as the surface of the beam divider that focuses the energy onto the detector.

*Narrow-band pyrometer* optics are designed to focus only energy in

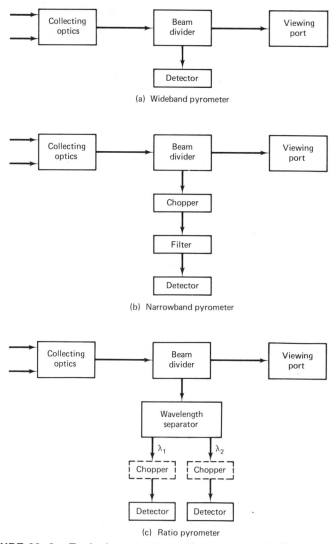

**FIGURE 20–3.** Typical pyrometer optical path block diagrams: (a) wide-band pyrometer; (b) narrow-band pyrometer; (c) ratio pyrometer.

the required narrow band of wavelengths on the detector. The detector itself is also chosen to respond best in that wavelength region. Any window used reflects this requirement. At least one narrow-band filter is placed into the optical path between the beam divider and the detector. The IR photon detectors used in such pyrometers are characterized by short time constants; hence, a chopper can be used to modulate the energy beam so that the detector output is quasi-ac, which is easier to amplify stably than is dc. A block diagram for this optical path is shown in Figure 20–3(b).

*Ratio pyrometer* optics are not only designed to pass a relatively narrow band of wavelengths; they must additionally separate energy in two extremely narrow wavelength bands which are very close to each other. The energy in each of these quasi-monochromatic bands is then focused onto a separate detector (Figure 20–3(c)). Choppers are often used in conjunction with the fast-response IR detectors.

Examples of the constituent elements of pyrometer optics are shown in Figure 20–4. The beam divider can be a *concave mirror* with a central aperture (sighting hole); in this case the detector is located between the mirror and the entrance aperture. More frequently the beam-dividing function is performed by a *beamsplitter,* typically a *dichroic mirror,* which reflects energy primarily in the IR region toward the detector but transmits energy primarily in the visible region toward an eyepiece. Most of the focusing functions are then performed by a *lens. Field stops* are used to delineate the beam and prevent any unwanted energy (e.g., scattering products) from being included in the beam. A window, whose material must be able

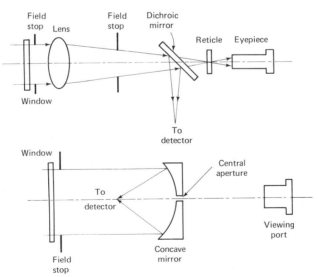

**FIGURE 20–4.** Examples of collecting optics and beam divider configurations.

to withstand ambient temperatures, which can be fairly high, is usually placed in front of the collecting lens. This window must especially be selected for spectral response characteristics (see Section 22.6.1). Most pyrometer optics are fixed in configuration, so that their focus is fixed. Some designs, however, provide for adjustable focusing.

### 20.4.2  Detectors

The wavelengths sensed by radiation pyrometers are generally in the *infrared* (*IR*) region of the electromagnetic spectrum. This region includes *near IR* (0.7 to 2 μm) and *far IR* (2 to 1000 μm). This breakdown is sometimes shown as near IR (0.7 to 1.2 μm), *middle IR* (1.2 to 7 μm), and far IR (7 to 1000 μm). However, even wide-band pyrometers usually do not respond to wavelengths significantly in excess of 20 μm, and this limit is adequate for essentially all remote temperature measurements. The choice of detector is governed primarily by spectral response requirements, and secondarily by time constant (response time) and detectivity (sensitivity). Wideband response is obtained from such thermal detectors as thermopiles (Section 22.4.6), thermistor bolometers (Section 22.4.7), and pyroelectric sensors (Section 22.4.8). When narrow-band response is required (including for ratio pyrometers), photovoltaic (Section 22.4.1) or, more commonly, photoconductive (Section 22.4.2) sensors are used. Among these, silicon, germanium, lead sulfide (PbS), and indium antimonide (InSb) detectors are probably most common in pyrometers. The response of silicon detectors extends only to about 1.1 μm; hence, such detectors are used mainly in narrow-band pyrometers when high temperatures must be sensed. The spectral response of germanium detectors extends to about 1.9 μm. PbS detectors are very often used in narrow-band and ratio pyrometers; they are usable at wavelengths up to about 2.8 μm. InSb detectors, with their response extending up to about 6.5 μm, are very useful when the temperature range to be measured is medium or low.

Semiconductor IR detectors, notably InSb, improve significantly in their sensitivity when they operate at lower temperatures. Such operation can be attained by use of a cooler; thermoelectric coolers are often used for this purpose. The mounting of a detector is very critical. It not only must be dimensionally stable over a range of temperatures, but must also provide good thermal isolation for the detector so that a minimum of the incident heat energy is lost by conduction to adjacent parts. A detector subsystem, then, includes the detector, its mounting, and any cooler that may be used.

### 20.4.3  Electronics

Pyrometer electronics can range from very simple to quite complex, depending on the type of detector used and the functions required. Very simple circuitry, consisting essentially of a reference-junction compensator (see

Section 19.6.1), has been used in conjunction with thermopile detectors. The pyrometer output is then in millivolts; it can be displayed directly on the type of millivoltmeter typically used for thermocouples, and it can be amplified as well for other display or data system uses. Thermistor bolometers require the signal-conditioning circuitry needed to convert the resistance change of the thermistor into a voltage change. The output from this circuitry is usually low enough to require amplification.

Electronic elements associated with IR detectors (radiation pyrometers using such detectors are often called ''IR radiation pyrometers'' or ''non-contact IR thermometers'') are somewhat more complex (see Figure 20–5). A *preamplifier* is usually located physically very close to the detector, which is characterized by a relatively high output impedance. When a chopper is used, it can also provide a synchronization (*sync*) signal, for example, by having the same slotted disk also interrupt a beam between a lamp and a light sensor whose output is then the sync signal. The chopper produces an ac (or pulsed dc) signal from the detector. After preamplification the signal is fed to an *amplifier*. Besides further amplifying the signal, this unit can also perform linearization, and it provides for such adjustments as *emissivity control* (gain adjustment to allow for known values of the target's emissivity) and *calibration setting* (adjustment of the output indication to a value obtained in a calibration mode). Many pyrometers include a *calibration source* in their optics, a lamp or heated black body that emits a known amount of heat energy toward the detector, which is blocked from seeing the target's radiation at that time.

The output of the amplifier is fed to a *demodulator,* which converts the ac or pulsed signal into dc suitable for display or as an input to a strip-chart recorder or a multifunction data system. Synchronous demodulation is desirable, and such demodulation can be obtained by using the sync signal derived from the chopper. A power supply converts the ac line voltage into

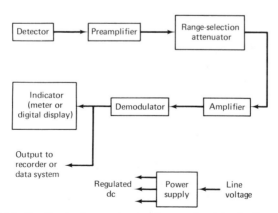

**FIGURE 20–5.** Typical pyrometer electronics block diagram (simplified).

one or more regulated dc voltages required by the electronics. Some pyrometers are battery operated rather than line-voltage supplied. Special functions such as "peak hold" are usually optional.

### 20.4.4  *Pyrometer Configurations*

Radiation pyrometers exist in a variety of configurations that make them suitable for different applications. A typical configuration is a sensing head that can operate in fairly severe environments connected by a cable to electronic components that are not exposed to those environments. Sensing heads connected to electronic components by fiber optics are gaining increasing popularity because of the increased immunity of the sensor and connecting cable to environmental effects. The electronics unit usually contains a digital display of measured temperature, as well as all provisions for adjustments and control, including the very important emissivity control. Output signals conditioned in the unit and available to ancillary equipment can be analog (dc voltage or dc current) or digital.

Several designs are portable and have their optics, detector, and electronics packaged in a single unit resembling a camera. A typical design is illustrated in Figure 20–6. This pyrometer is available in either of two ranges, 250 to 1000°C or 600 to 3000°C. It is basically a wideband pyrometer (1–20 μm). It can be used for fixed installations, mounted on a bracket or tripod, and powered through an ac adapter; or it can be used as a portable unit when the pistol-grip handle with nickel-cadmium rechargeable batteries is attached to it. The optics allow focusing on a target from any distance above 27 cm, and a built-in zoom feature permits focusing on very small targets (down to 1 mm in diameter). The built-in electronics system allows four display options: continuous, peak hold (display the highest reading of a temperature distribution), valley hold (display the lowest reading), and final hold (display the last of a series of readings). The ac adapter also furnishes a compatible digital output to ancillary computers or data processing systems.

A different design is shown in Figure 20–7. This unit is basically a wide-band radiation pyrometer which employs a thermopile detector and provides a thermocouple-type output in millivolts to a meter or strip-chart recorder. The wavelengths band and, hence, the usable temperature range (0.3 to 10 μm and approximately 250 to 1750°C, respectively) are established primarily by the calcium fluoride window. Related designs use fused silica and optical crown glass windows for lower wavelengths and high temperature ranges. This design has a 98% response time of 2 s. Related designs can be equipped with many of the optional provisions described (e.g., water cooling, various mounts). The design illustrated is intended for applications where installation space is limited.

A fiber optics pyrometer is shown in Figure 20–8. The sensor rod consists of a black-body cavity at the tip of a metal-sheathed single-crystal

**FIGURE 20-6.** Focusing radiation pyrometer with integral electronics. (Courtesy of Mikron Instrument Co., Inc.)

sapphire rod. It is optically coupled to a fiber optics cable whose other end is connected to the detector unit. The latter contains two-channel optics and detectors, thermistor-controlled ambient-temperature compensation, amplifiers, and multiplexers. The associated signal processor unit provides display and control functions as well as a digital interface with ancillary equipment. Different sensor-rod tip configurations allow remote temperature sensing in gases, liquids, or solids. Selection of either one optical path or both optical paths enables the instrument to be used as either a narrow-band or a two-color pyrometer. The signal processor unit provides numerous display, sampling rate, and output (to ancillary equipment) control and cali-

**FIGURE 20-7.** Miniature radiation pyrometer with thermopile detector (6 cm dia. × 7 cm long). (Courtesy of Honeywell Process Control Div.)

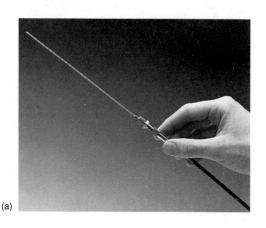

(a)

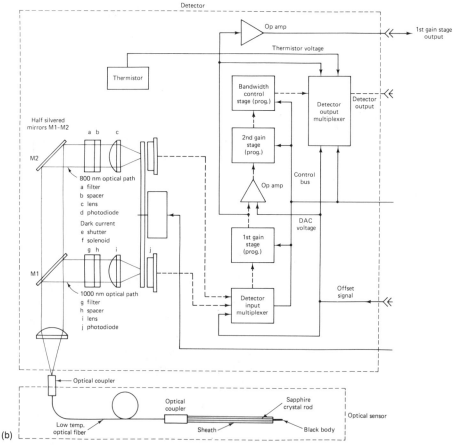

(b)

**FIGURE 20–8.** Multiple-mode optical fiber radiation pyrometer: (a) sensor rod; (b) pyrometer system block diagram (Courtesy of Accufiber, Inc.)

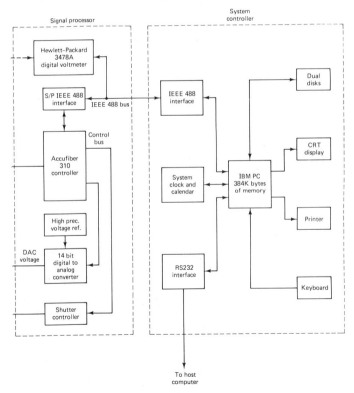

**FIGURE 20–8 (b).** *(Continued)*

bration functions in conjunction with a personal-type computer. The measuring range is 500 to 2000°C.

## 20.5 TRANSDUCER CHARACTERISTICS

A relatively small number of essential characteristics establish the primary criteria for selection and use of a radiation pyrometer. The *temperature (measuring) range* usually shows the upper and lower limits of temperature that will be sensed and displayed by a pyrometer. Ranges extend up to about 4000°C (in a few instances even higher) but, for most available designs, tend to be under 2800°C. The lower temperature limit is never zero; typically, it is between 5 and 20% of the upper limit. Many units have range selector switches and can provide full-scale displays of portions of the full range. The *time constant,* often stated as *response time* instead (usually meaning 98% response time), is dependent primarily on the type of detector used; typical values range from a fraction of a second to about 2 s. The type of detector used and the spectral response of the pyrometer (including optics) are very often stated in manufacturers' literature. Unless "wide-band" or "narrow-band" is stated elsewhere, this categorization can be obtained by looking at the callout for spectral response.

Since the proper operation of radiation pyrometers, particularly the wide-band and narrow-band types, requires that the target fill the field of view, the *minimum target diameter* (minimum diameter of the measured object) must be known. This minimum diameter is, of course, a function of the distance between the pyrometer lens and the object and is given by the characteristics of the optics used in the pyrometer. This optical characteristic is stated in various ways. In many specifications it is shown as the *resolution factor F*, which is equal to the ratio of the separation distance $D$ to the minimum acceptable target diameter $d$, or $F = D/d$. $F$ numbers can be anywhere between 3 and 500. Instruments having a relatively low $F$ number can be referred to as *wide angle*, whereas those having a relatively high $F$ number can be called *narrow angle*; however, there is no standard for the border line between those two categories. The resolution factor can also be shown as a *distance-to-size ratio* and is then expressed as, for example, "60:1" instead of the (equivalent) "$F = 60$." This ratio can also be expressed as an angle. If so, the corresponding resolution factor is equal to 60 divided by the angle (in degrees); for example, an angle of 4° corresponds to a resolution factor of 15. Some specifications simply show a table of separation distance (distance between lens and target) vs. minimum target diameter for a given pyrometer design. A statement of (minimum) *operating distance* is sometimes added.

*Accuracy* is shown in percent of full-scale output (usually meaning full-scale indication on an associated meter), or in percent of reading, or in units of (indicated) temperature; such tolerances are applicable when the instrument has been properly calibrated. Additionally, *repeatability* is often specified. When one or more outputs are provided (besides the usually provided meter), the characteristics and ranges of these outputs are stated. When special output holding features are included or optionally available, they are described. They may include "peak hold" (the peak reading over the period a button is depressed will be displayed), "valley hold" (same mechanization, but for the lowest reading), and "meter hold" or "reading hold" (any reading is held while a button is depressed). Other important characteristics include the *ambient temperature range* (sometimes accompanied by statements of window material used); the range of available *emissivity compensation* ("emissivity control range"); mass, size, and power consumption (sometimes type of power source); available options; and all "standard" as well as "optional" provisions for control functions.

## 20.6 APPLICATION NOTES

Radiation pyrometers are used primarily for high-temperature measurements, especially where the measured temperature or the environment at the point of measurement, or both, prevent any type of contacting high-temperature sensors, such as thermocouples, from being employed. The measurement of the temperature of molten metals and glasses is one of the most common applications. Similar applications are found in furnaces, fluid-bed reactors, and heat-treating equipment. Radiation pyrometers are also used for hot-gas measurements, such as those associated with gas turbines and in combustion research. There are many additional applications related to research as well as processing of a large variety of high-temperature materials.

Some radiation pyrometer designs are intended primarily for the remote sensing of moderate temperatures where contacting measurements are impractical. Examples are processes in the textile and paper industries involving moving webs whose temperature needs to be determined, and temperature measurements of bulk materials moving on conveyor belts.

# 21

# *Heat Flux Sensors*

## 21.1  BASIC DEFINITIONS

*Heat flux* is the amount of heat transferred across a surface of unit area per unit time. Section 19.1.1 sets out other heat-related definitions.

## 21.2  UNITS OF MEASUREMENT

### 21.2.1  *Units and Conversion Factors*

Heat flux is the time rate of transfer of heat energy. Energy is expressed in joules ($J$) and time in seconds ($s$); hence, heat flux is expressed in $J/s$ or watts ($W$), since the watt, the unit of power, is equal to J/s. Other units (non-SI units) have been used to express heat power as well, and their conversion factors are $1\ W = 1\ J/s = 9.498 \times 10^{-4}\ Btu/s = 10^7\ ergs/s = 0.2389\ g \cdot cal/s$. Conversion factors for the *Btu* (*British thermal unit*) are $1\ Btu = 8.139 \times 10^{-5}\ W \cdot s = 1055\ J = 252\ g \cdot cal = 1.055 \times 10^{10}\ ergs$.

Because heat flux is measured as heat flow to a surface, it is also expressed in units of (radiant, conductive, or convective) *heat flux per unit area*, that is, in $W/m^2$. Other (non-SI) units are still in popular use in the United States, and the most commonly used unit has been the $Btu/ft^2 \cdot s$ (Btu per square foot per second). Conversion factors for this unit are *1 Btu/*

$ft^2 \cdot s = 3600 \ Btu/ft^2 \cdot h = 0.00695 \ Btu/in^2 \cdot s = 0.271 \ cal/cm^2 \cdot s = 975$ $cal/cm^2 \cdot h = 1050 \ W/ft^2 = 7.3 \ W/in^2 = 1.136 \ W/cm^2 = 1.136 \times 10^4 \ W/m^2$.

Units for heat flux as well as the other heat-related quantities defined in Section 19.1.1 are tabulated as follows, together with their conversion factors:

| Quantity | U.S. Customary Unit[a] | Multiply by | to get SI Unit |
|---|---|---|---|
| Heat flux | Btu/ft$^2$·s | $1.136 \times 10^4$ | J/m$^2$·s |
| Heat flux | cal/cm$^2$·s | $4.184 \times 10^4$ | W/m$^2$ |
| Heat energy | Btu | $1.055 \times 10^3$ | J |
| Heat quantity | cal | 4.1868 | J |
| Thermal conductivity | Btu·in/s·ft$^2$·°F | $5.192 \times 10^2$ | W/m·K |
| Heat power | Btu/h | 0.293 | W |
| Specific heat | Btu/lb$_m$·°F | $4.1868 \times 10^3$ | J/kg·K |
| Heat-transfer coefficient | Btu/s·ft$^2$·°F | $2.044 \times 10^4$ | W/m$^2$·K |
| Thermal resistance | °F·h·ft$^2$/Btu | 0.1761 | K·m$^2$/W |
| Thermal diffusivity | ft$^2$/h | $2.58 \times 10^{-5}$ | m$^2$/s |

[a] Conversion factors for *British thermal units* (*Btu*) and *calories* (*cal*) are based on the International Tables; conversion factors for *thermochemical Btu* and *cal* differ from those shown by about 0.07%.

### 21.2.2 The Solar Constant

Since radiant heat flux measurements include measurements of the heat energy received from the sun (and such measurements have emerged from the theoretical to the practical in the many new solar energy applications), another unit needs to be introduced: the *solar constant*, defined as the energy per unit time (power) received from the sun per unit area, at the average sun—earth distance, and in the absence of the earth's atmosphere. The solar constant had been given the value of 1398 W/m$^2$ (frequently rounded off to 1400); however, more precise measurements were able to be made during the past decade, and the currently accepted value (as formally reported by M. P. Thekaekara, NASA/Goddard Space Flight Center, United States) is *1353 W/m$^2$. Hence, 1 solar constant = 1353 W/m$^2$ = 0.1353 W/cm$^2$ = 0.119 Btu/ft$^2$·s = 428.5 Btu/ft$^2$·h = 1.945 cal/cm$^2$·min = 125.66 W/ft$^2$.*

Note that this value is based on energy distribution with wavelengths ranging from 0.115 to 1000 μm, and the value for the average sun–earth distance (1353 W/m$^2$) is based on the value obtained at the earth's perihelion (closest distance to the sun), 1399 W/m$^2$, and the value obtained at aphelion (farthest distance from the sun), 1309 W/m$^2$. When measured on the earth's surface, this value is modified by the effects of the earth's atmosphere.

## 21.3   SENSING METHODS

Heat flux is measured as heat flow to a surface. For the purpose of the measurement, the surface is the active surface (of two surfaces) of the sensor.

Heat flux sensors can be grouped into three categories: *radiometers* sense radiant heat flux (and related instruments such as the *pyrheliometer* sense specific types of radiant heat flux, as explained in Section 21.4.2); *calorimeters* sense the combined radiant and convective heat flux; and *surface heat-flow sensors* sense conducted, convected, or radiated heat flux, or a combination of these, to or from a specific surface.

## 21.4   DESIGN, OPERATION, AND APPLICATIONS

### 21.4.1   *Calorimeters*

A calorimeter, as defined here, measures the heat flow into a system from the surroundings. Design variations of three basic types of calorimeter—the foil calorimeter, the slug calorimeter, and the thermopile calorimeter—are available.

A *foil calorimeter* ("membrane calorimeter," "Gardon gage") consists of a thin disk of one type of metal, bonded around its rim to a heat sink of a different type of metal. The two metals are so chosen that their junction, around the rim of the foil disk, forms a thermocouple junction. Since copper makes a good heat sink, and since Constantan with copper forms a Type T thermocouple, Constantan is commonly used for the foil disk and copper is used for the heat sink, which forms the body of the sensor. A thin copper wire is then welded to the exact center of the Constantan foil, forming a second thermocouple junction (Figure 21–1). The two junctions act as differential thermocouples, since the foil disk, when exposed to heat flux, is always hotter at the center than at the rim, where it is joined to the heat sink. The exterior surface of the foil disk is blackened, using a high-absorptivity coating, so that the calorimeter responds well to radiant heat flux as well as to convective heat flux. Heat absorbed by the foil disk is trans-

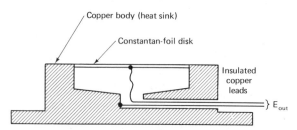

**FIGURE 21–1.**   Foil calorimeter.

ferred radially to the heat sink. The millivolt output of the sensor is proportional to incident heat flux. At constant heating rates the output provides a steady-state indication of the heat flux.

The *slug calorimeter* ("slope calorimeter") employs a circular metallic body ("slug") as thermal mass whose exterior surface is covered with a high-absorptivity (high-emissivity) coating (black). The mass is thermally insulated from its support in the housing of the sensor (Figure 21–2) A thin-wire thermocouple (or other very small temperature sensor) is attached to the center of the interior surface of the mass. Most of the incident heat flux will be retained as heat in the thermal mass, causing the temperature of the mass to rise. At a constant heating rate the slope of output voltage vs. time is indicative of heat flux; that is, the heat flux can be determined from the temperature history of the "slug." A second sensing junction can be placed at the inside of the housing, near the thermal barrier that insulates the thermal mass. The two junctions can then be connected as a differential thermocouple so that the output is independent of the heat rise in the housing.

Two disadvantages of the slug calorimeter are that it cannot be used for continuous measurements and that it tends to lose accuracy during a measurement, due to heat losses as the temperature rises.

The *thermopile calorimeter* consists basically of an insulating wafer around which are wound a series of thermocouples in such a manner that consecutive thermoelectric junctions fall on opposite sides of the wafer. Hence, the junctions on the upper surface of the wafer can all be considered "sensing junctions," whereas the junctions on the bottom side of the wafer can all be considered "reference junctions." This assembly is bonded to a heat sink to assure heat flow through the wafer/thermopile assembly. Heat is received on the exposed (upper) surface of the wafer and conducted through it to the heat sink. A temperature drop across the wafer is thus developed and is sensed by each of the junctions of the differential thermopile. As is typical for such thermopiles, the voltages produced by each of the junctions are additive, thereby amplifying the signal from one junction proportionally to the total number of junctions. The temperature drop across the wafer, and thus the output signal of the differential thermopile, is directly proportional to the heating rate. The appearance of the wafer with its thermopiles is somewhat similar to the thermopile shown in Figure 21–4, except

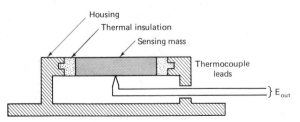

**FIGURE 21–2.** Slug calorimeter.

that only one set of junctions would be visible. The other set of junctions would be on the underside of the element; additionally, the junctions would be no bigger in size than the leads, since wire, rather than deposited metal, is used in the thermopile calorimeter.

Foil and thermopile calorimeters are used primarily for the measurement of transient heating rates or of sustained heating rates. Slug calorimeters are used mainly for the measurement of total heat quantity over a short period of time. A variety of calorimeter designs have been developed for different applications and different mounting methods; some of these are illustrated in Figure 21–3. Foil and thermopile calorimeters are often pro-

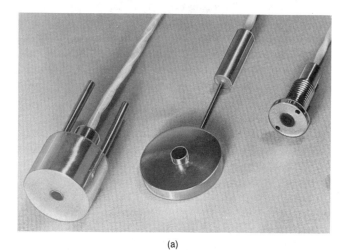

(a)

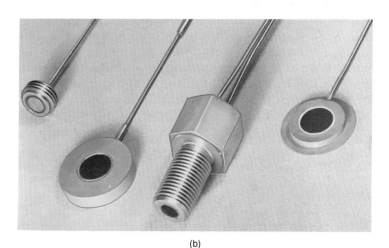

(b)

**FIGURE 21–3.** Typical calorimeter designs: (a) foil calorimeters; (b) slug calorimeters. (Courtesy of Hy-Cal Engineering, a unit of General Signal.)

vided with means for water cooling (e.g., the sensor at the left of Figure 21–3(a)). In some applications it is necessary to obtain a measure of the pure radiant constituent of heat flux; two calorimeters can then be used, one with the usual black high-absorptivity surface, the second with a polished gold (low-absorptivity) surface. When the two sensors are used simultaneously and placed close together, the "gold" calorimeter will provide an output of just the convective (and conductive, if any) heat flux, and the output of the "black" calorimeter can be corrected by subtracting this amount from its output. A number of different calorimeter designs have also been developed for special applications, and some designs have been tailored to utilize the advantages offered by miniature thermopiles in providing a higher output for lower values of heat flux.

### 21.4.2 Radiometers

Radiometers are related to radiation pyrometers (see Chapter 20) in that they respond to incident radiant heat flux; they differ from radiation pyrometers, however, not only by being calibrated in terms of heat flux (in W/cm$^2$ or equivalent units) but mainly by having a generally much wider viewing angle, typically 45 to 90° (although designs with other viewing angles exist) as compared to the angles ranging between a fraction of one degree to a few degrees typical for radiation pyrometers. This wider angle simplifies their design since collimating optics are not needed and viewing ports (requiring a beam splitter) are rarely used.

The simplest form of radiometer is a calorimeter with a window sealed over the sensing surface so as to isolate it from convective heat flux. Only when used in a vacuum (where there is no heat convection) can a windowless radiometer be used. The window tends to limit the spectral response; the thermal sensors used in radiometers, typically thermopiles but sometimes resistive sensors, have an inherently wide-band spectral response. Differential thermopiles, with their sensing junctions exposed to the incident heat radiation and their reference junctions shielded from it, have been used in radiometers. The example shown in Figure 21–4 provides a sensing area, along the center of its mount, which is 1.5 mm wide and 15 mm high for the 12-junction thermopile illustrated. Circular differential thermopiles are used similarly, when the (small) aperture is circular. The radiation–sensitive area is blackened with a high-absorptivity (low-reflectance) coating. Other thermal detectors include wire-wound thermopiles, pyroelectric detectors, and bolometers. Wide-band window materials used for radiometers include (with usable upper limit of transmittance shown in parentheses) quartz (4.5 μm), synthetic sapphire (6.5 μm), calcium fluoride (10 μm), silicon (about 200 μm, but nonuniform above 10 μm), and cesium iodide (over 50 μm); the latter is quite hygroscopic and should not be exposed to humidity above 45% RH.

Radiometers are often equipped with gas purge provisions, used to

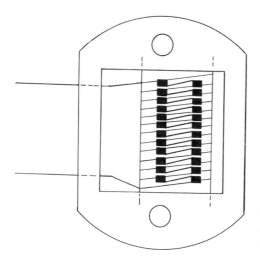

**FIGURE 21-4.** Thermopile sensing element for radiometer use. (Courtesy of The Eppley Laboratory, Inc.)

prevent window contamination, and water-cooling provisions for use in high-temperature environments. The unit shown in Figure 21-5 has such provisions; its recessed aperture has antireflection serrations, the diameter of the housing is 3.8 cm, and it uses a synthetic sapphire window and a thermoelectric detector.

Radiometers are used wherever radiant or radiated energy is to be measured as such and read out in terms of energy received, or received per unit time. Applications range from measuring the radiant energy from heating lamps, engine performance analyses, and rocket exhaust plume studies to high-temperature research and airborne radiometric measurements. An important application of radiometers has evolved with the increasing use of lasers. Laser radiometers are used to measure laser power and provide a direct readout in terms of energy, in joules. A wavelength band of about 0.3

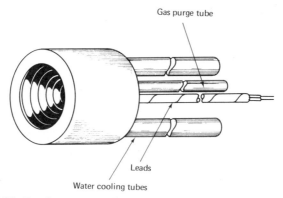

Gas purge tube

Leads

Water cooling tubes

**FIGURE 21-5.** Compact radiometer with air-purge and water-cooling provisions. (Courtesy of Hy-Cal Engineering, a unit of General Signal.)

to 1.1 μm handles most types of lasers and allows the use of fused-quartz windows in such radiometers; other windows can often be supplied when far-IR laser power (e.g., $CO_2$, 10 μm) needs to be measured.

Specialized types of radiometers are used to measure the energy due to, or related to, solar thermal radiation received on earth. A *pyrheliometer* measures direct solar radiation intensity at the surface of the earth and typically provides an output in solar constants. Pyrheliometers can be mounted on a servo-controlled solar tracker to measure radiation at normal incidence continuously. The *Angstrom pyrheliometer* has a detector composed of two blackened magnesium strips, one of which is shaded from the sun and heated by an electrical current while the other is exposed to the direct solar beam. The current through the heater is adjusted until the temperatures of the two strips are balanced. The solar intensity can then be determined at the null condition, using the instrument constant. This type of instrument is usable in good weather with steady sky conditions. The *absolute cavity pyrheliometer* absorbs radiation on a high-absorptance (0.997 or greater) conical receiver, uses a wire-wound toroidal thermopile detector, and determines heat flux by electrical substitution; this type of instrument can be self-calibrating.

A *pyranometer* (see Figure 21–6) measures the "global radiation," the combination of direct solar radiation and diffuse sky radiation incident on a horizontal surface. An important application of the pyranometer is in the evaluation of solar panels. Two designs of a pyranometer have been developed, the blackened-surface and the black-and-white receiver. The blackened-surface type has two domes (hemispheres); the inner one blocks the infrared radiation from the outer one. The hemispheres can be of clear optical glass or fused quartz, depending on spectral response requirements. A thermopile detector is used, with the hot junctions under the black receiver

**FIGURE 21–6.** Pyranometer. (Courtesy of The Eppley Laboratory, Inc.)

**FIGURE 21–7.** Surface heat-flow sensors. (Courtesy of Hy-Cal Engineering, a unit of General Signal.)

responding to the incident radiation and the cold junctions facing downward toward the interior of the instrument. In the black-and-white type the output signal is produced by a differential thermopile with the hot-junction receivers blackened and the cold-junction receivers whitened. A single optical-glass hemisphere provides uniform transmittance from 0.285 to 2.80 µm and protects the instrument from the weather. The instrument base is provided with a circular spirit level and adjustable leveling screws.

A *pyrgeometer* is a wide-band, far-IR radiometer used to measure (separately) incoming and outgoing unidirectional terrestrial radiation (measurements needed, e.g., for thermal balance studies). A typical design uses a silicon window with an interference filter vacuum-deposited on its inside surface to make the instrument solar blind and provide a spectral response from about 4 to 50 µm in daylight as well as darkness. Radiation emitted by the thermopile detector is automatically compensated for.

### 21.4.3  Surface Heat-Flow Sensors

The category of surface heat-flow sensors comprises thin surface-mounted sensors, typically employing a differential thermocouple or thermopile to sense the difference in temperature between their two surfaces due to heat conduction through the sensor. The heat flow measured by such sensors can be the flow through a wall across which a temperature gradient exists, or it can be the heat flow to or from any type of surface (e.g., walls, pipes, tanks). The heat flow sensed can be due to conduction, radiation, or convection, or combinations of these. The outside surface (when mounted) is blackened to provide good emittance when heat flow from a surface is measured, and to provide good absorptance when heat flow into a surface is measured (see Figure 21–7). The units can be mounted with adhesive or pressure-sensitive tape; some models are provided with mounting holes. The thickness of the sensors ranges between 0.7 and 1.5 mm, and up to 3 mm for some designs. Most units are thin enough to be able to be attached to a curved surface; precurved designs are also available.

# 22

# *Optical Detectors*

## 22.1 BASIC DEFINITIONS

*Optical detectors* (*photodetectors*, "photoelectric cells") are transducers which respond, with a usable electrical output, to radiant energy in the optical region of the electromagnetic spectrum (in the *light* spectrum).

Light is a form of radiant energy; it is electromagnetic radiation in that portion of the spectrum lying between 10 nanometers (nm) and 1 millimeter (mm) (1 mm = 1000 μm = $10^6$ nm). This definition of the light spectrum is used throughout this chapter. By strict definition, however, only *visible radiation* (visible light), whose spectrum extends from 380 to 780 nm, can be considered "light." The band of wavelengths between 10 and 380 nm is called *ultraviolet (UV) radiation* ("ultraviolet light"). The band of wavelengths between 780 and $10^6$ nm is called *infrared (IR) radiation* ("infrared light"). Within the IR band, the portion between 780 nm and 3 μm (3000 nm) is sometimes called *near IR,* and when it is, the band between 3 μm and 1000 μm is called *far IR.*

Most of the IR band overlaps the *heat radiation* band of electromagnetic radiation. Optical quantities can be stated in terms of either *visual* or *nonvisual* (thermally radiative) magnitudes, and optical detectors can be divided into two major groups: *quantum detectors* (photon sensors) and *thermal detectors* (thermal radiation sensors).

Spectral characteristics of light are most commonly shown in terms of wavelength. However, they can also be shown in terms of frequency, wave number, or photon energy. The interrelationship of these quantities, and their relationship to black-body temperature, are illustrated in Figure 22–1.

*Frequency* of light is related to wavelength by the speed of light in vacuum (2.99 793 × $10^{10}$ cm/s, normally rounded off to 3 × $10^{10}$ cm/s):

$$\nu = \frac{3 \times 10^{10}}{\lambda}$$

where $\nu$ = frequency of light, Hz
$\lambda$ = wavelength of light, cm

*Wave number* is the reciprocal of wavelength (in cm); hence, wave number is expressed in $cm^{-1}$.

*Photon energy* is related to the frequency of light by *Planck's constant* $h$ according to the equation

$$\mathscr{E}_p = h\nu$$

where $\mathscr{E}_p$ = photon energy, J (*Note:* 1 electron volt = 1.6022 × $10^{-19}$ joule)
$h$ = Planck's constant, 6.626 196 × $10^{-34}$ J·s
$\nu$ = frequency, Hz

*Black-body temperature,* the temperature at which a black body radiates energy such that the radiation has greatest intensity at a certain wavelength, is related to wavelength by *Wien's displacement law,*

$$\lambda_m T = 0.2897 \text{ cm·K (a constant)}$$

where $\lambda_m$ = wavelength at which radiant energy density is maximum, cm
$T$ = absolute temperature of black-body radiator, K

*Luminous flux* is the time rate of flow of light; the equivalent nonvisual quantity is *radiant flux.*

*Luminous intensity* is the luminous flux (emitted by a point source) per unit solid angle; the equivalent nonvisual quantity is *radiant intensity,* the radiant flux per unit solid angle.

*Illuminance* (illumination) is the luminous flux per unit area (of a uniformly illuminated surface on which this flux is incident); the equivalent nonvisual quantity is *radiant flux per unit area.*

*Luminance* ("brightness") is the luminous intensity of a (light-emitting) surface, in a given direction, per unit of (projected) area of the surface, as viewed from the given direction; the equivalent nonvisual quantity is *radiance,* the radiant flux per unit area, per unit solid angle.

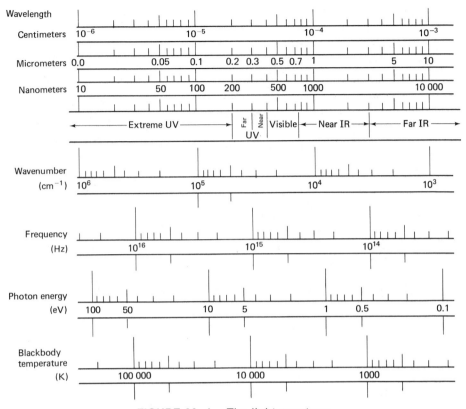

**FIGURE 22-1.**   The light spectrum.

*Luminosity* is the ratio of the luminous flux to the corresponding radiant flux.

A *photon* is the quantum of an electromagnetic field.

*Color* is a characteristic of visible light associated with its wavelength (see Figure 22–2) and also with purity and luminance.

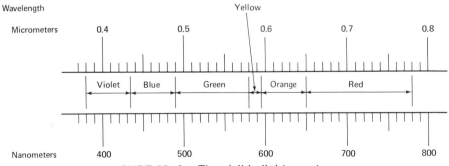

**FIGURE 22-2.**   The visible-light spectrum.

## 22.2  UNITS OF MEASUREMENT

*Luminous intensity (of a point source).* The unit of luminous intensity is the *candela (cd)*, one of the base units of the SI. It is defined as the luminous intensity, in the perpendicular direction, of a surface of 1/600 000 m² (1/60 cm²) of a black body at the temperature of solidifying (freezing) platinum under a pressure of 101 325 Pa. A proposed redefinition of the candela is the luminous intensity in a given direction of a source emitting a mono-chromatic radiation of frequency $540 \times 10^{12}$ Hz and whose radiant intensity in that direction is 1/683 W/sr.

An older unit, still popular, is the *candlepower (cp)*, defined as the luminous flux of one *candle* (see luminous flux, next) when viewed in the horizontal plane. There is a one-to-one relationship between the old and new units (1 cp = 1 cd).

*Luminous Flux.* The unit of luminous flux is the *lumen (lm)*, defined as the flux emitted within a solid angle of one steradian by a point source having a uniform intensity of 1 candela (*lm = cd·sr*). An older unit was the *candle*, the luminous flux from a spermaceti candle burning at the rate of 120 grains per hour.

*Illumination (of an area).* The unit of illuminance is the *lux (lx)*, the illu-mination of one lumen per square meter (*1 lx = 1 lm/m²*). An older unit was the *footcandle (fc)*, the illumination at a spherical distance of 1 foot from a 1-candle source (*1 fc = 10.76 lx*).

*Brightness (Luminance).* The unit of luminance is the *candela per square meter (cd/m²)* (of light-emitting area). An older unit is the *footlambert (fL)*, the luminance equal to 1/(4π) candle per square foot (*1 fL = 3.426 cd/m²*). Among other units that have been used are the *lambert (L)*, the luminance equal to 1/π candle per square centimeter (*1 L = 3183 cd/m²*) and the *stilb (sb)*, the brightness equal to 1 candle per cm² (*1 sb = 10 000 cd/m²*).

*Wavelength.* Wavelength in the UV and visible portions of the spectrum is expressed in *nanometers (nm)*, whereas it is usually expressed in *microm-eters (μm)* in the IR region. The micrometer has previously often been re-ferred to as a "micron (μ)." Another older unit, whose use is now dis-couraged, was the ångstrom (Å, or just A), which is equal to 0.1 nm (*10 Å = 1 nm*).

*Other Quantities.* Luminosity is expressed in *lumens per watt (lm/W)*.
Black-body temperature is expressed in *kelvins (K)*.
Photon energy is usually expressed in *electron volts (eV)* or *ergs;* the SI unit is the *joule (J)* (*1 eV = 1.602 × 10⁻¹⁹ J; 1 erg = 10⁻⁷ J*).
Wave number is expressed in *cm⁻¹*.

*Nonvisual magnitudes.* Radiant flux is expressed in *watts*, radiant intensity in *watts per steradian,* radiant flux per unit area in *watts per square meter,* and radiance in *watts per square meter per steradian* ($W\ m^{-2}\ sr^{-1}$).

## 22.3 SENSING METHODS

Optical detectors can be classified into two general categories: quantum detectors (or photon detectors) and thermal detectors. *Photon detectors* depend on effects produced when quanta of incident radiation (photons) react with electrons in a sensor material. *Thermal detectors* respond to total incident radiant energy and they are used primarily for IR sensing. Photon detectors employ photovoltaic, photoconductive, photoconductive-junction, photoemissive, or photoelectromagnetic transduction. Thermal detectors use thermoelectric, bolometric, or pyroelectric transduction methods. These transduction methods are illustrated in Figures 22–3 and 22–4 (where shown, $e$ = electrons, $hv$ = photons or radiant energy).

### 22.3.1 Photovoltaic Detection

Photovoltaic detectors are self-generating; that is, they require no external excitation power. Their output signal (voltage, current, or power) is a function of the illumination of a junction between two dissimilar materials (Figure 22–3(a)). Several types of pairs of materials exhibit the photovoltaic effect, such as iron/selenium (as in the classic selenium cell, the earliest "photocell") and copper/copper oxide. Modern photovoltaic detectors are made from semiconductor materials. Most of these are homojunction detectors (as explained in Figure 22–3(a)). Materials such as silicon, germanium, indium arsenide, or indium antimonide are made dissimilar by introducing different types of impurities (dopants) into opposite ends of the material. This makes one portion of the material *p-type* and the other *n-type*, and the junction between them, the *p-n junction*, then becomes the potential barrier. The *p on n* types are most common, with the *p layer* exposed to the incoming photons. An example of a heterojunction detector is the *n on p* type; it uses *n-type* glass deposited on *p-type* single-crystal silicon.

### 22.3.2 Photoconductive Detection

Photoconductive detectors (Figure 22–3(b)) are made of semiconducting materials which reduce their resistance in response to incident illumination. The material is contained between two conductive electrodes to which connecting leads can be attached. The change in conductance results from a change in the number of charge carriers created by the absorption of incident photon energy. Polycrystalline films, such as lead sulfide and lead selenide, as well as bulk single-crystal materials (doped germanium and silicon) are

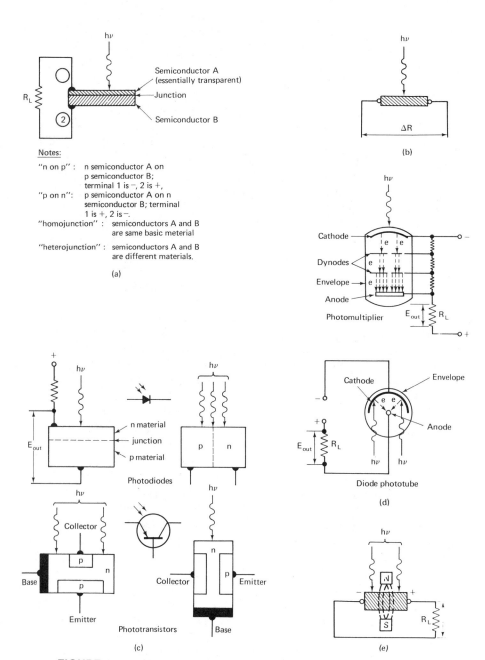

**FIGURE 22–3.** Basic light-sensing methods (photon detection): (a) photovoltaic; (b) photoconductive: (c) photoconductive-junction; (d) photoemissive; (e) photoelectromagnetic.

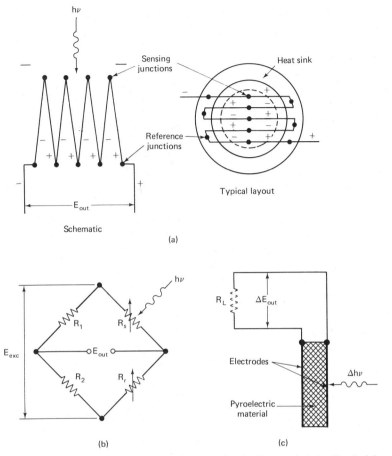

**FIGURE 22–4.** Basic light-sensing methods (thermal detection): (a) thermoelectric (thermopile); (b) bolometric; (c) pyroelectric.

used as photoconductors. The cadmium sulfide (CdS) photoconductive detector has often been used in cameras, for automatic exposure control. Some materials—for example, mercury cadmium telluride—are used in the photoconductive as well as the photovoltaic mode.

### 22.3.3  Photoconductive-Junction Detection

In the version of photoconductive detection called photoconductive-junction detection, the resistance across a junction between $p$ and $n$ semiconductor material changes as a function of incident light. The junction photocurrent (the leakage current varied by electron-hole pairs generated at or near the *depletion region*) increases with increasing incident photon flux. This principle is employed in *photodiodes* as well as *phototransistors* (Figure 22–3(c)).

### 22.3.4   Photoemissive Detection

Photoemissive detectors emit electrons from a cathode when photons impinge on it (Figure 22–3(d)). The electrons are ejected from the cathode surface when the energy of the radiation quanta is greater than the work function of the cathode material. This effect is utilized in the diode phototube (either evacuated or gas filled) as well as in the photomultiplier tube. In the diode phototube some of the electrons are collected by an anode which is at a positive potential with respect to the cathode. This causes current flow which can be used to produce an output voltage across a load resistor in series with the anode. In the photomultiplier tube additional electrodes (dynodes), at sequentially higher positive potential, are located between the cathode and the anode so as to amplify the electron current by means of secondary emission from the dynodes.

### 22.3.5   Photoelectromagnetic Detection

Photoelectromagnetic detection is a specialized detection method that is effected in a semiconductor that is acted upon by an external magnetic field (see Figure 22–3(e)). When photons are absorbed near the front surface of the semiconductor, the resulting excess of carriers at that surface and their relative absence at the rear surface cause a diffusion of carriers toward the rear. The force due to the application of a transverse magnetic field then deflects the holes toward one end of the semiconductor and the electrons toward the other end, thus causing an emf to be developed between the two end terminals.

### 22.3.6   Thermoelectric Detection

The thermoelectric (Seebeck) effect is explained in detail in Section 19.3.1. This effect is used in miniature multijunction thermocouple devices called *thermopiles*. The thermopiles, which consist of a number of thermocouples connected in series, produce an output voltage when the temperature of their sensing junctions is higher than the temperature of their reference junctions. Thermopiles used for radiant-flux sensing have their reference junctions in thermal contact with a heat sink, whereas their sensing junctions are blackened (to absorb heat radiation) and thermally isolated from the heat sink (Figure 22–4(a)). The reference ("cold") junctions remain relatively stable in temperature; they are also shielded so as not to receive any of the incident radiant flux. The sensing junctions are heated by the radiant flux. If the heat-sink temperature is known, the temperature difference between the reference and sensing ("hot") junctions, as indicated by the thermopile output voltage, is a measure of the incident radiant flux.

### 22.3.7 Bolometric Detection

Bolometers used for radiant-flux sensing generally consist of a pair of matched thermistors (or other temperature-sensitive resistive material such as silicon, indium antimonide, or single-crystal gallium-doped germanium) connected in a half-bridge or full-bridge circuit. One of the thermistors is blackened and mounted in such a manner that it senses radiant flux, while the second one is isolated from radiant flux and responds only to the temperature of the heat sink. A typical bolometer circuit is shown in Figure 22–4(b), where $R_s$ is the radiation-sensing thermistor, $R_r$ is the reference thermistor shielded from radiation, and $R_1$ and $R_2$ are a pair of matched, stable resistors, usually located some distance away from where thermal radiation is being sensed. When an excitation voltage $E_{exc}$ is applied to the circuit, the output voltage will be proportional to the difference in resistance between $R_s$ and $R_e$, which, in turn, is a measure of incident radiant flux. If this were a half-bridge circuit, $E_{exc}$, $R_1$, and $R_2$ would be replaced by two excitation voltages of exactly the same amplitude but opposite polarity, inserted in the place of $R_1$ and $R_2$.

### 22.3.8 Pyroelectric Detection

Pyroelectric detectors are composed of a ferromagnetic crystal material (e.g., triglycene sulfate, $LiTaO_3$) between two electrodes. The crystal exhibits a spontaneous polarization (electric charge concentration) which is temperature dependent. Changes in incident radiant flux, absorbed by the crystal, cause a change in the crystal temperature, resulting in a change in the potential difference (voltage) across the electrodes. This voltage is later neutralized by current flow through the internal leakage resistance and the external load resistance. Figure 22–4(c) illustrates the pyroelectric sensing method. Note that pyroelectric sensors are basically capacitive in nature.

### 22.3.9 Pressure-Actuated Photoelectric Detection

Pressure-actuated photoelectric detection is a transduction method that is employed in the *Golay-type IR detector* (*Golay cell*). A gas of low thermal conductivity (e.g., xenon) is enclosed in a cylinder capped by a blackened membrane on one end and a mirror-coated diaphragm at the other end. IR radiation incident upon the blackened membrane causes the gas to expand and the mirror-surfaced diaphragm to deform. The diaphragm is positioned in the optical path between a light source and a light sensor in such a manner that deformation of the diaphragm causes the light reflected by the mirrored surface toward the light sensor to increase with increasing pressure. The output of the light sensor is then proportional to incident IR radiation.

## 22.4 DESIGN AND OPERATION

### 22.4.1 *Photovoltaic Optical Detectors*

Silicon (Si) detectors, which are closely related to the Si solar cell used for solar-energy conversion, are the most commonly used photovoltaic detectors when good spectral response in the visible-light region is needed. Photovoltaic detectors made of other materials, notably indium arsenide, indium antimonide, and mercury cadmium telluride, are designed for use in the infrared region.

The *silicon photovoltaic detector* ("silicon cell") uses a junction between *p*-type silicon and *n*-type silicon to produce an output (current or voltage) proportional to incident illumination. Most of these detectors are of the *p on n* type. A typical design (see Figure 22–5) consists of an arsenic-doped, thin (about 0.5 mm) slice of silicon (*n*-type), with boron diffused into its upper surface to create a light-transparent layer of *p*-type silicon there. The *p-n* junction acts as a permanent electric field. When the active surface (*p*-type layer) is illuminated, the incident photons cause a flow of "holes" (positive charges) and electrons. The electric field at the junction directs the flow of holes toward the *p*-type silicon and the flow of electrons toward the *n*-type silicon. The resulting unbalance of charge carriers within the sensor causes an emf to be developed between the two surfaces. When a load resistance is connected across the surfaces, on which solder terminals have been formed on metallized (e.g., nickel-plated) areas, the hole and electron carriers flow through the circuit until a condition of balance is achieved. This current flow is then a function of the incident photon flux (illumination).

Some silicon cells use an *n-on-p* junction instead of the more common

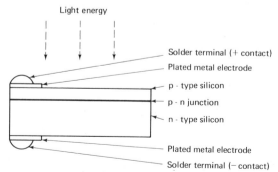

**FIGURE 22–5.** Silicon (*p*-on-*n*) photovoltaic sensor.

*p*-on-*n* junction. In a typical design this is accomplished by diffusing a very thin layer of phosphorus (*n*-type layer) into the surface of a boron-doped silicon slice (*p*-type). This tends to shift the spectral response curve of the sensor slightly toward the blue. The response peak of an *n*-on-*p* silicon cell is near 800 nm, whereas the response of *p*-on-*n* silicon cells (see Figure 22–6) has its peak around 900 nm. Some special silicon cell designs ("blue-enhanced," "blue-sensitive," "blue cell") have spectral response peaks as low as 560 nm, the peak of the human eye's response. The response time of silicon cells depends on load resistance, illumination level, and junction capacitance. Rise times and decay times are substantially equal for low illumination levels and are around 20 μs in typical circuits. At high illumination levels the rise time tends to be considerably shorter than the decay time. Special low-capacitance designs have rise and decay times as low as 2 μs.

Typical silicon photovoltaic sensor configurations used for light-sensing purposes are illustrated in Figure 22–7. Silicon ("solar") cells are also widely used for conversion of solar energy into electrical power. For such applications a large number of typically square, thin cells are mounted in close proximity to each other in the form of large "solar arrays." The cells are connected in series to provide the required output voltage, and series strings are connected in parallel to provide the required output current.

*Germanium photovoltaic sensors* are similar to silicon types in construction and operation. Their spectral response, however, peaks near 1.5 μm, and their operating temperature range tends to be significantly narrower.

*Photovoltaic infrared sensors*, used for IR sensing in the spectral region

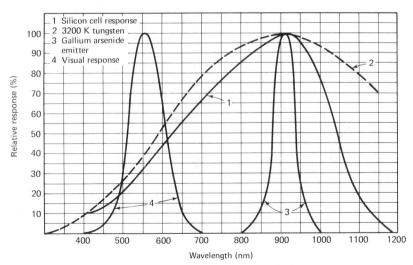

**FIGURE 22–6.** Relative spectral response of silicon cell compared with spectral characteristics of human eye, tungsten emitter, and GaAs (LED) emitter. (Courtesy of E G & G Vactec.)

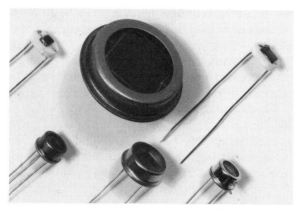

**FIGURE 22–7.** Typical silicon photovoltaic sensors. (Courtesy of E G & G Vactec.)

from 1 to 15 μm, exist in various designs and semiconductor material combinations. A common characteristic is their performance (detectivity) improvement with decreasing operating temperatures. They are most frequently used while mounted in a small cryogenic dewar, usually cooled with liquid nitrogen to 77 K, but they are sometimes used while mounted on a Peltier-effect (thermoelectric) cooler and cooled to somewhat higher temperatures. Included in this group are InAs, InSb, PbSnTe, and HgCdTe sensors. Figure 22–8 illustrates the relative spectral response of each of these types of sensors, which are described momentarily. Note that the detectivity at peak response differs for each of these sensors; the individual response curves are shown on the same graph only for the purpose of showing relative spectral characteristics.

*Photovoltaic indium arsenide sensors* consist of a small wafer of single-crystal indium arsenide (InAs) into which a relatively broad-area *p-n* junction is diffused. If intended to operate at room temperature (300 K), they are typically packaged in a "TO-18" (transistor-type) case, about 5 mm in diameter by 5 mm high, evacuated and sealed with a sapphire or quartz window. For operation (with improved detectivity) at dry ice (196 K) or liquid nitrogen (77 K) temperatures, they are integrally packaged within a small dewar whose liquid-nitrogen holding time is between 8 and 12 hours. For operation at relatively low temperatures without use of liquid cooling, they are integrally packaged with a thermoelectric cooler. As detectivity is improved with decreasing operating temperature, the output impedance of the sensor increases (from about 100 Ω at 300 K to about 10 MΩ at 77 K). InAs photovoltaic sensors have a time constant of about 1 μs.

*Photovoltaic indium antimonide (InSb) sensors* employ single-crystal InSb with a broad-area diffused *p-n* junction. Since they are intended to operate at 77 K, they are integrally packaged in a cooling assembly (coolant well) which either is used as dewar for liquid nitrogen or provides a volume sufficient for incorporation of nonliquid (e.g., thermoelectric) coolers. The sensing surface is always maintained with an evacuated area, sealed by a window typically made from 0.5-mm-thick sapphire. Complete sensor assemblies with provisions for flow-through cryogenic cooling are as small as 2 cm in diameter by 4 cm long. The sensing element itself can be circular or rectangular, with the circular unit ranging in size from 0.2 to 10 mm in diameter, with comparable dimensions for rectangular units. At 77 K, InSb sensors provide a detectivity $D^*$ (5 μm, 900, 1) (see Section 22.5.1 for a definition of detectivity) of approximately 40 to 75 × $10^9$ cm·$Hz^{1/2}$/W and a time constant around 1 μs, at an output impedance of 20 to 50 kΩ.

*Photovoltaic lead–tin–telluride (PbSnTe) sensors* are trimetal photovoltaic sensors which utilize intrinsic photoconductivity in single-crystal PbSnTe material but are operated in the photovoltaic mode. They are intended for operation at 77 K and are typically packaged integrally in a small dewar filled with liquid nitrogen. They are capable of providing a very short response time (down to 50 ns), and their spectral response is in the region from 8 to 12 μm. PbSnTe detectors have generally been phased out of production in favor of other detectors with spectral responses in the 8–15 μm region, notably HgCdTe detectors (see next).

*Photovoltaic mercury–cadmium–telluride (HgCdTe) sensors* are made from very thin (10 to 15 μm) slabs of this ternary alloy material and provide time constants down to about 10 ns. They are intended for operation at 77 K, but can operate at slightly higher temperatures (around 120 K) without a severe degradation of performance. One of their outstanding character-

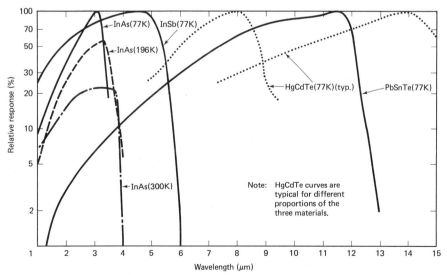

**FIGURE 22–8.** Relative spectral response of photovoltaic infrared sensors.

istics is that their spectral response curve can be adjusted to any desired relatively narrow band within the overall wavelength band of 2 to 14 μm by varying the proportions of the three materials in the alloy. Examples are shown by the two dashed curves in Figure 22–8. Photographic techniques have been successfully applied to the production of single-element sensors as well as multielement arrays.

*Heterojunction photovoltaic sensors* have been developed with a variety of semiconductor material combinations. *Heterojunctions* are junctions between two different semiconductor materials, whereas *homojunctions* are junctions between differently doped portions of the same material (e.g., *p*Si–*n*Si in the silicon photovoltaic sensors described previously). Although most of the research in the heterojunction area has been aimed at developing cheaper, lighter, and more efficient solar-energy converters, some of the results have shown the feasibility of producing light-sensing devices having certain desirable characteristics, such as sensitivity within certain broad spectral regions and, for special applications, a wavelength-dependent polarity change. The latter is peculiar to isotype heterojunctions (e.g., *n*Ge–*n*Si). Anisotype heterojunctions have been utilized in a photovoltaic sensor containing *n*-type special transparent material on *p*-type single-crystal material, resulting in a significantly improved blue response as compared to homojunction *n*Si–*p*Si sensors. Thin-film solar cells with a *p*Cu$_2$S–*n*CdS heterojunction were developed primarily for aerospace solar-power-generation purposes. Spectral response from about 0.5 to 2 μm has been reported for *p*Si–*n*CdS and *p*Ge–*n*GaP heterojunction sensors. Many other combinations of materials have been examined, and some of them show promising characteristics for different areas of applications, including as infrared sensors.

### 22.4.2  Photoconductive Optical Detectors

Light sensors employing bulk-effect photoconductors are resistive devices whose resistance decreases with increasing illumination. They have also been referred to as "photoresistors." The absolute resistance value of a photoconductive sensor, at any instant, depends on the photoconductive material chosen, the thickness, surface area, and geometry of the material, the geometry of its electrodes, the spectral composition of the incident light, the illumination level, the operating temperature, and the difference between present and previous light levels, as well as the exposure times at those levels (*light history effect*).

*Cadmium sulfide (CdS) and cadmium selenide (CdSe) photoconductive sensors* are the most popular types because their spectral response is close to the visible-light region and because they have relatively high sensitivity to changes in illumination level. Their sharply pronounced spectral response peak also enhances their usefulness in colorimetry. Typical spectral response curves for several types of CdS and CdSe photoconductors are shown in Figure 22–9. The ratio of dark to light resistance of these sensors is between 100:1 and $10^4$:1, depending on the specific material used. The measurand-to-output (illumination-to-resistance) relationship is generally nonlinear, and the nonlinearity becomes more pronounced with increasing illumination. Temperature coefficients of resistance are higher for CdSe than for CdS, and the polarity of the coefficients is usually negative for CdS and positive for CdSe. The coefficients also vary as an inverse function of light level. Response time is normally shorter for CdSe than for CdS and decreases with increasing illumination levels. Rise times are longer than decay times. At an illumination of 10 lx, typical rise times are 35 to 150 ms for CdS and 25 to 90 ms for CdSe on a time-constant basis (63% of final output). The active area geometry sometimes is simply rectangular, but more often is of a special

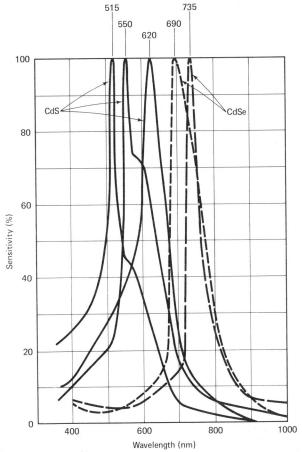

**FIGURE 22–9.** Spectral response of typical CdS and CdSe photoconductive sensors. (Curves courtesy of Clairex Corp.)

"wavy" shape (see Figure 22–10) obtained by evaporation techniques, which achieves a much larger active area for a given sensor size. Development efforts are directed mainly at improving linearity, expanding the usable spectral region, and reducing the light history effect. Research for this has been reported in combining CdS and CdSe in varying proportions in one photoconductive element, in treating the light-sensitive surface by the admixture of various chemical elements, and in studies of the interface between the (conductive) electrodes and the photoconductor.

*Lead sulfide (PbS) and lead selenide (PbSe) photoconductive sensors* are used for light sensing in the near-infrared region of the spectrum. The PbS spectral response is between 1 and 3 μm (at room temperature) and 1 and 4.5 μm (at 77 K). The PbSe spectral response is between 1 to 4.5 μm (at room temperature) and 1 to 6.8 μm (at 77 K). PbS sensors have a somewhat better detectivity than PbSe sensors; however, time constants are substantially shorter for PbSe than for PbS material (between 1 and 5 μs for PbSe, and between 40 and 1000 μs for PbS, at room temperature). Time constants increase with decreasing sensing element temperatures. Typical time constants at 77 K are 5000 μs for PbS and 80 μs for PbSe. Sensor configurations include the plate type, the hermetically sealed type with or without a built-in thermoelectric cooler, and the dewar type, which can be designed for either side or end-on viewing and allows the detector to be operated at the temperature of liquid nitrogen. Figure 22–11 shows a PbS (or PbSe) detector sealed in a TO-37 (a standard transistor configuration) can, with an integral thermoelectric cooler and a thermistor for close-loop temperature control. The actual size of PbS and PbSe detectors ranges from 0.05 to 10 mm.

*Gold-doped germanium [Ge(Au) or Ge:Au] and mercury-doped germanium [Ge(Hg) or Ge:Hg] photoconductive detectors* have been used for far-IR sensing, with a spectral response from 1 to 9 μm for Ge(Au) and from

**FIGURE 22–10.** CdS and CdSe photoconductive light sensors. (Courtesy of Clairex Corp.)

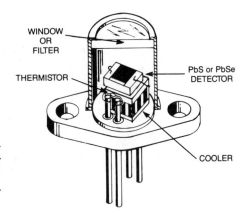

WINDOW
OR
FILTER

THERMISTOR

PbS or PbSe
DETECTOR

COOLER

**FIGURE 22–11.** Typical configuration of PbS or PbSe detector with built-in thermoelectric cooler. (Courtesy of Infrared Industries, Inc.)

4 to 14 μm for Ge(Hg). The use of cadmium, copper, and zinc as dopants has reportedly extended the spectral response to about 40 μm. Gallium-doped germanium photoconductors, which have also been used in bolometers, have a spectral response of about 10 to 100 μm when cooled to about 170 K. When operated at liquid-helium temperatures (3 to 4 K), their spectral response extends from 60 to 120 μm and they exhibit other favorable characteristics, except that their resistance (output impedance) is very high (around $10^{15}$ ohms). Such high output impedances dictate the use of a very closely coupled preamplifier which can operate at very low temperatures. Developments in doped silicon photoconductors have also been carried on. All such *extrinsic-semiconductor* detectors require cryogenic cooling, not necessarily to 3 or 4 K, but generally to below about 40 K.

*Mercury cadmium telluride (HgCdTe) detectors* can operate in the photovoltaic mode (see Section 22.4.1), but are more frequently used as photoconductors. Their spectral response can be adjusted to cover selected portions of the overall band from 2 to 20 μm by varying the ratio of the amounts of HgTe and CdTe in the material. They can be operated at room temperature, but their detectivity improves by about 1.5 orders of magnitude when operated at 77 K, and further reductions in operating temperature provide even higher values of detectivity. Their size ranges from 0.05 to over 10 mm. The detector shown in Figure 22–12 has a size of 4 mm × 4 mm.

HgCdTe detectors are increasingly used in multielement array form, chiefly in IR imaging. Single-line arrays have been designed with up to 640 elements. Figure 22–13 shows two 16-element HgCdTe photovoltaic arrays.

Some research has been carried on toward developing photoconductive sensors whose spectral response extends into the ultraviolet. Solid-solution *ZnCdS photoconductors* are reported to have the capability of providing spectral response peaks of 400 nm or below.

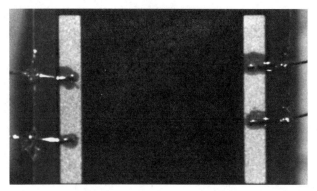

**FIGURE 22–12.**   HgCdTe photoconductive detector. (Courtesy of New England Research Center, Inc.)

### 22.4.3 *Photoconductive-Junction Optical Detectors*

The category of photoconductive-junction optical detectors encompasses phototransistors, avalanche photodiodes, and those types of solid-state photodiodes that are not operable, or not usually operated, in the photovoltaic mode.

*P-n junction photodiodes* have a built-in field enabling them to operate in the photovoltaic mode, but have better performance in the photoconductive (reverse-biased) mode. A typical design is illustrated in Figure 22–14. The relative responsivity, at the peak wavelength of about 950 nm, is only 60% of that obtained in the photoconductive mode. The light-sensitive por-

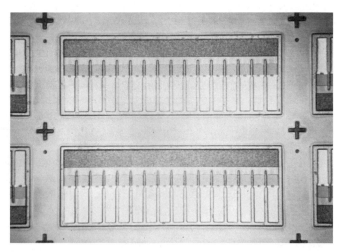

**FIGURE 22–13.**   Sixteen-element HgCdTe arrays. (Courtesy of New England Research Center, Inc.)

tion of a junction photodiode can be in the *p* region, the *n* region, or the depletion region at the junction. Typical spectral response is 400 to 1100 nm, and rise times are in the vicinity of 10 ns.

*PIN (or p-i-n) photodiodes* are characterized by an intrinsic layer (which defines the depletion layer) between the *p* and *n* regions. They, too, can operate in the photovoltaic mode, but are more commonly used in the photoconductive, reverse-bias mode. Their operation is based on the excitation of electron-hole pairs in the intrinsic layer by photons received at the surface of the *p* region. Spectral response is 400 to 1200 nm for silicon (200 to 1200 nm for "UV-enhanced" silicon), 500 to 1800 nm for germanium, and 1000 to 1700 nm for indium gallium arsenide. Rise times are extremely short, and the frequency response extends up to 1 or 2 GHz.

*Avalanche photodiodes* are depletion-layer photodiodes whose bias voltage can be increased to the point where electron-hole pair multiplication by collision (*avalanche effect*) takes place in the depletion region. The avalanche effect produces significant amplification of the photocurrent with negligible noise. The operation of avalanche photodiodes is similar to that of photomultiplier tubes (see Section 22.4.4). Amplification factors of over 200 times have been achieved. The spectral response is typically 400 to 1100 nm for silicon, 800 to 1800 nm for germanium, and 900 to 1700 nm for indium gallium arsenide. Frequency response is only slightly lower than for PIN photodiodes using these materials.

*Photofets* are photodiodes combined, in one small package, with a field-effect-transistor-based high-impedance (but low-output-impedance) amplifier.

*Phototransistors* are light-sensitive *pnp* or *npn* junction transistors which provide inherent amplification of the photocurrent. *Photodarlingtons* are planar epitaxial phototransistors with a Darlington-connected second

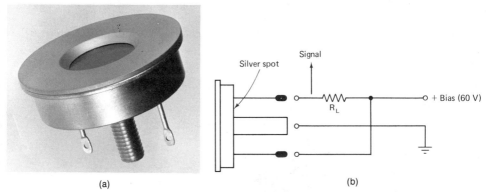

(a)          (b)

**FIGURE 22–14.** Silicon *p-n* photodiode operated in photoconductive (reverse-biased) mode: (a) photodiode; (b) basic circuit diagram. (Courtesy of Thorn EMI Electronics Ltd.)

transistor in a single device for additional amplification. Both types of devices have spectral responses that peak between 800 and 850 nm.

### 22.4.4  Photoemissive Optical Detectors

The category of photoemissive optical detectors comprises phototubes (vacuum or gas-filled photodiodes) and photomultiplier tubes. Since gas-filled phototubes are no longer in common use, the subcategory of phototubes is now limited to vacuum photodiodes.

*Vacuum photodiodes* are cold-cathode diode vacuum tubes. They consist of a photocathode, an anode, and sometimes a guard ring within an envelope having an entrance window. When a potential is applied between the photocathode and the anode, electrons emitted from the photocathode, due to photon incidence, are attracted toward the anode, and a current (photocurrent) flows through the circuit. Vacuum photodiode characteristics depend primarily on the photocathode and window materials. The spectral response is typically 200 to 1100 nm, but some vacuum photodiodes have been designed especially for UV measurements and have a spectral response to below 150 nm. Vacuum photodiodes exist in a variety of configurations. The design shown in Figure 22–15 has 13 connecting pins, only two of which are used (anode and cathode). Quite different electrode geometries and connections are used in the two biplanar vacuum photodiode configurations shown in Figure 22–16. Among many other designs are those having sturdy plug-in sockets and those containing a long, concave cathode with a central slim anode rod.

*Photomultiplier tubes (PMTs)* are photoemissive devices differing from photodiode tubes in that additional electrodes (*dynodes*) are placed between the photocathode and the anode. The dynodes are connected into a voltage-divider network in such a manner that, starting with the dynode nearest the photocathode, each successive dynode has a higher potential applied to it. The dynode nearest the anode is at a potential close to the high voltage applied to the anode itself. When photons strike the cathode, free electrons

**FIGURE 22–15.** Vacuum photodiode. (Courtesy of ITT Electro-Optical Products Div.)

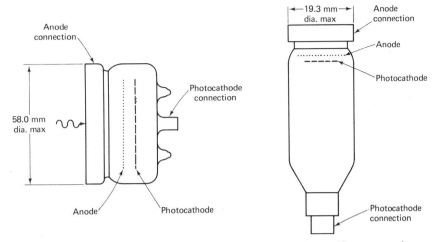

**FIGURE 22–16.** Vacuum photodiode configurations. (Courtesy of ITT Electro-Optical Products Div.)

are liberated and drawn to the first (lowest-voltage) dynode because its potential is positive with respect to the cathode. Several electrons are liberated at the first dynode for each electron emitted by the cathode. These secondary-emission electrons are drawn to the next dynode because its potential is positive with respect to the previous dynode. Several electrons are liberated from the second dynode for each electron emitted by the first dynode. This process is multiplied by each successive dynode (hence the term "photomultiplier"). All electrons resulting from this multiplication process are collected by the anode. The amplification of the cathode photocurrent, due to the multiplication process, is typically between $10^5$ and $10^8$ (*current gain*), depending primarily on the number of dynodes in the photomultiplier tube. Three basic dynode structural arrangements are used: the circular cage, venetian-blind cage, and in-line cage (see Figure 22–17). Typical photomultiplier configurations are shown in Figure 22–18.

Spectral response characteristics of photomultiplier tubes depend on the material of the photocathode surface as well as on the *window* material (the portion of the tube's envelope directly in front of the photocathode). Many different types of photocathode materials have been developed to meet a variety of application requirements. The spectral response characteristics of some of these materials are illustrated in Figure 22–19. Some of the materials are identified by a "spectral response" (S-) number, such as AgO–Cs (S-1), or BiO–Ag–Cs, semiopaque (S-10). Others are identified by the material mixture (e.g., GaAs, GaInAs, GaAsP) or by a generalized material term such as bialkali (Sb–Na–K) or multialkali (Na–K–Cs–Sb), including the "ERBA" (extended-range bialkali) and "ERMA" (extended-range multialkali) photocathode materials. Additional materials are often designated

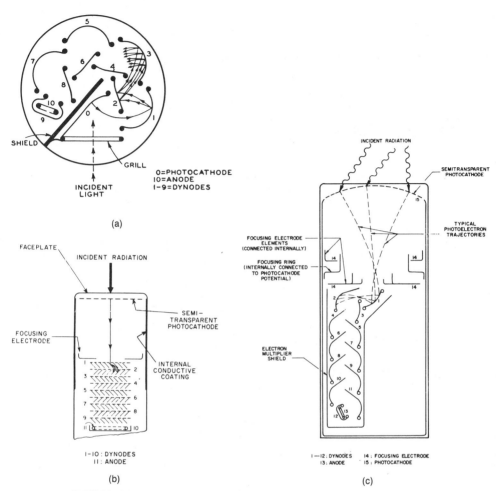

**FIGURE 22–17.** Schematic arrangements of photomultiplier tube internal structures; (a) circular cage; (b) venetian-blind cage: (c) in-line cage. (Courtesy of Burle Industries, Inc.)

just by a manufacturer-assigned number or name. Among window materials, lime glass and fused silica are most frequently used. Other materials are used for obtaining specific overall spectral characteristics, such as LiF or MgF for response in the UV portion of the spectrum (see Figure 22–23). Certain photocathode-window material combinations have resulted in the development of "solar blind" photomultipliers, i.e., those having a spectral response only in the UV, with a cutoff below 350 nm. Multialkali, GaAs, GaInAs, or GaAsP photocathodes are used when response in the IR region of the spectrum is required. Photomultipliers have very short response times, in many cases less than 10 ns, but they require the provision of a very stable

**FIGURE 22-18.** Typical photomultiplier tubes. (Courtesy of Thorn EMI Electronics Ltd.)

high-voltage (600 to 1500 V) supply. A 1% change in this voltage can produce a 10% variation in current gain. Some photomultiplier tubes are available in an integrated package together with their voltage-divider network.

*Crossed-field detectors* are a special version of a photomultiplier in which very fast response time is obtained by the addition of a magnet such that the electric and magnetic fields cross each other.

### 22.4.5  *Photoelectromagnetic Optical Detectors*

A relatively small number of photoelectromagnetic optical detectors have been built for IR sensing, first using indium antimonide, and later mercury cadmium telluride, as semiconductor material. A permanent magnet is usually employed to establish the required magnetic field. Photoelectromagnetic IR detectors offer two advantages: they are fully operable at room temperature, and they have a very low output impedance. Characteristics reported for a photoelectromagnetic HgCdTe detector include a spectral response of 3 to 12 μm, an output impedance of 50 ohms, a size of about 1 mm, a responsivity between 2 and 10 mV/W, and a frequency response from dc to 0.8 GHz.

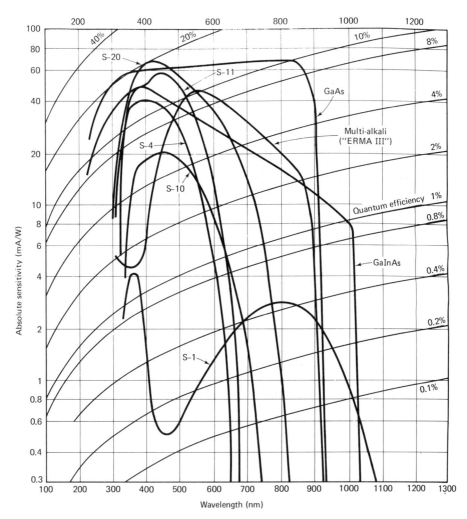

**FIGURE 22–19.** Typical photocathode spectral-response characteristics. (Courtesy of Burle Industries, Inc.)

### 22.4.6 *Thermoelectric Optical Detectors*

Thermal detectors using thermopiles (multijunction thermocouples) are self-generating (i.e., they need no external excitation power). Their design is similar to that of the thermopile radiometer described in Section 21.4.2. The sensing junctions that constitute the active area are blackened for better response to incident radiant flux. The reference junctions are in good thermal contact with a usually annular heat sink, but are electrically isolated from it. The case of the detector is provided with an aperture, usually sealed by a window, which is small enough to admit radiant energy only to the active area. The window material can be selected so as to provide a spectral response narrower than the inherently wide-band (0.2 to about 30 $\mu$m) response of the thermopile.

Thin-film techniques have been successfully applied to the production of very small multijunction thermopiles. Between 8 and over 300 junctions can be contained within an area less than 1 cm × 1 cm. To increase the number of junctions, however, less thermocouple material must be used, and this increases the overall resistance of the detector. Still, detectors with enough junctions to provide a responsivity on the order of 50 V/W have a resistance not exceeding about 10 kilohms. Most thermopile detectors are sealed, evacuated, and then back-filled with an inert gas, such as argon or xenon, to preserve the delicate active area. This prolongs detector life substantially, but at the cost of a reduction in sensitivity. Thermopile detectors cannot be used for rapidly fluctuating radiant energies, their frequency response does not usually extend above 10 Hz, and further, they cannot be used for long-term steady-state measurements where "dc response" would be needed, because the reference-junction temperature will gradually increase to the same value as the sensing-junction temperature. When this happens, the output will be zero.

### 22.4.7 *Bolometer Optical Detectors*

Bolometer detectors (*bolometers*) are categorized as thermal detectors. Semiconductors such as silicon and indium antimonide have been used in specialized bolometers, which are cooled by liquid helium and provide a spectral response into the millimeter range. Most bolometers, however, employ a closely matched pair of thermistors. An active (radiation-sensing) thermistor flake and a compensating flake, shielded from radiation, are typically cemented to a block of solid material (e.g., sapphire) which acts as heat sink. In an "immersed" design, the active flake is attached to the inside surface of a lens and the compensating flake is mounted in the base of the sensor. The two thermistors are electrically connected as two arms of a bridge circuit. Two voltage sources providing equal voltage, but of opposite polarity, form the other two arms of the bridge circuit. A difference in resistance of the closely matched thermistors, due to incident radiant flux on the active flake, then causes an output voltage of the bridge circuit. Thermistor bolometers are designed to operate at room temperature. Their inherent spectral response extends from 0.25 to about 500 $\mu$m; however, they are usually equipped with a window that not only forms a seal but also limits the spectral response to a selected band of IR wavelengths. For example, a germanium window provides a response in the 2 to 20 $\mu$m band, and a silicon window limits the response to a band from 1 to 16 $\mu$m. The time constant of a thermistor bolometer is generally between 1 and 5 ms, and typical values for responsivity are near 300 V/W.

### 22.4.8  Pyroelectric Optical Detectors

Pyroelectric detectors are thermal detectors utilizing the pyroelectric effect provided by certain materials whereby an output is generated by the rate of change of the temperature of the sensor due to the rate of change of incident radiant flux. Their performance characteristics depend primarily on the sensing material, the preparation and geometry of the electrodes, the use of absorbing coatings, the thermal design of the structure of the sensor, and the nature of the electronic interface. Since the output is rate-of-change dependent, the radiant flux incident upon the sensor must be chopped, pulsed, or otherwise modulated. The sensors are normally operated at room temperature or other ambient temperatures, without artificial cooling.

The spectral response of pyroelectric sensors extends from below the far (vacuum) UV to beyond the far IR, but is often limited to a band of wavelengths in the IR region by means of a window. Rise times are between less than 1 ns to about 200 ns. Detectivities are in the vicinity of $10^8$ cm·Hz$^{1/2}$/W.

Figure 22–20 shows a pyroelectric detector, including details of the sensing head. This design is intended for bench tests. Other available configurations include small TO-5 (transistor standard) cans and other plug-in or lead-mounted detectors with pyroelectric elements between 1 and 10 mm in size.

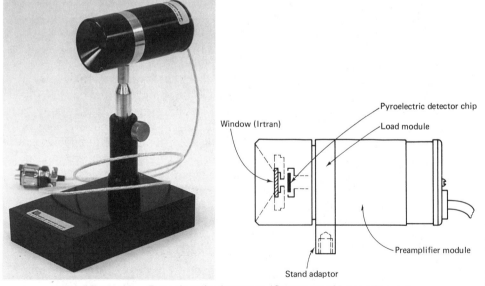

**FIGURE 22–20.**   Pyroelectric detector. (Courtesy of Laser Precision Corp.)

Developments in pyroelectric sensors have been primarily in the area of the pyroelectric materials. These are, in general, single crystals, ceramics, or plastics. Triglycine sulfate (TGS) was originally the most widely used material. Although it is characterized by a very good responsivity, it was found difficult to manufacture, it is hygroscopic, and it has a very low Curie point (49°C), the temperature at which depolarization occurs and pyroelectric properties are lost. Some of these difficulties were later overcome by improvements in the material. Lithium tantalate ($LiTaO_3$) was found to have comparable pyroelectric properties, but with a much higher Curie point (610°C). Research in plastic pyroelectric materials has led to the production of polyvinylidene fluoride and $PVF_2$ (polyvinyl fluoride) detectors which, in some cases, could provide detectivities as high as $10^9$ cm·$Hz^{1/2}$/W, although detectivities of plastic pyroelectrics in general tend to be lower than this. A significant advantage of plastic pyroelectric sensors is their producibility and relatively low cost.

### 22.4.9  Golay-Type Detectors

Pressure-actuated photoelectric transduction of radiant energy, primarily in the IR portion of the spectrum, is employed by the Golay-type sensor (*Golay detector, Golay cell*). The basic sensor was invented by M. J. E. Golay in 1947 to satisfy a requirement for a relatively sensitive detector having a flat spectral response curve over a wide portion of the infrared spectrum.

Figure 22–21 illustrates a Golay-type sensor and its operating principles. The sensing system consists of a chamber filled with a gas of low thermal conductivity and a flexible mirror. Incident radiation passes through a window (A) made from a material chosen for a specific response in the IR region of the spectrum and reaches a thin energy-absorbing film (B) which warms the gas in the chamber. The resulting gas expansion causes a deflection of the mirror membrane (C) which seals the other end of the chamber. A fine leak (D) compensates for changes in ambient temperatures. The incident radiation is periodically interrupted by an external chopper (not shown) so that radiation pulses at a repetition rate of between 5 and 25 Hz (selectable) are incident on the absorbing film, which causes the deflections of the flexible mirror to occur at the same rate. However, the amplitude of each deflection is given by the amount of radiation in each pulse. The flexible mirror receives a light beam originating at the light source (E) via a condenser lens (F), a grating (G), and a meniscus lens (H). The latter lens focuses the beam so that an image of one part of the grating is superimposed on another part of the same grating when the flexible mirror is not deflecting. When the flexible mirror deflects, a corresponding change in the relations of the line image and grid occurs. This varies the intensity of light reflected from the flexible mirror through lenses H and F (and grating G) toward a stationary mirror (L) and a light sensor (K). In the case of the design illustrated, the

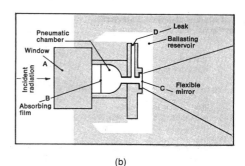

(b)

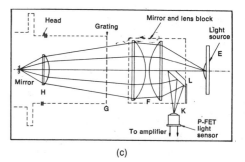

(c)

(a)

**FIGURE 22–21.** Golay-type IR detector: (a) complete sensor package; (b) sensing system (pneumatic chamber); (c) transduction system (photoelectric). (Courtesy of Oriel Corporation.)

light source is a light-emitting diode and the light sensor is a photo-FET whose output is fed to a solid-state amplifier contained in the same sensor package.

The spectral response of the sensor is essentially flat from 0.5 to 1000 μm. The choice of windows both determines the specific spectral response of a sensor within this broad band of wavelengths and affects the sensitivity, since different windows have different values of transmittance. Typical detector-package sensitivities are $3 \times 10^6$ V/W with KBr, KRS-5, or CsI windows (spectral response below about 50 μm) and $1.5 \times 10^6$ V/W with diamond, quartz, or Si windows (spectral response above 50 μm to between 350 and 900 μm). The 90% response time is around 20 ms, followed by a longer recovery time.

## 22.5   DETECTOR CHARACTERISTICS

### 22.5.1   *Performance Characteristics Terminology*

Since many of the essential performance characteristics of light sensors are substantially different from those normally specified for most other types of sensing devices, these characteristics are grouped separately here.

*Sensitivity* is generally defined as the ratio of the change in output (of a sensing device) to a change in the measurand. This characteristic is often specified for photoemissive light sensors, primarily photomultiplier tubes, as cathode (or anode) luminous (or radiant) sensitivity. *Cathode luminous sensitivity* is the ratio of photoelectric emission current (from the photocathode) to the incident luminous flux (under specified conditions), expressed in *A/lm*. *Cathode radiant sensitivity* is the ratio of the photoelectric emission current to the incident radiant flux (at a given wavelength and under specified conditions of irradiation), expressed in *A/W*. *Anode luminous (or radiant) sensitivity* is defined similarly, except that the sensitivity is referred to the anode instead of the cathode.

*Quantum efficiency* is the average number of photoelectrons (or similar electronic charges) emitted from the light-sensitive portion of a light sensor (e.g., a photocathode) per incident photon, expressed in *percent (%)*; it is shown either for a given wavelength or, as *peak quantum efficiency,* for the wavelength of peak radiant sensitivity of the light-sensitive surface.

*Responsivity* is the ratio of the sensor's output amplitude to the incident radiant flux (at a stated wavelength), expressed in *V/W* or *A/W*, depending on the type of sensor.

The following three characteristics are specified primarily for radiant-energy (IR) sensors; they relate incident radiant power to signal-to-noise ratio, the ratio of the signal voltage to the noise voltage produced across the output terminals of the sensing device.

*Noise equivalent irradiance* is the minimum radiant flux density necessary to provide a signal-to-noise ratio of 1 when the noise is normalized to unit bandwidth; it is expressed in $W/Hz^{1/2}{\cdot}cm^2$.

*Noise equivalent power (NEP),* at a given wavelength, is the minimum radiant flux necessary to provide a signal-to-noise ratio of 1 when the noise is normalized to unit bandwidth; it is usually expressed in $W/Hz^{1/2,}$ but can also be expressed in *W* at a specified bandwidth (greater than 1 Hz).

*Detectivity* is usually shown as *D\** (*D-star*), a "figure of merit" originally introduced to remove the dependence of NEP on the radiation-sensitive area of the sensing device; it is the actual detectivity normalized to unit bandwidth and unit area and is a measure of the signal-to-noise ratio. Detectivity (*D\**) is given by

$$D^* = \frac{S/N(A{\cdot}\Delta f)^{1/2}}{W}$$

where *S/N* = signal-to-noise ratio (ratio of signal voltage to noise voltage)
   *A* = sensitive area of sensing device, $cm^2$
   $\Delta f$ = noise equivalent bandwidth of amplifier in test setup, Hz
   *W* = rms value of radiant power incident on sensitive area, W

It is usually expressed as "$D^*(\lambda, f_c, \Delta f)$ = _____ $cm{\cdot}Hz^{1/2}{\cdot}W^{-1}$"; that is, the value of $D^*$ is given at a specified wavelength ($\lambda$), chopper frequency of the test setup ($f_c$), and noise equivalent bandwidth ($\Delta f$). In some specifications the absolute temperature of a black-body source (e.g., 500 K) is substituted for $\lambda$. $D^*$ is related to NEP by NEP $= A^{1/2}/D^*$.

*Wavelength cutoff* is that value of wavelength, reached with increasing wavelengths, at which the responsivity is reduced to 30% of the peak spectral responsivity. It is usually expressed in $\mu m$.

*Dark current* is the current flowing in a light sensor in the complete absence of any incident luminous or other radiant flux. An equivalent term, sometimes applied to photoconductive sensors, is *dark resistance.*

*Field of view* is the solid angle, or the angle in a specified plane, over which radiant energy incident on the sensitive area is measured within specified tolerances.

*Spectral response* is the band of wavelengths over which the sensor performs within specified tolerances under stated conditions. It is frequently shown as a graph of percent relative response (with peak response = 100%) vs. wavelength, or some other measure of sensor response (e.g., cathode radiant sensitivity, quantum efficiency, $D^*$, etc.) vs. wavelength. It can also be shown as the two wavelength values bracketing the peak response and at which the sensor response is reduced to a stated fraction or percentage of peak response. Sometimes the wavelength of peak response is shown additionally.

Other light-sensor performance characteristics are of the type frequently specified for many other types of sensing devices and are, therefore, not discussed here in detail. They include response time (time constant, rise time, decay time), power dissipation, and various electrical, mechanical, and environmental characteristics (e.g., electrical connections, dimensions, mounting provisions, operating temperature range, etc.). Note that certain light-sensor designs exhibit a marked temperature sensitivity, and some types (notably some IR sensor designs) are intended to operate at very low temperatures.

### 22.5.2  Specification Characteristics

In this section the design, performance, and environmental characteristics that should be specified for light sensors are listed and described. Since some fundamental differences exist between sensor types, not all characteristics are specified for all sensor types. For purposes of brevity, the following abbreviations will be used: PV, photovoltaic; PC, photoconductive; PCJ, photoconductive-junction; PED, photoemissive diodes; PMT, photomultiplier tubes; TP, thermopiles; BOL, bolometers; PYR, pyroelectric.

*Design characteristics* always include the overall configuration with all necessary dimensions; mounting means and mounting dimensions; type and dimensions of external electrical connections; location and dimensions of any integrally packaged components, such as windows, lenses, coolers, and preamplifiers; degree of sealing, if any; and the nature of the internal atmosphere (vacuum or a specific gas) if the sensor is sealed. The viewing configuration (side viewing or end viewing) is usually specified for PMT and for some other sensor types if they are contained within a cooling jacket. The light-sensitive material and the dimensions of the light-sensitive area (active area) are shown for all types except for most BOL; in the case of PED and PMT the size and type of photocathode are stated. When the sensor employs a combination of two or more sensing materials, all materials are stated. The location of the sensitive surface is frequently defined (e.g., by its distance from a lens or window). The material and thickness of an integrally packaged window is always specified. The number of dynodes, the dynode structure, and, sometimes, the dynode material are shown for PMT. For TP the number of junctions is stated. The field of view is normally shown only for sensors with integrally packaged optics. Many light sensors are purchased "bare" and then connected with other components by the user.

Electrical design characteristics include the (element or output) impedance (or resistance) for most types of sensors; the resistance at a specified illumination for PC; the dark resistance for PC; the capacitance (between electrodes) for PV, PCJ, PED, and PYR; the maximum supply voltage for PC, PED, and PMT; a typical, nominal, or recommended supply (or bias) voltage for PC and PMT; and the maximum electrical-power dissipation for PCJ and PMT. For phototransistors, the breakdown voltage rating (collector-to-emitter and emitter-to-collector) and the collector-emitter saturation voltage are stated. Sensors having a high output impedance require a very closely coupled preamplifier to avoid noise and interference pickup in the interconnecting leads.

*Performance characteristics* always include the spectral response of the sensor, usually shown in graphical form but sometimes stated in numerical form, together with the wavelength of peak spectral response. For thermal sensors (TP, BOL, PYR) the spectral response is dictated primarily by the material of any integrally packaged window.

"Sensitivity" is specified in one or more ways for the various sensor types. Responsivity (usually in terms of current, sometimes in terms of voltage) is specified for all sensors except PMT. For BOL, a graph of responsivity vs. bias voltage is typical. Detectivity ($D^*$) and noise equivalent power (NEP) are stated for all except photoemissive devices. For PMT it is customary to show the anode and/or cathode radiant and/or luminous sensitivity. Current or voltage sensitivity is specified for PV and PCJ sensors. Specifications for PV sensors usually show the open-circuit output voltage as well as the short-circuit current, at a specified illumination, and the dark reverse current. Quantum efficiency is sometimes stated, particularly for PCJ and PED. For some applications it is important that the sensitivity be uniform over the entire active surface. In such cases, a tolerance for variations in sensitivity is specified and special techniques are used to map the active area in order to establish any nonuniformities in sensitivity.

Response time (sensor response to a step input in luminous or radiant flux) is shown in various ways for all sensors. It is sometimes shown in graphical form, in other cases as 90% (or some other percentage) response time, as rise time (and fall time), or as time constant. Additionally, frequency response is stated for some types of sensors. PMT specifications usually show time-response characteristics as anode pulse rise time, anode pulse FWHM (full width at half magnitude), and transit time.

Additional performance characteristics, shown only for certain types of light sensors, include linearity (PV and PCJ), dark current (PCJ and PMT), current amplification (PMT), gas amplification (gas-filled PED), maximum anode current (PED, PMT), and maximum as well as minimum detectable radiant-power density for some thermal detectors.

*Environmental characteristics* normally include the operating temperature range (or one or more recommended operating temperatures) and thermal effects on sensitivity, responsivity, or detectivity. For some applications requiring ruggedized sensors, the shock and vibration environments to be withstood by the sensor are also specified, either as operating or as non-operating conditions.


## 22.6   APPLICATION NOTES


### 22.6.1   *Optical Detection Systems*

Although there are some relatively simple applications in which optical detectors are used with little or no ancillary elements (see Section 22.6.2), the majority of applications call for elements to be added to the detector. Such elements include those which modify the incident light, those which modify the detector output, those that furnish electrical excitation to the detector, and those that lower the temperature at which the detector operates. Together with the detector, these elements constitute an optical detection (or light-sensing) system. A basic, typical light-sensing system is illustrated in the block diagram of Figure 22–22. Not all of the elements are needed in all such systems. On the other hand, some of the more sophisticated sensing systems may include additional elements to further modify the incident light and to condition and process detector output signals.

*Choppers* are used to interrupt the incoming light periodically. This causes the detector output to be ac rather than dc and is used because ac is generally easier to amplify than dc. Also, some detectors perform better with chopped (or pulsed) light, and certain types have no "frequency response down to dc," which means that they do not respond unless the light

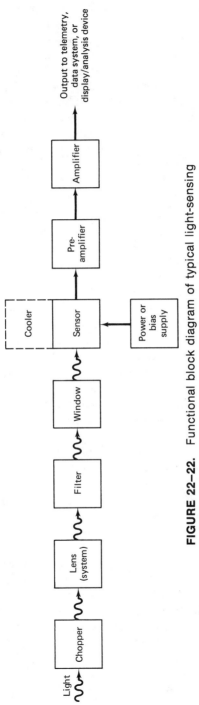

**FIGURE 22–22.** Functional block diagram of typical light-sensing system.

is chopped (at the detector) or pulsed (at the source). Finally, chopping provides brief periods during which the sensor is "dark," which facilitates the correction of sensor output signals for any shifts or drifts not related to the light being sensed (e.g., thermally caused shifts, random drifts, aging).

Most choppers are rotary electromechanical devices in which a rotating blade alternatingly covers and uncovers a hole through which the light beam passes. Various blade shapes are used, such as a butterfly-shaped blade (which interrupts the light beam twice per revolution) and a disk with a hole in the same position as the hole through which the light beam passes. Some sensing systems employ a chopping mirror, electromechanically driven so that it moves back and forth through a small angle and reflects the incident light beam either to the sensor or away from the sensor. Variable-speed controls are used to select the angular speed of the blade or mirror and, hence, the *chopping frequency*, which is typically between 5 and 400 Hz but can be 4 kHz or higher. The blades or mirror drives are so designed that the chopped light beam has a specific wave shape (e.g., square wave, sinusoidal). An angular-speed sensor (e.g., reed switch, optoelectronic) is usually built into the chopper to provide an output usable for synchronization of the sensor-output electronics.

*Lenses* are used to focus the light beam onto the active area of the sensor. The characteristics of lenses of various shapes (e.g., convex, biconvex, concave, biconcave, meniscus, etc.) can be found in any basic optics text and will not be described here in detail. Various glasses are used as lens materials. Some lenses are made from materials described as window materials (e.g., Ge, Si, KRS-5, $BaF_2$) and can then perform the additional spectral-response controlling function otherwise provided bv windows. Lens systems, consisting of two or more lenses, are employed in some light-sensing systems.

*Filters* are used to limit the spectral response of the sensor (often in combination with a window) to a desired band of wavelengths. They are square or circular in shape, with parallel surfaces, and usually less than 1 cm thick. Two or more filters can be used in combination. The filter is made either from one of a large variety of carefully selected materials or material mixtures, or from a glass or quartz substrate coated with a thin film of such materials or mixtures.

A variety of spectral-response characteristics can be obtained by proper selection of a filter. *Band-pass filters* have a high transmittance over a selected band of wavelengths and a transmittance asymptotic to zero for wavelengths above and below this band. *Narrow-band filters* are a version of band-pass filters having a very narrow passband and a very sharp peak. *Short-pass filters* transmit wavelengths below a specified wavelength. *Long-pass* filters transmit wavelengths longer than a specified value. Filters are often classified by the portion of the spectrum in which they transmit light (i.e., UV, visible, or IR filters). *Interference filters* (usually used as narrow-band filters) consist of several parallel and partially reflecting surfaces, each such surface being the interface between two thin layers of transparent dielectric materials having different refractive indices. The principle of interference is used so that some transmitted wavelengths are reinforced whereas other wavelengths are weakened. *Neutral density filters* are used to attenuate the incident light by specified amounts. They are designed either for the visible portion of the spectrum or for a broader range (UV to near IR). The different levels of attenuation are attained by different thicknesses of one or two metallic coatings on either glass or quartz. Standard densities range from "0.1" (79.5% transmission) to "4.0" (0.01% transmission) in eight standardized steps. Continuously variable band-pass as well as neutral-density filters are used for some applications, as are filter wheels which place one of several selectable filters into the optical path.

Among other optical components sometimes used in optical detection systems are *right-angle prisms* (which deflect a light beam by 90°) and other prisms; *beam splitters* (prisms or *dichroic mirrors*), which reflect a portion of the light beam 90° while permitting the remaining portion of that beam to travel straight through; and *coatings* (usually for antireflection purposes, sometimes to provide reflection). *Monochromators* are continuously variable very narrow band-pass devices. *Field stops* are opaque disks with a small hole; they can be placed in front of a light sensor to assure that only an appropriately narrow beam of the light to be measured will reach the sensor and that all other light (or radiation) will be excluded.

*Windows* are those optical components of a light-sensing system that are placed closest to the sensitive area of the sensor. In photomultiplier tubes they are an integral part of the tube's envelope and are directly in front of the photocathode. In many other sensor types they are used as the transparent seal which protects the sensing material from the ambient atmosphere. The light-transmission range of window materials (see Figure 22–23) is the most essential window characteristic. Some materials are usable for UV (e.g., LiF, sapphire); others are most useful in relatively narrow or relatively wide portions of IR. Many sensor designs are available with a choice of materials for their window to provide the desired spectral response. The spectral-response characteristics of the light-sensing material itself must, of course, be compatible with the window transmission range. Other important characteristics of window materials which influence design and selection are the refractive index (specified at a stated wavelength or wave number), the thickness, reflection losses, solubility (NaCl and KBr are water soluble), the hardness, and any tendency to cold flow. Some sensors employ more than one window.

*Coolers (detector cooling devices)* are used for sensors whose performance improves significantly as their operating temperature is reduced below room temperature. They are necessary accessories, in particular, for many types of IR sensors, some of which are commonly operated at cry-

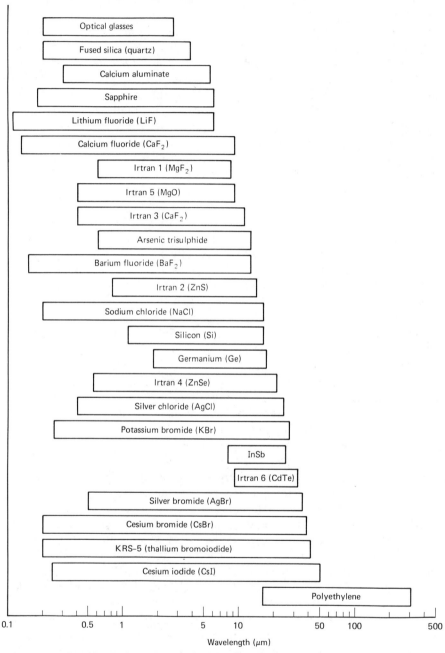

**FIGURE 22–23.** Transmission ranges of window materials.

ogenic temperatures (between 4 and 77 K). Very low sensor temperatures are attained by means of a dewar, a cryostat, or a Joule–Thomson cooler. A *dewar* (dewar flask) is a double-walled "vacuum bottle," a vessel in which the region between the two walls has been evacuated and sealed. The wall surfaces in contact with the evacuated space are metallized (e.g., silvered) for heat-rejection (reflection) purposes. A liquefied gas stored in a dewar can maintain its temperature for varying periods of time. A *cryostat* is a vessel holding liquefied gas whose temperature is maintained over relatively long periods of time by externally controlling the pressure in the vessel. The sensor is mounted within the dewar (or cryostat), which is provided with a window. Detector temperatures down to 4 K have been obtained by means of dewars and cryostats. Temperatures below 4 K have been achieved by the use of superfluid helium in a dewar.

A *Joule–Thomson cooler* relies on the adiabatic expansion of a pressurized gas through a small orifice at which the temperature of the escaping gas becomes much colder than the temperature upstream of the orifice. The sensor is mounted so that it remains in good thermal contact with the point at which the gas flows from the orifice. Such coolers have been used to provide temperatures down to about 23 K, with temperatures considerably below that considered achievable.

*Refrigerators (cryorefrigerators)*, using the expansion-engine principle to expand a gas from a high pressure to a low pressure in closed-cycle operation, provide another alternative for detector cooling. Such mechanical refrigerators have cooled IR detectors to about 13 K in some applications. Development efforts on absorption (thermal) refrigeration systems and on magnetic refrigeration systems promise potential temperature reductions of detectors to below 1 K. Mechanical refrigerators tend to consume fairly large amounts of power (between tens of watts to over a kilowatt), a disadvantage in power-constrained applications.

*Thermoelectric coolers* employ the Peltier effect, by which heat is absorbed by a thermocouple junction when current of the correct polarity is passed through the junction. These devices are quite popular for applications where sensors must be cooled to 273 K (0°C) or slightly below this point, since they are small in size and consume little electrical power (15 to 25 W). Multistage thermoelectric coolers can provide lower temperatures, down to about 170 K, at a somewhat higher power consumption (around 50 W).

When optical-quantity sensors, especially IR sensors, are flown on satellites, spacecraft, or even high-altitude research balloons, passive cooling can be applied, using techniques by which heat from the detector region is radiated out toward space. In some applications of this type, detector temperatures down to 90 K have been achieved by using passive cooling techniques (usually by means of *radiator plates* or otherwise configured radiators).

Optical components used in conjunction with cooled sensors frequently require cooling as well, so as to reduce local background radiation effects and increase the contrast between target and background.

*Power supply or bias supply* circuitry is needed for all except self-generating (photovoltaic, thermoelectric, pyroelectric) sensors. Resistive (photoconductive, bolometric) sensors require a well-regulated bias supply. Photoconductive-junction diodes require a bias supply. Phototransistors require a power supply of a type typical for transistors in general. Photoemissive diodes require a power supply to provide the potential difference between anode and cathode. A basic circuit is shown in Figure 22–24; the picoammeter in this circuit can be replaced by a load resistance across which a variation in output voltage can be detected and fed to a preamplifier and amplifier. Photomultiplier tubes require a well-regulated high-voltage supply. Either the positive or the negative terminal of the power supply is grounded, depending on the application, and a carefully designed voltage-divider network is connected between anode and cathode to provide the proper voltage to each of the dynodes (see Figure 22–25).

*Preamplifiers* are used with most types of light sensors which typically provide relatively small output signals and sometimes have a very high output impedance. The preamplifier is located physically very close to the sensor to keep the length of the interconnecting leads as short as possible. In some designs the preamplifier is packaged integrally with the sensor in a single ''sensing head'' (''optical head''). Preamplifiers must be matched to the sensor output impedance (which can be between less than 100 and more than $10^{10}$ $\Omega$) and should provide a high and stable gain while minimizing electrical noise. Their output impedance should be low, which helps to minimize noise pickup in the cable connecting the preamplifier to an amplifier or other electronic components.

*Amplifiers* are often required to further amplify the sensor output signal and to present a signal of the appropriate level and characteristics to the display, analysis, or telemetry equipment. When the light detected by the

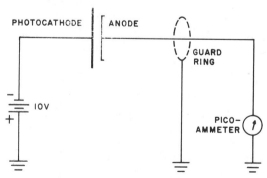

**FIGURE 22–24.** Typical photodiode operating circuit. (Courtesy of ITT Electro-Optical Products Div.)

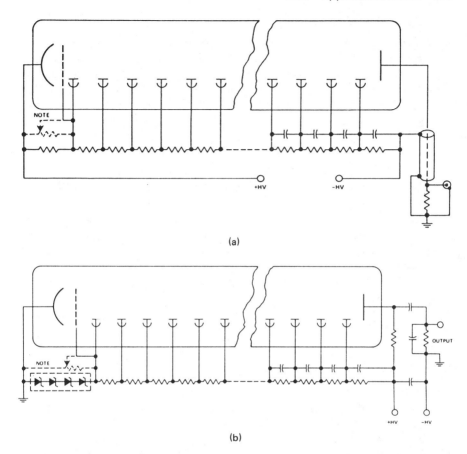

(a)

(b)

Note: In modern photomultipliers, the focusing electrode is normally connected to dynode No. 1.
In older tube types, the focusing electrode may be connected to the arm of a potentiometer,
between cathode and dynode No. 1, to permit adjustment for maximum anode current.

**FIGURE 22–25.** Typical voltage-divider arrangements for photomultiplier tubes: (a) anode return at ground potential (for fast-pulse-response and high-peak-current systems); (b) photocathode at ground potential (for scintillation counting systems). (Courtesy of Burle Industries, Inc.)

sensor is chopped, a demodulator is used to remove the chopper-induced modulation from the signal. The demodulator is synchronized with the chopper by means of synchronization pulses generated by the chopper.

### 22.6.2  *Typical Applications*

Relatively simple applications of optical detectors are in *photometry*, the measurement of illumination (or brightness). A well-known version of a photometer is the photographic light meter, a close relative of which is the light meter used to check on illumination in the work place. Many cameras now have made light meters unnecessary by incorporating a photodetector for automatic exposure control. Photometers usually operate in the visible-light region, but can be designed for UV or near-IR operation. In the far IR, radiance (or irradiance) is measured by radiometers (thermopiles, pyroelectric detectors, bolometers, with Golay cells typically used for laboratory calibration). A frequent use of these devices is the measurement of laser output power.

Another relatively simple group of applications of photodetectors includes all their uses for detection of the presence or absence of light. Objects moving along a conveyor belt can be counted, safety devices can be activated, doors can be opened, and many other control functions can equally be initiated when a light beam, normally seen by a photodetector, is interrupted. Some of these applications use visible light, and even more of them (such as intrusion detection) use IR beams. When light-beam interruption must be sensed in the presence of strong illumination (at wavelengths the detector would normally respond to), the beam can be modulated at the source; detector output conditioning circuitry is then designed to respond only to the modulation frequency used. Another application in the area of detecting the presence or absence of light is the reading of bar codes (in supermarkets and other retail stores, in banks, and even on freight cars), typically by scanning, using reflected light and either a single detector or a detector array. Light spot sensing (often using small reflectors) is also employed for positioning control, e.g., in robotics.

In electronics, photodetectors in conjunction with light-emitting diodes (LEDs) are used for circuit isolation. Such photoisolators (or optical couplers) are usually required to provide a high frequency response to transfer high-frequency data from the circuit while still isolating the circuit electrically. Similar LED-photodetector combinations are used on fiber-optics communications, with the LED at one end and the detector at the other end of an optical fiber. A high frequency response is usually of extreme importance since digital data have to be transmitted over such lines at data rates in the hundreds of MHz (and even in the low GHz region).

For certain applications, such as flame detection, "solar-blind" detection systems can be used; these use a combination of detector and window materials with filters to prevent response of the detector in those spectral regions where solar illumination is significant.

Photodetectors are also widely used in analysis equipment such as spectrometers, colorimeters, refractometers, turbidimeters, opacity detectors (including smoke detectors), and nephelometers. Their use in transducers (e.g., for measuring displacement, speed, and liquid level) is covered in other chapters of this book.

# 23

# *Nuclear
Radiation Detectors*

## 23.1 DEFINITIONS AND RELATED ITEMS

### *23.1.1 Basic Definitions*

*Nuclear radiation* is the emission of charged and uncharged particles and of electromagnetic radiation from atomic nuclei. *Charged particles* include alpha and beta particles and protons. *Uncharged particles* are typified by the neutron. Gamma rays and X-rays are forms of (nuclear) *electromagnetic radiation* (see Figure 23–1).

*Elementary particles* comprise electrons, positrons, neutrons, protons, various mesons, hyperons, neutrinos, and photons.

*Alpha particles* (α-particles) are nuclei of helium atoms; an alpha particle consists of two protons and two neutrons and has a double positive charge.

*Beta particles* (β-particles) are negative electrons or positive electrons (positrons); they are emitted when *beta decay* occurs in a nucleus, a radioactive transformation by which the atomic number is changed by +1 or −1 while the mass number remains unchanged.

*Gamma rays* (γ-rays) are electromagnetic radiation quanta resulting from quantum transitions between two energy levels of a nucleus.

*X-rays* are quanta of electromagnetic radiation originating in the extranuclear part of the atom. Note from Figure 23–1 that *hard X-rays* are more energetic than *soft X-rays* and, therefore, have a greater penetrating power. Also see bremsstrahlung, subsequently.

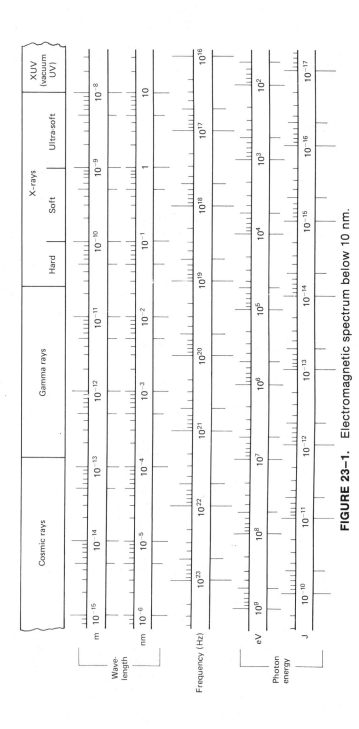

**FIGURE 23–1.** Electromagnetic spectrum below 10 nm.

*Cosmic rays* are high-energy charged particles from outer space. *Primary cosmic rays* are almost entirely composed of positively charged atomic nuclei, and about 92% of these are protons. *Secondary cosmic rays* are formed by collisions of primary cosmic rays with air nuclei in the earth's upper atmosphere and contain mostly gamma rays, electrons, mesons, and neutrinos.

*Protons* are positively charged elementary particles of mass number 1.

*Neutrons* are uncharged elementary particles of mass number 1. They can exist only briefly in a free state. The energy ranges are up to $10^8$ eV for *slow neutrons*, $10^3$ to $10^5$ eV for *intermediate neutrons*, and above $10^5$ (to about $5 \times 10^7$) eV for *fast neutrons*. *Thermal neutrons* are slow neutrons having velocities in equilibrium with the velocities of thermal agitation of the molecules in the medium in which they are situated; the energy of thermal neutrons is in the lowest portion of the slow neutron energy range (0.03 eV or less).

*Ionization* is the process of formation of atoms with a positive or negative charge (*ions*).

*Scintillation* is the emission of light energy (*photons*) by a photoluminescent material (*phosphor*) due to the incidence of ionizing radiation upon the material.

*Bremsstrahlung* is a continuous spectrum of X-radiation produced by the acceleration or deceleration applied to a high-velocity charged particle when it is deflected by another charged particle. The literal translation of this German word is "braking radiation," from *Bremse* (brake).

*Cerenkov radiation* is the radiation emitted by a high-energy charged particle when it passes through a medium in which its speed is greater than the speed of light. The index of refraction of such a medium must be greater than unity.

*Compton scattering* is the elastic scattering of photons by electrons. A *Compton electron* is an electron set in motion by interaction with a photon.

*Neutron density* is the number of neutrons per unit volume.

*Neutron flux* is the product of neutron density and speed (for neutrons of a given energy).

*Half-life* is the time required for the disintegration of half the atoms of a radioactive substance. *Biological half-life* is the time required for a living biological organism to eliminate one-half of a radioactive substance that was introduced into it by means of biological processes. The *effective half-life* is the half-life of a radioactive isotope in a biological organism, expressed as the product of radioactive and biological half-life divided by the sum of radioactive and biological half-life.

*Dose* is a quantity of ionizing radiation to which a biological sample is exposed or which it absorbs (*exposure dose, absorbed dose*).

### 23.1.2  *Related Fundamental Constants*

*Speed of light* in vacuum: $c = 2.997\ 925 \times 10^8$ m/s (round off to $3 \times 10^8$)

*(Unified) atomic mass unit (amu):* $u = 1.660\ 565 \times 10^{-27}$ kg (round off to $1.66 \times 10^{-27}$)

*Electron rest mass:* $m_e = 9.109\ 534 \times 10^{-31}$ kg (round off to $9.11 \times 10^{-31}$)

*Proton rest mass:* $m_p = 1.672\ 648 \times 10^{-27}$ kg (round off to $1.673 \times 10^{-27}$)

*Neutron rest mass:* $m_n = 1.674\ 954 \times 10^{-27}$ kg (round off to $1.675 \times 10^{-27}$)

*Electron charge:* $e = 1.602\ 189 \times 10^{-19}$ C (round off to $1.602 \times 10^{-19}$)

*Planck's constant:* $h = 6.626\ 176 \times 10^{-34}$ J·s (round off to $6.626 \times 10^{-34}$)

$\quad = 4.135\ 673 \times 10^{-15}$ eV·s (round off to $4.136 \times 10^{-15}$)

### 23.1.3  *Related Laws*

*Special Theory of Relativity*

$$E = mc^2 \qquad m = \frac{m_0}{\sqrt{(1 - v^2/c^2)}}$$

where $E$ = energy
$\quad m$ = mass (in motion)
$\quad m_0$ = mass (at rest) (*rest mass*)
$\quad c$ = velocity of light
$\quad v$ = velocity of mass

*Planck's Law*

$$\mathscr{E} = h v$$

where $\mathscr{E}$ = photon energy, J     or     $\mathscr{E}$ = photon energy, eV
$\quad v$ = frequency, Hz              $\quad v$ = frequency, Hz
$\quad h$ = Planck's constant, J·s       $\quad h$ = Planck's constant, eV·s

*Radioactive Decay Law*

$$N = N_0 e^{-\lambda t}$$

where $N$ = number of atoms (of a radioactive species) at time $t$
$N_0$ = number of atoms present at $t = 0$
$t$ = elapsed time
$e$ = base of natural logarithm ($\approx$2.71828)
$\lambda$ = decay constant

## 23.2 UNITS OF MEASUREMENT

*Radioactivity* (activity of a radionuclide) is expressed in becquerels. The *becquerel* ($Bq$) is the activity of a radionuclide decaying at the rate of one spontaneous transition per second ($Bq = s^{-1}$). The becquerel is the SI unit that replaces the curie. The *curie* ($Ci$) equals the quantity of any radioactive material in which the number of disintegrations per second is $3.7 \times 10^{10}$. *Conversion: 1 Ci = 3.7 $\times$ 10$^{10}$ Bq.*

*Absorbed dose* is expressed in grays. The *gray* ($Gy$) is the absorbed dose when the energy per unit mass imparted to matter by ionizing radiation is 1 joule per kilogram ($Gy = J/kg$). The gray is the SI unit that replaces the rad. The *rad* is the absorbed dose when the energy per unit mass imparted to matter by ionizing radiation is 100 ergs per gram. *Conversion: 1 rad = 10$^{-2}$ Gy.*

*Exposure dose* (X-ray and gamma-ray exposure) is expressed in coulombs per kilogram. The *coulomb per kilogram* ($C/kg$) is the exposure when X rays or gamma rays produce in free air 1 coulomb of electrical charge per kilogram of dry air. The C/kg is the SI unit that replaces the roentgen. The *roentgen* ($R$) equals the quantity of X-ray or gamma-ray radiation whose associated secondary ionizing particles produce ions, in air, carrying one electrostatic unit of charge (of either sign) per 0.001 293 g of air. *Conversion: 1 R = 2.58 $\times$ 10$^{-4}$ C/kg.*

*Dose equivalent* (in radiobiology) is expressed in sieverts. The *sievert* ($Sv$) is the dose equivalent when the absorbed dose of ionizing radiation multiplied by the dimensionless factors $Q$ (quality factor) and $N$ (product of any other multiplying factors) stipulated by the International Commission on Radiological Protection is 1 joule per kilogram ($SV = J/kg$). The sievert is the SI unit that replaces the rem. The *rem*, a unit whose name was derived from "*roentgen equivalent man*," is the unit of *relative biological effectiveness dose* (*RBE dose*). It equals the absorbed dose (in rads) times an agreed conventional value of the RBE. It was originally defined as the absorbed dose that will produce the same effect in human tissue as that produced by one roentgen. *Conversion: 1 rem = 10$^{-2}$ Sv.*

Energy (electromagnetic energy, or *photon energy*, since the photon is a quantum of electromagnetic energy) is expressed in joules. The *joule*

(*J*) is the SI unit that replaces the *electron volt* (*eV*). *Conversion: 1 eV = 1.602 × 10⁻¹² erg = 1.602 × 10⁻¹⁹ J (1 erg = 10⁻⁷ J).*

*Neutron flux* is expressed in *neutrons per square meter-seconds* (*n/m²·s*); the use of the centimeter instead of the meter is fairly common (n/cm²·s).

The *window thickness* of a radiation detector is commonly expressed in *milligrams per square centimeter* (*mg/cm²*).

## 23.3   SENSING METHODS

The methods of sensing nuclear radiation are generally based on the interaction of the radiation with a substance contained in the detector. The detection of electrons usually involves multiplication of the incoming electrons within the detector.

### 23.3.1   Radiation Detection Using Ionization

The ionization radiation method of detection relies upon the production of an ion pair in a gaseous or solid material and the separation of the positive and negative charges by an electric field to produce an output signal.

Charged particles, such as alpha and beta particles and protons, can exert sufficient electromagnetic forces on the outer electrons of atoms which they pass at high velocity to separate one of the electrons. When this occurs, an ion pair is formed. The ions move to the electrodes of opposite sign, and an ion current is created. One particle can cause such ionization to take place several times before its energy is expended. Electrons that are released in the ionization process can be accelerated sufficiently to produce additional ions.

Neutrons and other uncharged particles can produce ionizing particles by transferring some of their energy to nuclei with which they collide. X-rays and gamma rays can remove secondary electrons from atoms with which they interact, and these electrons then produce ion pairs.

Radiation detection by ionization in different substances is illustrated in Figure 23–2. The separation of positive and negative ions is due to the electric field between two electrodes (cathode and anode), or between the positively and negatively charged material of a semiconductor, formed by connecting them across a dc power supply. In the *current mode* of operation, the increase in current caused by the flow of charges to the electrodes of opposite polarities (the *ionization current*) can be monitored as the average *IR* drop across the load resistor ($R_L$). In the *pulse mode* of operation, the ionization is measured as a single event. The output is a train of voltage pulses, each generated by the ionization due to one particle. The output pulses are conveniently taken through a coupling capacitor.

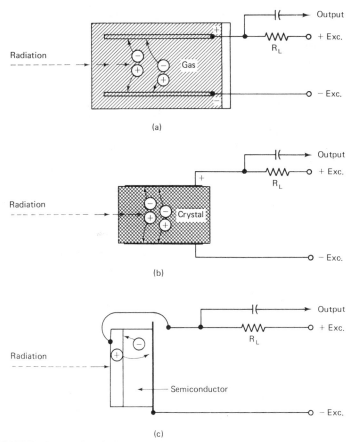

**FIGURE 23–2.** Radiation detection using ionization (shown for output in pulse mode; output is taken directly across $R_L$ in current mode): (a) ionization in gas; (b) ionization in solid crystal; (c) ionization in solid semiconductor.

Different particles cause different amounts of ionization; hence, the output pulse amplitude can be indicative of the types of incident particles.

The ionization current will also be greater when some of the same type of particles have higher energy than others of that type. The pulse amplitude then becomes a measure of particle energy. The average ionization current resulting from a steady radiation flux can be used as a measure of the average magnitude of this flux.

Ionization due to nuclear radiation occurs in certain gases and solids to various extents. Figure 23–2(a) shows ion pairs formed in a gas such as argon or krypton. Ionization in solid crystals, such as diamond or silver chloride, has also been used for radiation detection. Charge separation and flow of ionization current are effected by two electrodes of opposite polarity, on opposite crystal surfaces, connected to a dc power supply, as shown in Figure 23–2(b). Solid-crystal detectors have been largely replaced by semiconductor detectors. Semiconductor materials, such as germanium and silicon, can produce positive/negative charge pairs (electron-hole pairs) as a result of incident radiation in the presence of an electric field established by the junction between *p*- and *n*-type material when dc excitation is applied, as illustrated in Figure 23–2(c)

### 23.3.2   Radiation Detection Using Scintillation

Certain materials have been found to convert nuclear radiation into light with a relatively high conversion efficiency. Such *scintillator* materials are used in conjunction with a photon detector in radiation detectors. Photomultipliers are used for the detection of the photons produced in the scintillator (see Figure 23–3). The scintillator is optically coupled very closely with the photocathode of the multiplier. Due to their light emitting action, scintillator materials act as *phosphors*. These materials can be organic or inorganic, solid, liquid, or gaseous.

A different effect, unrelated to scintillation, also produces light in response to nuclear radiation. The *Cerenkov effect* (first observed by Cerenkov in 1934 in the USSR) occurs in pure, transparent, nonluminescent material (e.g., certain glasses) whose refractive index is such that the velocity of the radiation through the material is greater than the velocity of light through the material. Cerenkov radiators are used in conjunction with photon detectors in detectors known as *Cerenkov counters*, for the detection of very high-energy radiation.

### 23.3.3   Electron Detection

The detection of electrons, as such, can be performed by a very high input-impedance *electrometer tube* or by other electrometers such as the quartz-fiber and vibrating-reed types. The most commonly used devices, however, are *electron multipliers,* in which the incoming electron releases secondary electrons in increasing numbers (*avalanche effect*) from a number of dynodes or one continuous dynode. The dynodes are placed ahead of the anode, and the output signal is a change in anode current due to total electron flow toward the anode.

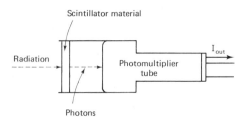

**FIGURE 23–3.** Radiation detection using scintillation.

## 23.4 DESIGN AND OPERATION

Emphasis in this section is placed on those types of radiation detectors that are readily available commercially. Some additional types are described briefly as well. Categorized by transduction principle, ionization chambers, proportional counters, and Geiger counters are gas-filled ionizing sensors, semiconductor detectors are solid-material ionizing sensors, scintillation counters use photoelectric sensors, and electron detectors are related to photoemissive sensors (see Section 22.4.4).

### 23.4.1 Ionization Chambers

The usual configuration of an ionization chamber ("ion chamber") is cylindrical (Figure 23–4). The outer metallic cylinder is the cathode, while an internal rod or wire running along the axis of the cylinder is the anode. The outer surface of the cathode is usually uninsulated since it is normally kept at ground potential. Electrical connections are simple and are not necessarily peculiar to the configurations illustrated. The *end-window* ionization chamber (Figure 23–4(a)) mates with a socket that provides spring contacts for the cathode and an insulated female socket for the anode pin. The *side-window ionization chamber* (Figure 23–4(b)) has leads attached to the cathode and anode terminals. The *windowless ionization chamber* (Figure 23–4(c)) is shown with two connector pins of different diameters. The "pancake" style (Figure 23–4(d)) is a special configuration of compact design. Cathodes are usually made from a corrosion-resistant metal such as stainless steel. Earlier designs have a glass envelope surrounding cathode and anode.

Although all gases will produce some ionization, those used in ionization chambers are selected for optimum ionization potential and energy per ion pair. Typical fill gases are argon, krypton, neon, xenon, helium, hydrogen, nitrogen, and certain compounds such as barium trifluoride ($BF_3$) and methane ($CH_4$). Favorable operation can be obtained by admixing a small amount of a different gas, for example, neon with a small amount of argon, xenon or argon with a small amount of nitrogen. The pressure of the gas within the chamber can be below, at, or above atmospheric pressure. When very thin windows are used, it must be close to atmospheric pressure to minimize any pressure differential across the window.

Window material and thickness are governing factors of the response of an ionization chamber to different types and energy levels of radiation to be measured. If the mean range of a heavy particle or a particle of very low energy is exceeded by the window thickness, the particle will be stopped by the window and will not be detected. Hence, very thin windows are required for alpha-particle detection. Mica windows having a thickness of less than 4 mg/cm$^3$ are typically used to meet this requirement. Thin metal windows are used for beta, gamma, and X-rays when no alpha radiation

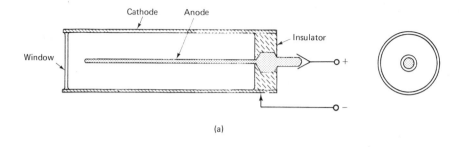

(a)

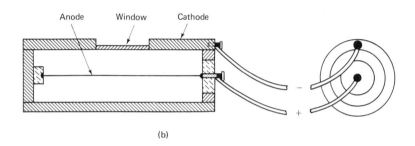

(b)

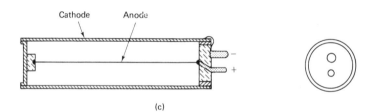

(c)

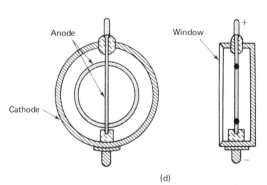

(d)

**FIGURE 23–4.** Typical configurations of ionization chambers: (a) with end window; (b) with side window; (c) without window; (d) pancake style, with window.

needs to be measured. Windowless thin-wall chambers have a nearly spherical field of view. Their cylindrical case, typically 30 to 50 mg/cm³ thick, acts as a window as well as cathode. When narrow fields of view are required, the chamber can be placed in a thick-walled enclosure (*shield*) having a small aperture. For gamma-ray detection, windows should be made of a metal having a large gamma-ray absorption coefficient (e.g., tantalum or tungsten). Stainless steel has been used as a compromise material for windows and thin-wall-chamber cathodes when both beta and gamma rays are to be detected. Anodes are typically made of thin tungsten wire.

### 23.4.2 Proportional Counters

When the excitation voltage across the electrodes of a gas-filled chamber is close to zero, most ion pairs will be lost due to recombination. As the voltage is increased, the recombination loss is negligible and ionization-chamber operation is achieved. As the voltage is further increased, the chamber operates in the *proportional region* where *gas amplification* occurs: electrons released in the primary ionization are accelerated sufficiently to produce additional ionization and thus add to the collected charge. The output pulse is proportional to the total collected charge, which is proportional to the energy of the incident particle that causes the primary ionization. As the voltage is increased still further, the output pulse size becomes increasingly independent of the primary ionization, until a region called the *Geiger plateau* is reached (see Section 23.4.3) in which the collected charge is entirely independent of the amount of initial ionization.

In gas amplification (gas multiplication) the additional ions multiply themselves by an avalanche effect. Each secondary electron ionizes additional gas molecules, producing more secondary electrons, and so on. The avalanche effect starts at a threshold value of anode-to-cathode potential and increases when the potential is raised above that value. Threshold values and the gas amplification factor (which can vary between 1 and $10^6$) depend on excitation voltage, detector geometry, and the fill gas. Gas mixtures such as xenon/nitrogen, argon/$CO_2$ and hydrogen/$CH_4$ are often used in detectors which operate in the proportional region and are, therefore, called *proportional counters*. The admixture of gases such as $CO_2$ and $CH_4$ (methane) greatly reduces the ionization caused by ultraviolet radiation created during the avalanche process; the absorption of this UV radiation by the gas additives allows higher anode voltages to be used and larger gas amplification factors to be realized (see Figure 23–5) while maintaining stable counter operation.

Since a proportional counter produces a pulse for each ionizing event, and since the amplitude of the output pulse is proportional to the energy of the incident radiation (as long as the counter operates within the truly proportional region), this type of detector can be used to differentiate between

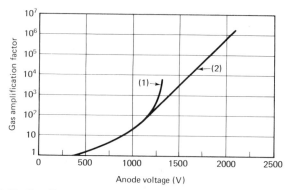

**FIGURE 23–5.** Gas amplification factor as a function of anode voltage for (1) pure argon and (2) 95% argon/5% $CO_2$.

various particles (e.g., to measure alpha particles in the presence of beta and gamma radiation). This is accomplished by pulse-height discrimination in the output conditioning circuitry.

The windowless *gas-flow counter* is a special version of the proportional counter. (It can also be a special version of the Geiger counter.) It is very usable for the detection of alpha radiation and weak beta radiation such as emitted by certain commonly used radioisotopes (e.g., $C^{14}$, $S^{35}$, $Ca^{45}$, and $H^3$). The device consists of an ionization chamber hood, equipped with inlet and outlet hose connections, that is slipped over and sealed against a sample holder (see Figure 23–6). A gas (-mixture) supply is connected to the counter, and the required voltage is applied to it while the gas flows through the counter. The quick make-and-break seal allows the rapid insertion and removal of radiation-emitting samples. Continuous replacement of the gas also extends the life of the counter substantially. Very thin replaceable windows (about 0.25 mg/cm³) are sometimes used in gas-flow counters to reduce possible effects of static charges, accidental contamination from a loosely packed sample, and vapor effects from moist samples.

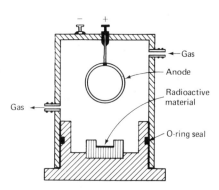

**FIGURE 23–6.** Gas-flow counter.

### 23.4.3 Geiger Counters

As explained at the outset of Section 23.4.2, increasing the anode-to-cathode potential of a proportional counter above the proportional region causes it to operate in a different region, called the *Geiger-Mueller (G-M)* region, in which the total collected charge, greatly increased by gas amplification, is independent of the energy that initiated the primary ionization. If the anode potential is increased still further, the discharge region is reached. Discharge can, however, be kept from becoming continuous (*quenched*) by admixing a *quenching agent* to the gas in the chamber. The quenching vapor, typically alcohol or a *halogen* such as bromine or chlorine, quenches the discharge by preventing the production of secondary electrons by positive ions at the cathode. A counter that operates in this region and contains a quenching agent is called a *Geiger-Mueller tube*, or *G-M tube*, or *Geiger counter*.

There is, of course, a limit to the anode potential that can be applied before discharge becomes continuous, even in the presence of a quenching agent. The anode-to-cathode potential must be kept within a *plateau* bounded by minimum and maximum voltage limits (see Figure 23–7). The quenching action begins to fail when the voltage exceeds that maximum, and operation reverts to the proportional mode when the voltage decreases below the minimum. Hence, the anode voltage is set, and regulated about, a value equivalent to the center of the plateau. Note that the absence of such a plateau characteristic in proportional counters (and some ionization chambers) poses a much more severe requirement for the regulation of their power supplies.

Since some of the quenching fluid is dissociated during each discharge, alcohol-quenched Geiger counters have a limited life (about $10^8$ counts). Halogen quenching gas, however, tends to recombine after dissociation and does not impose a constraint on the life of the counter. Gas-flow proportional counters can be operated as Geiger counters by admixing a quenching agent to the gas and raising its anode voltage into the Geiger region.

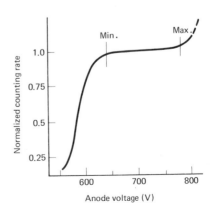

**FIGURE 23–7.** Plateau of typical Geiger counter.

The output pulse of a Geiger counter is characterized by a fast rise time (less than 1 μs) and a duration of several μs before its decay due to quenching following the discharge. The decay of the pulse is followed by a *dead time* of 50 to 150 μs. During a subsequent *recovery time*, usually shorter than the decay time, the counter gradually becomes capable of producing a full-height output pulse again. The counter is inoperative during the dead time, and any particle that could produce an ionizing event would not be detected during this time. A particle causing an ionizing event during the recovery time would create an output pulse of less than full height.

Dead time can be reduced by pulsing the anode voltage rather than applying it steadily. The pulses extend from a quiescent voltage at which no ionization occurs, or even from a slightly negative bias level, to the voltage required for operation in the Geiger region. The counter is then operative only while the Geiger-region voltage pulse is applied. The pulse on and off times are selected as much shorter than the normal dead time and recovery time. As a result, the counter will respond to a larger number of particles per unit time.

### 23.4.4  *Semiconductor Detectors*

Semiconductor detectors tend to be significantly more efficient than gas-filled ionization detectors: the material is about $10^3$ times denser, its average threshold energy for electron-hole pair production is roughly 10 times lower, its carrier mobilities are higher, and the difference between the mobilities of positive and negative charges is less.

*Intrinsic semiconductors* are pure crystals; the concentration of charge carriers is characteristic of the material itself rather than of the impurity content or any imperfections within the crystal. Intrinsic semiconductor detectors have been used for gamma-ray and X-ray measurements as well as light measurement. Charge carriers (electron-hole pairs) are produced in intrinsic semiconductor material (typically Si or Ge) by the interaction between incident photons and the atoms of the material as a result of (1) the elastic scattering of photons by electrons, which sets other electrons into motion (*Compton effect*); (2) the ejection of a bound electron from an atom with absorption of all of the photon's energy (*photoelectric effect*); and (3) the conversion of a photon, traversing a strong nuclear electric field, into an electron and a positron (*pair production*). When an excitation potential is applied across the two electrodes on the crystal, incident radiation will create electron-hole pairs, with the electrons and holes flowing toward their corresponding electrodes; the resulting current can then be detected as an IR drop across the load resistor $R_L$ and becomes the detector's output voltage (see Figure 23–8).

The output pulse produced by each event is characterized by a short rise time and a somewhat slower decay time because the electrons have a

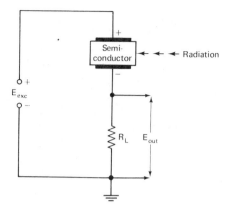

**FIGURE 23–8.** Intrinsic-semiconductor radiation detector.

mobility typically three times that of the holes. Although noise current is reduced by using materials with high bulk resistivity, the thermal noise level of intrinsic semiconductor detectors at room temperature tends to be very high. Far more efficient operation is obtained by cryogenic cooling of the detector.

*High-purity* (*HP*) detectors are intrinsic semiconductors. *HP Ge* detectors have been used for gamma-ray detection as well as gamma-ray spectroscopy. Although they can be exposed to room temperature while nonoperational, they are used while cooled to cryogenic temperatures, effected by a cryostat or dewar (typically filled with liquid nitrogen to keep the detector at a temperature of 77 K). Preferably, HP detectors should operate in a vacuum, and the same dewar can be used for cryosorption pumping to maintain a vacuum for the detector.

*Extrinsic semiconductors* are those whose properties are dependent on controlled impurities added to the crystal. They exist in two major types. The *surface-barrier* type (Figure 23–9(a)) usually consists of *n*-type single-crystal silicon on one surface of which a *p*-type layer of silicon dioxide is

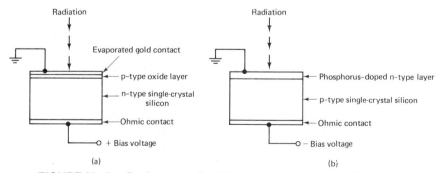

**FIGURE 23–9.** Basic types of extrinsic-semiconductor radiation detectors: (a) surface-barrier type; (b) diffused-junction type.

formed. This thin (often monomolecular) layer is typically covered with an evaporated gold film which serves as contact (electrical ground). The back electrode, to which the bias voltage is applied, is also formed by evaporation; gold or aluminum is typically used. A few designs use p-type silicon with an n-type layer; here the ground contact (entrance contact) is at positive potential, and a negative bias voltage is applied to the back electrode. The *diffused-junction* type (Figure 23–9(b)) is commonly made by a shallow diffusion of phosphorus to form an n-type layer in p-type single-crystal material; p-type diffusion in n-type material has also been used in a few designs.

N-type silicon has an excess of electrons (negative charges), and p-type silicon has an excess of holes (positive charges). An electric field (potential gradient) exists in the space-charge region (*depletion layer*) close to the junction, where the net charge density of donors and acceptors is not neutralized by the mobile-carrier density and is, hence, substantially different from zero. It is in this *depletion region* that electron-hole pairs are produced by incident radiation. The depth of the depletion region (the *depletion depth*) is governed by the resistivity (bulk resistivity) of the silicon wafer and by the externally applied bias voltage (voltage applied across the two electrodes). The depletion depth must be chosen on the basis of the maximum energy of the particle to be detected. The *range in silicon* (determined from particle type and energy) required to stop that particle then equals the required depletion depth. Once this depth has been established, the appropriate combination of bulk resistivity and bias voltage can be calculated (or read off a nomogram). In *partially depleted* detectors the depletion depth is some fraction of the silicon wafer thickness. In *totally depleted* detectors the depletion depth is equal to the wafer thickness.

Between the sensing surface of the detector and the active region there exists a layer in which some of the incident energy is dissipated without producing electron-hole pairs collectable by the electrodes (*dead layer*). The dead layer of surface-barrier detectors is usually smaller than that of diffused-junction detectors. This advantage is somewhat offset by the location of the barrier right at the surface, which makes surface-barrier detectors more subject to mechanical damage.

In the *p-i-n junction* radiation detector the n- and p-regions are separated by an intrinsic region to increase depletion depth.

Besides Si and Ge, a few other semiconductor materials are used in radiation detectors. Although some of these are still in various stages of development, *cadmium telluride (CdTe)* detectors have been in production for some time. They can be used, at room temperature, for beta, X-ray, and gamma detection. CdTe surface-barrier-type detectors are characterized by the nearly intrinsic resistivities of their large single crystals.

The implementation of recent technology developments in semiconductor manufacture can be expected to show increasing use of ion-implantation and epitaxy techniques.

### 23.4.5 Scintillation Counters

In a scintillation counter the scintillator (phosphor material in the counter's enclosure) acts as the sensing element of a transducer in that it produces a burst of light quanta (photons) in response to an incident radiation particle; the light sensor, usually a photomultiplier tube (see Sections 22.4.4 and 22.6.1), acts as the transduction element, converting light into an output current (or, as IR drop across a load resistor, an output voltage). The term "counter" should really be applied to the entire detection system, but is commonly applied to the detector itself. Figure 23–10 illustrates a scintillation-counter system. There are two basic groups of scintillators: Cerenkov scintillators, which are used in specialized high-energy-particle detection and are made of transparent solids such as glasses, and fluorescent scintillators, which are made of organic or inorganic crystals or plastic or liquid phosphors (solution scintillators).

Inorganic crystals are impurity activated; the impurity content is less than 0.1%. In scintillator nomenclature the symbol of the activator is placed in parentheses behind the symbol of the compound of the crystal. Thallium-activated sodium iodide, *NaI(Tl)*, is the most frequently used material; it produces the highest light output, and its wavelength of maximum emission is 410 nm, well within the spectral response of many photomultipliers. Other thallium-activated alkali halides used as scintillators are CsI(Tl), CsBr(Tl),

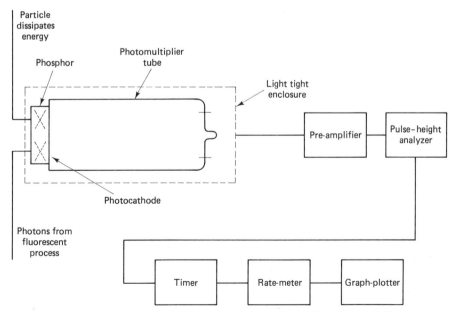

**FIGURE 23–10.** Scintillation counter system. (Courtesy of Thorn EMI Electronics, Ltd.)

and KI(Tl), with additional such materials used less frequently. CsI(Tl) has a wavelength of maximum emission between 420 and 570 nm, but its light output (relative pulse height) is only about 25% of that of NaI(Tl). There are other activators besides thallium. CsI(Na) provides a relative pulse height about 60% that of NaI(Tl), and LiI(Eu) provides a relative pulse height about 30% that of NaI(Tl). The decay constant is shortest for NaI(Tl), about 0.25 $\mu$s. The decay constants of other inorganic crystals, including $CaF_2(Eu)$, are between 0.9 and 1.4 $\mu$s.

Organic crystals are typically grown from coal-tar derivatives such as naphthalene ($C_{10}H_8$) and anthracene ($C_{14}H_{10}$), which have conjugate double bonds in their benzene-ring structures. Other organic scintillator materials include naphthacene, pentacene, and trans-stilbene. The relative pulse height of these materials is 50% or less compared to NaI(Tl); however, their decay constants tend to be in the nanosecond region.

Solution scintillators are usually either solid plastics (plastic phosphors) or solid materials dissolved in a liquid such as toluene. An example of a plastic scintillator is a polymerized solution of *p*-terphenyl and tetraphenylbutadiene in styrene. These scintillator materials are characterized by very short decay constants, on the order of a few nanoseconds; their relative pulse height tends to be between 10 and 20% compared to NaI(Tl). Gaseous scintillators (e.g., xenon) are infrequently used.

Photomultiplier tubes, normally used in conjunction with scintillators, are discussed in Section 22.4.4. The optical coupling between scintillator and photocathode must be very close. It can be a thin layer of silicone oil, or a glass or plastic "light pipe" which may be relatively long. The assembly comprising scintillator, optical coupling, and photomultiplier tube (and, often, its dynode voltage-divider resistors) is covered by an enclosure that is lightproof (except at the scintillator sensing surface) and frequently acts also as a magnetic shield. The emission spectra of scintillator and photocathode should be closely matched. Scintillator peak-emission wavelengths tend to be in the blue region. Certain compounds (*wavelength shifters*) can be added to some scintillator materials to shift their spectral response toward the red, if required by photocathode characteristics.

### 23.4.6  Electron Multipliers

Electron multipliers are used for the detection of electrons, as such, or as a detection element in other instruments such as mass spectrometers. They are either of the multiple-dynode type or of the continuous-dynode type. Multiple-dynode electron multipliers contain typically between 10 and 14 dynodes and are very similar to the dynode assemblies of photomultiplier tubes (see Section 22.4.4); the similarity extends to their operation in a vacuum. A typical design is shown in Figure 23–11.

**FIGURE 23–11.** Fourteen-dynode electron multiplier, with in-line cage structure. (Courtesy of Burle Industries, Inc.)

The operating principle is the avalanche effect due to secondary electron emission. The anode is operated at a high potential (typically between 2000 and 4500 V dc) referred to electrical ground. A resistance network is connected between each of the dynodes and the anode so that voltage increments are equal between successive dynodes. Incoming electrons impact the first dynode, where they release secondary electrons. Because of the higher potential on the second dynode, these electrons are all accelerated toward it. Upon impact with the second dynode additional secondary electrons are released by the already multiple electron stream, and this process continues until a vastly multiplied electron stream reaches the anode. By use of this avalanche effect, current amplifications between $10^5$ and $10^6$ can be attained. However, it must be noted that the current of the electrons entering the entrance aperture of the electron multiplier (the "radiation opening") is on the order of some picoamperes; hence, the total anode current is usually on the order of one or a few microamperes. It is, therefore, important that the preamplifier to which the electron multiplier (*EM*) is connected be coupled to it as closely as possible. Copper-beryllium is most commonly used as dynode material, and many designs use only high-temperature metal and ceramic materials in their construction. This permits bake-out at temperatures between 450 and 600°C, as required for use in ultra-high-vacuum systems. Besides the in-line cage structure illustrated, the other cage structures used in photomultipliers (venetian blind, circular cage, etc.) are also employed in electron multipliers.

Electron multipliers of the multiple-dynode type as well as those of the continuous-dynode type described next are generally usable for the detection of positive and negative electrons, positive and negative ions, protons, and electromagnetic radiation in the soft X-ray and vacuum ultraviolet (XUV) regions.

The continuous-dynode *channel electron multiplier* (*CEM*) is a hollow glass cylinder with a resistive secondary-emission coating on its inside surface. The electron multiplication (typically around $10^7$ to $10^8$) is a result of cascade action in which incoming (primary) electrons collide with the secondary-emission-coated inner wall and cause several secondary electrons to be emitted. These secondary electrons become primary electrons for the next collision with the wall surface farther along the cylinder (*channel*) wall. A high-voltage potential is applied across the two ends of the channel, setting up the axial electric field that accelerates the continuously multiplying electrons from the entrance-aperture end to the exit end of the channel. The cascade and multiplication process continues down the channel length (see Figure 23–12(a)) until the end is reached or gain saturation occurs. For CEMs having an *l/d* ratio between 60 and 100, gain saturation results from the device having a net secondary-emission gain of unity in the last portion of the channel length. This gain saturation results in output pulses which are all of nearly constant amplitude.

CEMs exist in a number of curved configurations. The most common configuration is shown in Figure 22–12(b). Cones of various sizes can be used to expand the input aperture diameter, but are not used on all designs. The anode (collector) can be user furnished, spaced about 1 mm away from the channel exit, and supplied with a voltage of about $+250$ V dc, referenced to the high-voltage (2 to 3 kV) supply's positive terminal, which is then connected directly to a contact at the exit end of the channel. The collector can also be furnished in the form of a thin metallic cap, bonded to the exit end with conductive epoxy and electrically connected as shown in Figure 23–12(b); however, this reduces the gain significantly and reduces the bakeout temperature from 300°C to about 100°C. The configuration shown has a subtended arc of 270°. Other configurations include helical channels subtending arcs of 840°. Electron gain varies with excitation voltage (high voltage across channel ends), with this relationship becoming nonlinear above a gain of about $10^6$ and then becoming asymptotic to a gain of around $10^8$. The diameter of the circular-arc configuration illustrated is 4.6 cm.

The Spiraltron® CEM is used where configurational constraints impose a preference for a straight-line, rather than a curved, design. This detector has a conical entrance aperture, a straight channel "preamplifier" section, and a section containing six thin channels which have been twisted together. Spiraltron® electron multipliers (SEMs) can be connected (with a collector bias of around $+200$ V, referenced to the "+ H.V." terminal) either in a two-terminal mode, with "+ H.V." at the exit end and "− H.V." at the (grounded) aperture end, or in a three-terminal mode, with an intermediate terminal located at the junction of the preamplifier and twisted-channel sections. A typical SEM is 5.3 cm long and is illustrated in Figure 23–13.

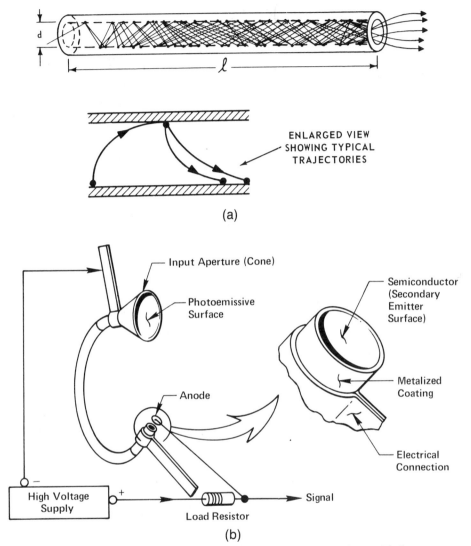

**FIGURE 23-12.** Continuous-dynode channel electron multiplier (CEM): (a) principle of CEM operation; (b) Channeltron® electron multiplier. (Courtesy of Galileo Electro-Optics Corp.)

### 23.4.7 *Neutron Detectors*

Neutrons, which are uncharged particles, are detected by detecting charged particles emitted from a material in which a nuclear reaction is caused by the incident neutrons. The reaction can be caused by the interaction of the neutrons with atomic nuclei, or by recoils of charged particles from collisions

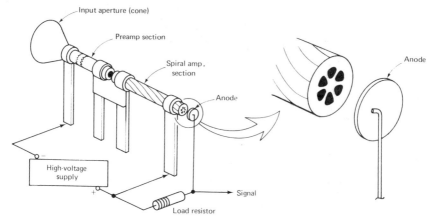

**FIGURE 23–13.** Spiraltron® electron multiplier. (Courtesy of Galileo Electro-Optics Corp.)

with neutrons. The materials (*conversion materials*) used for this purpose are usually stable isotopes of elements having a high absorption cross section in the energy range of interest. They can be employed either as a fill gas or as an internal coating, or both, in an ionization chamber or proportional counter; they can be used as a constituent of a scintillator material; or they can be the material of a film, or on a foil, placed in front of the sensitive area of a charged-particle or gamma-ray detector (depending on the reaction used and its products).

Lithium and boron isotopes are often used as conversion materials for *slow-neutron* detection. In the $Li^6(n,\alpha)T$ reaction, in which alphas and tritons ($T$) are produced when neutrons react with $Li^6$, the most penetrating particle (besides the neutron) is the triton. In the $B^{10}$ $(n,\alpha)Li^7$ reaction, in which alphas and $Li^7$ nuclei are produced as charged particles, the $\alpha$ particle has the greater range. $B^{10}$ is commonly used in ionization chambers and proportional counters in the form of $B^{10}$-enriched $BF_3$ (boron trifluoride) gas. In semiconductor detectors the conversion material is deposited on thin (0.5-mm-thick) metal (typically aluminum or stainless steel) foil, with a plastic light-tight cap providing the retainer for the foil as well as a cover for the detector. In another design, the conversion material is vacuum-evaporated onto the active surface of a semiconductor detector, which is then mounted face to face with another such detector. The resulting "sandwich" geometry allows simultaneous counting of the two reaction products in the two detectors.

*Fast neutron* detection requires different conversion materials. $He^3$, used as fill gas in proportional counters, and pressurized at between 1 and 10 atm, is usable from the thermal through the fast neutron range. $Li^6F$ (solid) can also be used for neutron energy ranges from slow to fast. $H^1$ has been used as a polyethylene film for fast-neutron conversion. Semiconductor detectors can be provided with a layer of hydrogenous material for fast-neutron detection; recoil protons, resulting from ($n,p$) scattering, are then the detected particles. An alternative is to use a $Li^6F$ or $B^{10}$ layer and surround the entire detector in paraffin; fast neutrons are moderated in the paraffin and detected as slow neutrons. Heavy charged particles can also be produced, as *fission fragments*, by neutron reactions with materials such as $Pu^{239}$, $U^{235}$, $Np^{237}$, and $Th^{232}$.

## 23.5 DETECTOR CHARACTERISTICS

*Mechanical characteristics* such as outline dimensions, dimensions and locations of mountings and electrical connections and of gas connections (for gas-flow counters), case material and sealing, identification markings, and mass ("weight") are described for radiation detectors in the same manner as for other types of transducers. Additionally, window material and thickness are shown, for gas-filled ionization-type counters the cathode material is described (if different from the case or envelope material), and shown as well are anode thickness and material, effective cathode dimensions, and type and composition of the fill gas and its pressure (and type of quenching agent for Geiger counters). For scintillation counters, descriptions are included of the scintillator dimensions and material and the type of optical coupling, and either the model number of a commercial photomultiplier (or CEM with photocathode) is given, or the complete characteristics of an integrally packaged scintillation detector are provided. For semiconductor detectors, the material, sensitive area and wafer thickness (or volume, for some types of detectors), and depletion depth are stated.

    *Electrical characteristics* include the range of operating excitation voltages (anode voltages or equivalent), maximum allowable voltage, and separate bias voltage, if any. For gas-filled detectors, sufficient details must be given to establish output impedance and load characteristics, such as the recommended range of external series resistances, internal capacitance, and conductor and insulator materials of electrical connections. Maximum anode current is shown for scintillation counters and electron multipliers, and plateau characteristics are shown for Geiger counters. For semiconductor detectors, bulk resistivity, internal capacitance, bias voltage, and depletion depth are shown.

    *Performance characteristics* list the types of radiation to be measured by the detector as well as the characteristics of wanted outputs and those of unwanted outputs. The latter comprise the maximum *background counts*, or *dark counts*, per second or per minute, maximum leakage current (for semiconductors), and maximum *dark current* (for photomultiplier scintillation counters) at a nominal operating voltage. Output pulses are described by rise time and decay time.

Transfer characteristics are shown as detection efficiency for given types of radiation (in percent), as charge collection efficiency for some types of semiconductor detectors, or as counting efficiency. Timing characteristics are often shown, mainly to facilitate coincidence-circuit design. Energy resolution is shown as *full width at half maximum* (*FWHM*) for semiconductor detectors; this is the width, expressed in energy units, of a monoenergetic peak, as measured at an amplitude corresponding to 50% of the peak height. For some detectors the *full width at one-tenth maximum* (*FWTM*) is additionally specified. When FWHM is shown for scintillation and proportional counters, the term refers to the width of an energy distribution curve (which somewhat resembles a Gaussian distribution curve) measured at one-half the height of the peak of this curve. Energy resolution is the smallest difference in energy between two particles (or photons) that can be discerned by the detector.

Geiger-counter specifications should show dead time as well as recovery time. The effects of photosensitivity and the hysteresis, if any, should also be stated. For scintillation counters, the wavelength of maximum scintillator emission (or a curve descriptive of the complete emission spectrum) is usually shown in specifications. Where detector life is limited (e.g., by radiation damage, or by use of an organic quenching fluid in a Geiger counter), the detector life is shown, as maximum obtainable counts.

Environmental characteristics are normally limited to thermal characteristics; only in some cases are capabilities of withstanding shock, vibration, and acceleration specified. The operating temperature range and the effects of temperature variations within this range on detector output should be stated. More complete temperature characteristics must be specified for those types of semiconductor detectors that are intended to be operated at cryogenic temperatures; in this case the effects on the detector of exposure to room temperature should also be stated. Special considerations are also necessary for detectors that are intended for operation in a vacuum.

## 23.6   APPLICATION NOTES

### 23.6.1   *Signal Conditioning and Display*

A number of special design considerations apply to signal-conditioning circuitry and to display equipment for nuclear radiation detectors.

Low-level output signals are typical for many detectors. Ionization chambers, for example, produce output currents as low as $10^{-15}$ A which must be conditioned and displayed. To obtain a measurable IR drop across the load resistor, such resistors can have values up to $10^{13}$ $\Omega$, and special design and manufacturing precautions must be taken with such resistors.

Electrometer tubes characterized by low grid current, or quartz-fiber electrometers, have been used to match the high output impedances of many types of detectors and avoid loading errors. Integration of the instantaneous ionization currents over long time periods during which the weak current charges a capacitor has provided another means for measuring currents below 1 pA. FET preamplifiers are now often used for a high-output-impedance match and amplification of weak signals.

Since detector outputs are typically in pulse form, the associated amplifiers are specifically designed as *pulse amplifiers*, with large gain adjustment ranges (e.g., 50 to 50 000) and the ability to amplify pulses with a variety of rise times and decay times. Except for spectrometry, the usual output indication is in terms of either total counts (over a given time interval) or count rate (e.g., counts per minute). Electronic counters can display total counts at fairly high rates; when the rates are too high to be handled by such a counter, a *scaler* is used. This produces and displays one output pulse for a specified number of input pulses. Counting-rate meters (*ratemeters*) indicate the time rate of occurrence of pulses averaged over a specified time interval. Selection of this interval can be accomplished by changing the circuit constants of an integrating network, thus changing its time constant, in known discrete increments. Analog indications of counting rate can be displayed on a panel meter.

The *pulse-height analyzer* furnishes displays of radiation energy spectra. It can indicate either the number of pulses falling within one or more specified amplitude ranges or the rate of occurrence of such pulses. Each channel of such an analyzer is adjusted to pass only those pulses whose amplitude is between a threshold voltage $V$ and an upper limit $V + V_w$, where $V_w$ is the energy *window width*, and the range of energies passed by such a *single-channel amplifier* (*SCA*) is called the *energy window*. Both the threshold voltage and the window width are usually adjustable. A spectral plot can be obtained by adjusting the window width to a fixed value and then *sweeping* the threshold voltage at a known rate to provide a time correlation for each energy level analyzed.

*Coincidence circuits* are used to assure that a given indication is due only to a specific particle (or several related particles). The circuit obtains separate inputs from at least two detectors and uses logic electronics to produce an output pulse only when a specified number or a specified combination of input terminals receive pulses within a specified short time interval. A *delayed coincidence circuit* is actuated by two input pulses (from generically related events), one of which is delayed by a specified time interval with respect to the other. An *anticoincidence circuit*, on the other hand, produces an output pulse only when one of two input terminals receives a pulse and the other receives no pulse. Such a circuit is useful in reducing unwanted *background counts* (counts caused by ionizing events other than those that are supposed to be detected).

Some semiconductor detector designs are available with integrally packaged hybrid-electronics circuitry including not only a preamplifier, but also a shaping amplifier and single-channel amplifier (SCA) with provisions for window selection.

### 23.6.2  Selection Criteria

The selection of a specific detector design can often be based on known radiation-emission characteristics of materials used in a given process or procedure. The alpha, beta, and gamma energies of radioactive isotopes as well as their half-lives are generally well established. The detector design should be optimized for the required interaction of the radiation with the sensitive material of the detector. When two or more types of radiation can be incident simultaneously on a detector, response to only one selected type of radiation can sometimes be accomplished by appropriate detector design; it can also be accomplished by the design of the associated counting circuitry, including the use of coincidence or anticoincidence circuits.

Heavy charged particles such as alphas and protons of low or medium energy lose energy rapidly in passing through matter, without appreciable scattering. Their mean range decreases with increasing absorber density. Alpha particles can be detected with scintillation counters using silver-activated zinc as the scintillator material, with proportional and Geiger counters, with windowless or thin-window semiconductor detectors, and with windowless gas-flow counters.

Very high-energy charged particles are usually measured with fluorescent or Cerenkov scintillation counters.

Beta and gamma radiation can be detected with virtually all basic types of radiation detectors, although gamma rays are not charged particles and must first interact with matter to produce ions. Weak beta radiation detection requires a very sensitive detector such as a windowless or ultra-thin-window gas-flow counter. Thin beryllium windows have been found usable in low-energy X-ray and gamma-ray spectrometry. Window material and thickness can often be selected to exclude beta particles and provide an output only for gamma radiation when both are present. Since gamma rays and X-rays differ only in their energy level, those types of detectors that are suitable for gamma-ray measurements are often usable for X-ray measurements as well. Where particularly high performance, including fine energy resolution, is required from gamma-ray detectors, especially in spectrometry, the cryogenically cooled HP Ge detectors have been found most suitable. CdTe gamma-ray detectors have also found applications, primarily in nuclear medicine, health physics, and general radiation monitoring, where compact detectors, capable of operating at room temperature, are required.

### 23.6.3 Typical Applications

The major applications of nuclear radiation detectors are in the areas of nuclear power generation, weapons testing, biology and medicine, and environmental monitoring. Additional applications are found in transducer systems for other measurands, such as liquid level and density (see the respective relevant chapters).

Nuclear radiation detectors are almost invariably used for monitoring rather than for closed-loop control. In nuclear power stations the detectors are used to monitor radiation at various levels in all areas of the facility. There are many applications in medicine, both for diagnostics (using radio-isotope tracer materials) and for treatment (primarily for cancer). Environmental radiation monitoring is important not only in the vicinity of a nuclear power station, a weapons test site, or a medical or research facility using radioactive isotopes or X-rays, but also for the detection of naturally occurring radioactive material such as radon gas. Many monitoring sites measure and report on the radiation level in the atmosphere, due to either man-made reactions or natural causes. There are also many applications in research in chemical and metallurgical analysis, using gamma-ray spectroscopy, and, in astronomy, locating and measuring the intensity of X-ray emitters. The applications even extend to archaeology, in which dating on the basis of the abundance of an isotope of carbon ($C^{14}$ dating) is an important tool. Gamma and neutron detection is used by the petroleum industry for oil-well logging.

Combinations of a radiation source (usually a gamma-ray emitter) and a detector are used for thickness measurements, as well as for liquid (or quasi-liquid) level measurements and density measurements, where such a system offers the advantage of being entirely nonintrusive, i.e., where both source and detector can be attached to the outside wall of a pipe, duct, or vessel.

# 24

## Transducers for Electricity and Magnetism

### 24.1 BASIC DEFINITIONS AND UNITS OF MEASUREMENT

The definitions shown in this section are intended to present a brief overview of the most common electric and magnetic quantities, together with their units of measurement and the derivation of these units; the latter relationships are shown in brackets. Note that only the ampere is a basic unit of the SI. All other units are either derived from the basic SI units for mass, length, and time, or further derived from such derived units.

*Current* is the flow of charge (see shortly) per unit time. The unit of electrical current is the *ampere* $(A)$, which is defined in the SI as that value of constant current which, if maintained in two straight parallel conductors of infinite length, of negligible cross section, and placed 1 meter apart in a vacuum, would produce between those conductors a force equal to $2 \times 10^{-7}$ newton per meter of length.

*Power* is the time rate of change of energy; the unit of electrical power is the *watt* $(W)$ $[W = J/s]$. Energy is based on force; the unit of energy is the *joule* $(J)$ $[J = N \cdot m]$. Force is given by the basic SI units for mass, length, and time; the unit of force is the *newton* $(N)$ $[N = kg \ m \cdot s^2]$. (*Note:* AC power also has a reactive component that is measured in *vars* (from *v*olt *a*mpere *r*eactive).) The ratio between power and the product of rms voltage and rms current is called the *power factor*. A *root-mean-square* (*rms*) value is equal to the square root of the time average of the square of the quantity;

in practice, for an ac voltage, it is the effective value equivalent to a similar dc voltage in ability to do work in a resistive load. When ac voltages are measured as rms values, the corresponding peak and average voltages can be calculated as $V_{peak} = 1.414V_{rms}$ and $V_{avg} = 0.901V_{rms}$. These relationships assume a sinusoidal wave shape.

*Charge* (electric charge) is a quantity of electricity, and current is the flow of charge per unit time. The unit of charge is the coulomb ($C$), defined as the quantity of electricity transported in 1 second by a current of 1 ampere; hence, $[C = A{\cdot}s]$.

*Voltage* refers to either *electric potential difference* or *electromotive force* (*emf*); the unit for both is the *volt* ($V$), defined as the difference in electric potential between two points of a conductor carrying a constant current of 1 ampere when the power dissipated between these points is equal to 1 watt; $[V = W/A]$.

*Frequency* is expressed in *hertz* (*Hz*); the hertz is a frequency of 1 cycle per second.

*Resistance* (electrical resistance) is expressed in *ohms* ($\Omega$). The ohm is the electrical resistance between two points of a conductor (which is not the source of any electromotive force) when the constant potential difference of 1 volt, applied between these two points, produces in this conductor a current of 1 ampere; $[\Omega = V/A]$.

*Conductance* is the reciprocal of resistance ($1/R$); it was formerly expressed in reciprocal ohms ("mhos"); the SI unit for conductance is the *siemens* ($S$); $[S = 1/\Omega = A/V]$.

*Capacitance* (electric capacitance) is expressed in *farads* ($F$); the farad is the capacitance of a capacitor between the plates of which there appears a potential difference of 1 volt when it is charged by a quantity of electricity equal to 1 coulomb; $[F = C/V = A{\cdot}s/V]$.

*Inductance* (electric inductance) is expressed in *henrys* ($H$); the henry is the inductance of a closed circuit in which an emf of 1 volt is produced when the electric current in the circuit varies uniformly at a rate of 1 ampere per second; $[H = V{\cdot}s/A]$.

The *impedance Z* of an ac circuit is comprised of a *real* part, *resistance* ($R$), and an imaginary part, *reactance* ($X$); thus, $Z = R + jX$. The reactance can be *capacitive* ($X_C$) or *inductive* ($X_L$); where both exist in a circuit, $Z = \sqrt{R^2 + (X_L - X_C)^2}$. Impedance as well as reactance are expressed in *ohms*.

Similarly, the *admittance Y* of an ac circuit is the reciprocal of its impedance, i.e., $Y = 1/Z$. Admittance consists of a *real* part, *conductance* ($G$), and an *imaginary* part, *susceptance* ($B$); thus, $Y = G + jB$. Admittance as well as susceptance are expressed in *siemens*. (They were formerly expressed in reciprocal ohms, or "mhos.")

*Resistivity* ($\rho$), also called *specific resistance*, is the resistance offered by a unit cube of a substance to the flow of current; thus, $\rho = R{\cdot}A/l$, where

$R$ is the resistance of a uniform conductor, $A$ is the cross-sectional area, and $l$ is its length. Resistivity is expressed in *ohm-meters* ($\Omega \cdot m$) or its submultiple, ohm-centimeters.

*Conductivity* ($\sigma$) is the ratio of the current density to the electric field in a material. *Current density* is the current per unit cross-sectional area of a conductor (or conductive substance); *electric field intensity* or *electric field strength* is the electric force (vector) per unit positive (test) charge, expressed in volts per meter. Electrical conductivity is expressed in *siemens per meter* ($S/m$) or per centimeter.

*Magnetic flux* is expressed in *webers* ($Wb$); the weber is the magnetic flux which, linking a circuit of one turn, produces in it an emf of 1 volt as it reduces to zero at a uniform rate in 1 second; [$Wb = V \cdot s$].

*Magnetic flux density* is expressed in *teslas* ($T$); the tesla is the magnetic flux density given by a magnetic flux of 1 weber per square meter; [$T = Wb/m^2$].

The unit of *magnetomotive force* is, strictly speaking, the *ampere* ($A$), but the *ampere-turn* is more commonly used; a magnetomotive force of one ampere-turn may be the result of a current of 1 ampere flowing in 1 turn of wire, or the result of a current of 0.01 ampere flowing in 100 turns of wire.

The unit of *magnetic field strength* is the *ampere per meter* ($A/m$).

Non-SI units for electrical and magnetic quantities, and their conversion factors, are listed as follows in alphabetical order:

1 abampere = 10 A

1 abcoulomb = 10 C

1 abfarad = $10^9$ F

1 abhenry = $10^{-9}$ H

1 abmho = $10^9$ S

1 abohm = $10^{-9}$ $\Omega$

1 abvolt = $10^{-8}$ V

1 electron volt (eV) = $1.6 \times 10^{-19}$ J

1 faraday (physical) = $9.6522 \times 10^4$ C

1 gamma = $10^{-9}$ T (= 1 nT)

1 gauss = $10^{-4}$ T

1 gilbert = 0.7958 ampere-turn

1 horsepower (electric) = 746 W

1 maxwell = $10^{-8}$ Wb

1 oersted = 79.58 A/m

1 unit pole = $1.25664 \times 10^{-7}$ Wb

## 24.2 SENSING METHODS, TRANSDUCERS, AND OTHER SENSING DEVICES

### 24.2.1 Voltage Transducers

The most commonly used device for sensing ac voltages, particularly in electrical power distribution, is the *voltage transformer* (*potential transformer*), which steps down the high voltage into a low-voltage, low-current signal. This signal is then usually rectified and conditioned so that the voltage transducer provides a dc output. An example of such a transducer is shown in Figure 24–1; the unit illustrated has an output of 0 to 1 mA dc for an input range of 0 to 150 V ac. Another version of this design is used in three-phase, four-wire systems, where the same output is provided, for the same input, between each of the three "'hot'" wires and the neutral wire.

In such applications as power supplies handling relatively low ac and dc voltages a magnetic amplifier (amplifier using saturable reactors) can be

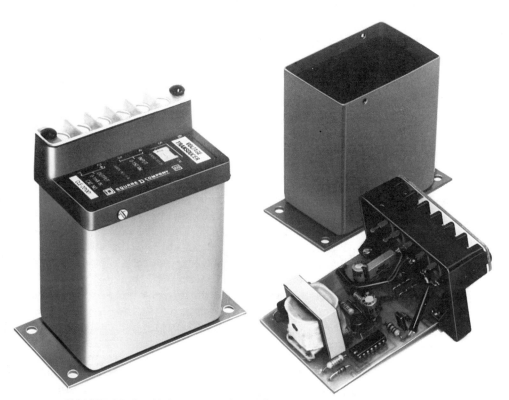

**FIGURE 24–1.** Voltage transducer: in housing (left); removed from housing (right). (Courtesy of Square D Company).

used, with one of the saturable-reactor windings connected across the voltage to be measured and acting as control winding. When an ac voltage is to be measured, it is first rectified into dc before being applied to the *voltage-controlled magnetic amplifier*.

Voltage dividers are also used for sensing (usually low-power) voltages. A voltage divider consists of two resistors connected in series, with their end terminal connected across the voltage source and the signal taken between the center connection and the ground line of the source. Unlike the two sensors just described, a basic voltage divider does not provide electrical isolation between the voltage source and the data system. However, isolation can be provided by connecting a differential amplifier across the signal terminals. The resistor across which the signal is taken should always have a lower resistance than the other resistor, low enough so that a short circuit across it does not affect the voltage source. AC voltages sensed by a voltage divider are usually rectified as part of the signal-conditioning process.

Discrete (switch-type) outputs can be obtained from a *voltage comparator*, which compares the voltage sensed to a set-point voltage and actuates a relay when the voltage increases above, or decreases below, the set point. The set point can be adjustable over a fairly wide range, and the switch-point accuracy can be maintained within very close tolerances.

*Hall-effect devices* are used increasingly in voltage transducers. They are the same type of device as explained for current transducers in the next section, except that the voltage is the input to the transducer and corresponding changes in current are the source for the output signal. *Modified D'Arsonval meters*, again as explained for current transducers, have also been used for voltage sensing.

### 24.2.2   Current Transducers

Basic current sensors are the shunt for dc currents and the current transformer for ac currents. A *shunt* is a section of conductor having a low resistance (typically in the form of a short bus bar) inserted in series with the conductor in which the current is to be measured. The resistance of the shunt is so selected that a specified current will produce a known amount of *IR* drop across it. The potential difference across the shunt is then measured as a voltage. A *current transformer* is a transformer which has a primary winding, consisting of a relatively small number of turns of heavy-gage wire, which is connected in series with the current-carrying conductor, and a secondary winding, consisting of a relatively large number of turns of small-gage wire, from which the signal voltage is taken. The turns ratio is so selected that a specified current through the primary winding will produce a known voltage in the secondary winding. Current transducers using current transformers commonly use rectification and conditioning to produce a dc output signal (as previously explained for voltage-transformer types of voltage transducers). When relatively high currents are to be measured, a clamp-on current sensor (see next paragraph) instead of a series connection is used to obtain the input for the current transformer. Single-phase as well as three-phase current transducers are available.

*Clamp-on current sensors* are used for measurements of relatively high currents without using a live connection to the conductor. These sensors can be installed permanently, or they can be clamped around the conductor temporarily to obtain a sample current reading and then removed. They typically contain a circular ferromagnetic (usually ferrite) core that is cut in half but provides a continuous core when the two halves are clamped together. The sensor is opened, slipped around the wire, and then clamped together. It then acts as a current transformer, with the current-carrying conductor itself acting as the primary and a multiturn winding around the core acting as the secondary which produces an output voltage proportional to ac current.

When isolation is required between the power supply and the data system and relatively low dc currents are to be measured, a *current-controlled magnetic amplifier* can be used. One of the saturable-reactor windings is the control winding, which is connected in series with the current-carrying conductor. A single-turn control winding is usually sufficient.

*Hall-effect current sensors* can be used for dc as well as ac measurements. The current is made to cause a change in the magnetic field in which a *Hall device* (Figure 24–2) is located. This device is a semiconductor of a type specifically selected for the stated use (one having a high Hall constant, or Hall coefficient). It produces an electric field which is both transverse to an excitation current passing through it and transverse to the magnetic field. The output voltage created by the electric field is proportional to the magnetic field, at constant excitation current. The magnetic field is usually generated by a ferromagnetic, gapped, toroidal core through which the current-carrying conductor runs. Most sensors of this type provide a circular opening through which the conductor must run. The opening is in the sensor housing,

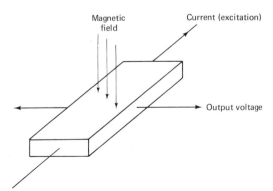

**FIGURE 24–2.** Operating principle of Hall-effect device.

which is made from an insulating material. The core is placed immediately inside of this opening. Four terminals are usually provided, two for the dc excitation of the Hall device and two for the output signal. Split-core, clamp-on versions are also available. The output voltage is of the same form as the input current. That is, a dc input current produces a dc output voltage and an ac input current produces an ac output voltage of the same wave shape. Depending on the frequency response of the sensor (which can extend from dc to over 1 MHz), transients in the measured current will be repro-duced in the output signal. The excitation current must be closely controlled. When a dc output is required for ac current measurement, the output signal can be rectified. (Circuitry for this can be located within the sensor housing.) Other forms of signal conditioning can also be provided within the sensor housing, and additional terminals may then be provided to supply external power to such circuitry. Some designs include a relay and control circuitry so that a switch output is provided at a set current level.

Very small currents (down to $10^{-12}$ A) can be measured by an *elec-trometer tube* (a high-vacuum, high-input-impedance electron tube), by a vibrating-reed capacitor, or by amplifiers using varactor diodes, metal-oxide-semiconductor field-effect transistors (MOSFETs), or junction FETs. In some low-power applications (e.g., scientific instruments) the current drawn from a high-voltage secondary can be derived from a calibration that pro-vides correlation with a conventional measurement of the current through the low-voltage primary winding of a transformer.

*Modified D'Arsonval meters* are also used for current sensing. Such meters contain an axially pivoted coil of rectangular configuration through which the measured current flows. A pointer is integral with the coil assem-bly. The coil rotates between the pole pieces of a permanent magnet. The angle of rotation is proportional to current through the coil. The meter is modified by having a transduction element (e.g., capacitive) attached to the pointer so that an output signal proportional to the angular displacement of the pointer is produced. The meter can also be modified by attaching a contact to the pointer and one or more contacts to the meter face so that one or more discrete outputs are produced as the pointer rotates.

### 24.2.3 Power Transducers

DC power is simply the product of voltage and current. The multiplication can be performed by electronic circuitry that is now usually digital and that can be incorporated into a dc watt transducer. However, it is often performed by the computer of the data system, which also handles other data manip-ulation tasks. This sort of a scheme allows voltage and current as well as the computer-derived power to be displayed.

AC power is the vectorial sum of *real* power ($V_{rms}I_{rms} \cos \phi$) and *imaginary* power ($V_{rms}I_{rms} \sin \phi$), where $\phi$ is the *phase angle* (and $\cos \phi$ is the *power factor*). Real power is expressed in watts, whereas imaginary power is expressed in *vars*, a contraction of "volt-ampere reactive." AC power transducers are, therefore, known as either *watt transducers* or *var transducers*, depending on whether they are designed to measure real or reactive (imaginary) power.

Some power transducer designs are capable of providing outputs for both watt and var. An example is illustrated in Figure 24–3. This watt/var transducer uses pulse techniques to multiply voltage and current and incorporates circuitry that provides a var output based on phase angle. The transducer can provide an output of either $\pm 1$ ma dc or $\pm 10$ V dc, depending on the part number selected; similarly, the part number designates use in one-phase, three-phase/three-wire, or three-phase/four-wire circuits. Watt/var measuring ranges are 500, 1000 or 1500, again depending on the part number selected.

Another frequently used type of power transducer uses a Hall-effect device, similar in operating principle to the Hall-effect current transducer described in Section 24.2.2. Hall-effect devices in such power transducers are used as *Hall multipliers*. The output voltage of a Hall-effect device is dependent on the excitation current through the device as well as on the magnetic field to which the device is exposed. In current sensors, the excitation current is held constant and the measured current is made to vary the magnetic field. In power transducers, which have to respond to both

**FIGURE 24–3.** Watt-and-var transducer. (Courtesy of Square D Company)

voltage and current, the voltage is typically connected to a potential transformer in the transducer and the output of the transformer is conditioned into a proportionally varying excitation current through the Hall-effect device. At the same time, the current is sensed by passing the conductor through a gapped toroidal core and the magnetic field produced by this electromagnet is applied to the Hall-effect device; the field varies proportionally to the current sensed. The Hall-effect device thus multiplies the voltage and the current sensed. Such Hall-effect transducers are available for use on single-phase, three-phase/three-wire or three-phase/four-wire lines. They can be used for power ranges having an upper limit between 50 W and 800 kW (single-phase) or 1.5 MW (three-phase) and with power-line frequencies of 50 or 60 Hz as well as with power supply frequencies extending to about 10 kHz (for the lower power ranges).

Hall-effect as well as electronically multiplying power transducers are also available specifically to perform as *power factor transducers* or as *phase-angle transducers*. They typically provide a dc output current or voltage for phase angles from $\pm 45$ to $\pm 72.5°$ and/or power factors from 0.3 (lead or lag) and unity to 0.7 (lead or lag) and unity.

At very high frequencies (e.g., microwave range), electrical power is typically measured by using a portion of the power to heat a temperature sensor. In this method, the temperature rise is measured and correlated to power by a calibration curve.

### 24.2.4  Electrostatic-Charge Sensors

Electrostatic charge accumulation on surfaces and electrostatic fields can be measured by several types of potential gradient sensors. Most designs are capable of measuring electrostatic voltage (typically in kV) or electrostatic fields (in V/m or V/cm) without physical contact between the sensor and the measured surface. The measuring range is given not only by a range selector switch on the instrument, but also by the distance between sensor and surface, which is typically between 5 and 30 cm.

One design uses an electromagnetically driven oscillating electrode located behind a small circular aperture in a metallic plate. The electrode is in a feedback loop which neutralizes the net field at the electrode by feeding a dc signal back to the aperture plate. The oscillation of the electrode provides an ac error signal proportional to the electrostatic field sensed. The output of the device is a dc signal given by the amplitude of the feedback voltage. Another design uses a small radioisotope source in conjunction with a high-resistance voltage sensor for transduction of the electrostatic charge or voltage measured. Typical configurations include a hand-held instrument, with a pistol grip, that is pointed at the measured surface from a known distance, and small probes connected by a cable to a control/display unit. A related design uses a dissipation array, in fixed or mobile installations, to detect charge buildup in the atmosphere. This device can provide warning of lightning.

### 24.2.5  Frequency- and Time-Sensing Devices

Time, as well as frequency, which is the inverse of time, is measured by reference to a periodic phenomenon, one whose periodicity is known within an amount of error (uncertainty) adequate for a given measurement. The earliest periodic phenomenon used for time determinations was the pendulum, which is still in use in some clocks, which, when properly designed and adjusted, can provide time displays with very low error. More recently the quartz-crystal oscillator was developed and refined; it is widely used in electronic equipment as a reference for frequency and time. The use of countdown circuitry allows obtaining exact submultiples of the frequency at which the crystal oscillates. The most recently developed frequency standards are atomic in nature. The resonance of the cesium atom has been used to define the unit of time, the second (as adopted by the CGPM in 1967): "The second is the duration of 9 192 631 770 periods of the radiation corresponding to the transition between the two hyperfine levels of the ground state of the cesium-133 atom."

Other atomic frequency standards include the hydrogen-atom maser and the rubidium gas cell; however, the primary frequency standard is the cesium beam standard, and continuing improvements in equipment design have enhanced the usability and stability of this device.

An unknown frequency is measured by comparing it with a known frequency. Within a limit on very high frequencies, this can be done by using a reference frequency to establish a known time interval (e.g., exactly 1 second), converting the wave shape of the unknown signal into pulses, and then digitally counting the number of pulses over the time interval.

A method also usable at extremely high frequencies is to mix the unknown frequency with a standard frequency (*beat* one against the other) and then measure the frequency difference (*beat frequency*) by a device such as a frequency counter. Various types of circuits are used in *frequency-to-dc converters*, which convert a specified range of frequencies into a dc voltage. The simplest of these is the FM discriminator, which produces zero output voltage at a preset center frequency and increasing dc voltage at frequency deviations from this center frequency, that is linear within a certain band of such frequency deviations; the polarity of the output voltage depends on whether the measured frequency is above or below the center frequency. *Frequency-to-digital converters* convert the count of an absolute or beat frequency into coded form (e.g., binary, binary-coded decimal). *Frequency-to-voltage conversion* techniques also include resistive-capacitive (*RC*) networks in feedback loops of operational amplifiers in conjunction with diodes or with a saturable transformer/rectifier. Among related devices is a frequency transducer using pulse techniques to provide an output of either 0–1 mA dc or 0–10 V dc. Full-scale output is obtained for frequency ranges of 45–55 Hz, 55–65 Hz, or 375–425 Hz.

### 24.2.6  Magnetic-Flux-Density Sensors

Magnetic-flux-density sensors include devices known as gaussmeters, magnetic field probes, and magnetometers. Since the SI unit for the quantity measured is the *tesla* (*T*), the term "gaussmeter" is now obsolescent (*1 gauss = 0.1 mT*). The weak magnetic fields sensed by magnetometers have been measured in *gammas* and are now measured in *nanoteslas* (*1 gamma = 1 nT*).

The most commonly used magnetic-flux-density sensors are those using Hall-effect devices. As explained for Hall-device current and power sensors, these devices (Figure 24–2) produce an output voltage proportional to the flux density of a transverse magnetic field when a constant excitation current is passed through them. The Hall-effect device is usually contained near the tip of a probe which is connected to an excitation- and signal-conditioning and display unit by a cable. Such probes (*Hall probes*) are designed for responding to either an axial or a transverse magnetic field. *Axial probes* are generally circular in cross section, whereas *transverse probes* are blade-shaped in most designs. Some probe designs include a flexible stem between the handle and the probe tip. Probes are also available for simultaneous measurements of an axial and a transverse magnetic field (*two-axis probe*) or an axial and two mutually orthogonal transverse fields (*Z-*, *X-*, and *Y*-axes, hence a *three-axis probe*). The excitation current can be dc or ac; the latter effects chopping of the sensed field and provides an ac output voltage which can then be amplified and synchronously demodulated. Two-axis probes contain two (and three-axis probes contain three) mutually orthogonally placed Hall-effect devices in their tip.

*Inductor probes* consist of an air-core inductor in a nonmagnetic (e.g., brass) housing. This type of probe responds to the rate of change of an ambient magnetic field or, generally, to varying (not steady-state) magnetic fields. It is quite usable for ac fields. The flux changes induce a voltage in the coil. The inductor probe is the sensor in a *search-coil magnetometer*. It is also used as the sensing coil in a *fluxmeter,* which can be used to measure steady-state magnetic fields by spinning the coil through the field so that an ac voltage, given by the spin rate, is induced in the coil.

The *nuclear magnetic resonance (NMR) magnetometer* (or magnetic field sensor) is based on proton precession (*proton-precession magnetometer*). The resonance frequency (in the RF range) of certain substances varies exactly and linearly with magnetic flux density and manifests itself by a sharp dip in applied RF power due to absorption as the frequency of an RF generator is swept through a given range of frequencies. The resonance frequency shift of a proton ($H^1$) is 42.5776 MHz/T (4257.76 Hz/gauss). The resonance frequency shift for lithium ($Li^7$) is 16.5461 MHz/T, and for deuterium ($H^2$ or D) it is 6.54 MHz/T. The sensing probe contains the substance in a small enclosed sample chamber which includes the RF coil system. The probe is connected to an oscillator and frequency counter in an excitation/display unit.

The *flux-gate magnetometer* operates on the basis of permeability changes, employing a core saturated in both directions (or two saturated cores whose windings are mutually opposed in polarity). The saturable reactor(s) are connected into an ac impedance bridge circuit which is balanced in the absence of any magnetic field. A second-harmonic signal is generated when a magnetic field changes the core saturation. In a *vector flux-gate magnetometer* three such flux-gates are mounted so that they are mutually orthogonal, and the combined sensor can then provide information as to the strength as well as direction of a magnetic field. Closed-loop servo control has been applied to this type of device, where the amount of current needed to restore balance is a measure of the magnetic field. Sensitive magnetometers exemplified by the flux-gate magnetometer can detect fields of 1 or a few nT (gammas). Other very sensitive magnetometers include the *helium magnetometer* and the *rubidium vapor magnetometer*, whose operation is based on optical pumping of monochromatic light and transition changes in resonance levels due to changes in the magnetic field.

# Appendix

**Table A–1. TEMPERATURE SCALE CONVERSION AND SELECTED SIGNIFICANT TEMPERATURES**[a]

| Celsius (°C) | Fahrenheit (°F) | Kelvin (K) | Celsius (°C) | Fahrenheit (°F) | Kelvin (K) |
|---|---|---|---|---|---|
| **Absolute Zero** | | | **Boiling Point of Hydrogen (d)** | | |
| −273.15 | −459.67 | 0 | −252.87 | −423.2 | 20.28 |
| | | | | | |
| −272.0 | −457.6 | 1.1 | −252.0 | −421.6 | 21.1 |
| −270.0 | −454.0 | 3.1 | −250.0 | −418.0 | 23.1 |
| | | | −248.0 | −414.4 | 25.1 |
| **Boiling Point of Helium** | | | | | |
| −268.9 | −452.0 | 4.2 | **Boiling Point of Neon (d)** | | |
| | | | −246.05 | −410.9 | 27.10 |
| −268.0 | −450.4 | 5.1 | | | |
| −266.0 | −446.8 | 7.1 | −245.0 | −409.0 | 28.1 |
| −265.0 | −445.0 | 8.2 | −244.0 | −407.2 | 29.1 |
| −264.0 | −443.2 | 9.1 | −242.0 | −403.6 | 31.1 |
| −262.0 | −439.6 | 11.1 | −240.0 | −400.0 | 33.1 |
| −260.0 | −436.0 | 13.1 | | | |
| | | | −238.0 | −396.4 | 35.1 |
| | | | −236.0 | −392.8 | 37.1 |
| **Triple Point of Hydrogen (d)** | | | −235.0 | −391.0 | 38.1 |
| −259.34 | −434.8 | 13.81 | −234.0 | −389.2 | 39.1 |
| | | | −232.0 | −385.6 | 41.1 |
| −258.0 | −432.4 | 15.1 | −230.0 | −382.0 | 43.1 |
| −256.0 | −428.8 | 17.1 | | | |
| −255.0 | −427.0 | 18.1 | −228.0 | −378.4 | 45.1 |
| −254.0 | −425.2 | 19.1 | −226.0 | −374.8 | 47.1 |

**Table A–1.** (*continued*)

| °C | °F | K | °C | °F | K |
|---|---|---|---|---|---|
| − 225.0 | − 373.0 | 48.1 | − 178.0 | − 288.4 | 95.1 |
| − 224.0 | − 371.2 | 49.1 | − 176.0 | − 284.8 | 97.1 |
| − 222.0 | − 367.6 | 51.1 | − 175.0 | − 283.0 | 98.1 |
| − 220.0 | − 364.0 | 53.1 | − 174.0 | − 281.2 | 99.1 |
| Triple Point of Oxygen (*d*) | | | − 172.0 | − 277.6 | 101.1 |
| − 218.79 | − 361.8 | 54.36 | − 170.0 | − 274.0 | 103.1 |
| | | | − 168.0 | − 270.4 | 105.1 |
| − 218.0 | − 360.4 | 55.1 | − 166.0 | − 266.8 | 107.1 |
| − 216.0 | − 356.8 | 57.1 | − 165.0 | − 265.0 | 108.1 |
| − 215.0 | − 355.0 | 58.1 | − 164.0 | − 263.2 | 109.1 |
| − 214.0 | − 353.2 | 59.1 | − 162.0 | − 259.6 | 111.1 |
| − 212.0 | − 349.6 | 61.1 | − 160.0 | − 256.0 | 113.1 |
| − 210.0 | − 346.0 | 63.1 | − 158.0 | − 252.4 | 115.1 |
| − 208.0 | − 342.4 | 65.1 | − 156.0 | − 248.8 | 117.1 |
| − 206.0 | − 338.8 | 67.1 | − 155.0 | − 247.0 | 118.1 |
| − 205.0 | − 337.0 | 68.1 | − 154.0 | − 245.2 | 119.1 |
| − 204.0 | − 335.2 | 69.1 | Boiling Point of Krypton | | |
| − 202.0 | − 331.6 | 71.1 | − 152.3 | − 242.1 | 120.1 |
| − 200.0 | − 328.0 | 73.1 | − 152.0 | − 241.6 | 121.1 |
| − 198.0 | − 324.4 | 75.1 | − 150.0 | − 238.0 | 123.1 |
| − 196.0 | − 320.8 | 77.1 | − 148.0 | − 234.4 | 125.1 |
| Boiling Point of Nitrogen | | | − 144.0 | − 227.2 | 129.1 |
| − 195.8 | − 320.4 | 77.3 | − 140.0 | − 220.0 | 133.1 |
| | | | − 136.0 | − 212.8 | 137.1 |
| − 195.0 | − 319.0 | 78.1 | − 135.0 | − 211.0 | 138.1 |
| − 194.0 | − 317.2 | 79.1 | − 132.0 | − 205.6 | 141.1 |
| − 192.0 | − 313.6 | 81.1 | − 128.0 | − 198.4 | 145.1 |
| − 190.0 | − 310.0 | 83.1 | − 124.0 | − 191.2 | 149.1 |
| Triple Point of Argon | | | − 120.0 | − 184.0 | 153.1 |
| − 189.3 | − 308.8 | 83.8 | − 116.0 | − 176.8 | 157.1 |
| Freezing Point of Argon | | | − 115.0 | − 175.0 | 158.1 |
| − 189.2 | − 308.6 | 83.9 | − 112.0 | − 169.6 | 161.1 |
| − 188.0 | − 306.4 | 85.1 | − 108.0 | − 162.4 | 165.1 |
| − 186.0 | − 302.8 | 87.1 | Boiling Point of Xenon | | |
| Boiling Point of Argon | | | − 107.1 | − 160.8 | 166.0 |
| − 185.7 | − 302.3 | 87.4 | − 104.0 | − 155.2 | 169.1 |
| − 185.0 | − 301.0 | 88.1 | − 100.0 | − 148.0 | 173.1 |
| − 184.0 | − 299.2 | 89.1 | − 96.0 | − 140.8 | 177.1 |
| | | | − 95.0 | − 139.0 | 178.1 |
| Boiling Point of Oxygen (*d*) | | | − 92.0 | − 133.6 | 181.1 |
| − 182.96 | − 297.3 | 90.19 | − 88.0 | − 126.4 | 185.1 |
| | | | − 84.0 | − 119.2 | 189.1 |
| − 182.2 | − 296.0 | 90.9 | − 80.0 | − 112.0 | 193.1 |
| − 180.0 | − 292.0 | 93.1 | − 78.9 | − 110.0 | 194.3 |

**Table A–1.   (continued)**

| °C | °F | K | °C | °F | K |
|---|---|---|---|---|---|
| Sublimation Point of Carbon Dioxide | | | 20.0 | 68.0 | 293.1 |
| −78.5 | −109.3 | 194.7 | 24.0 | 75.2 | 297.1 |
| | | | 24.4 | 76.0 | 297.6 |
| −76.0 | −104.8 | 197.1 | 28.0 | 82.4 | 301.1 |
| −75.0 | −103.0 | 198.1 | 28.3 | 83.0 | 301.5 |
| −72.0 | −97.6 | 201.1 | 32.0 | 89.6 | 305.1 |
| −68.0 | −90.4 | 205.1 | 32.2 | 90.0 | 305.4 |
| −64.0 | −83.2 | 209.1 | 36.0 | 96.8 | 309.1 |
| −60.0 | −76.0 | 213.1 | 36.1 | 97.0 | 309.3 |
| −56.0 | −68.8 | 217.1 | 40.0 | 104.0 | 313.1 |
| −55.0 | −67.0 | 218.1 | 44.0 | 111.2 | 317.1 |
| −52.0 | −61.6 | 221.1 | 44.4 | 112.0 | 317.6 |
| −48.0 | −54.4 | 225.1 | 48.0 | 118.4 | 321.1 |
| −44.0 | −47.2 | 229.1 | 52.0 | 125.6 | 325.1 |
| −43.3 | −46.0 | 229.8 | 56.0 | 132.8 | 329.1 |
| −40.0 | −40.0 | 233.1 | 60.0 | 140.0 | 333.1 |
| −38.9 | −38.0 | 234.3 | 64.0 | 147.2 | 337.1 |
| −36.0 | −32.8 | 237.1 | 68.0 | 154.4 | 341.1 |
| −35.0 | −31.0 | 238.1 | 72.0 | 161.6 | 345.1 |
| −32.0 | −25.6 | 241.1 | 76.0 | 168.8 | 349.1 |
| −31.1 | −24.0 | 242.0 | 80.0 | 176.0 | 353.1 |
| −28.0 | −18.4 | 245.1 | 84.0 | 183.2 | 357.1 |
| −27.2 | −17.0 | 245.9 | 88.0 | 190.4 | 361.1 |
| −24.0 | −11.2 | 249.1 | 92.0 | 197.6 | 365.1 |
| −23.3 | −10.0 | 249.8 | 96.0 | 204.8 | 369.1 |
| −20.0 | −4.0 | 253.1 | | | |
| −16.0 | 3.2 | 257.1 | | | |
| −15.6 | 4.0 | 257.6 | Boiling Point of Water (d) | | |
| −12.0 | 10.4 | 261.1 | 100.0 | 212.0 | 373.15 |
| −11.7 | 11.0 | 261.5 | 104.0 | 219.2 | 377.1 |
| −8.0 | 17.6 | 265.1 | 108.0 | 226.4 | 381.1 |
| −7.8 | 18.0 | 265.4 | 112.0 | 233.6 | 385.1 |
| −4.0 | 24.8 | 269.1 | 116.0 | 240.8 | 389.1 |
| −3.9 | 25.0 | 269.3 | | | |
| | | | 120.0 | 248.0 | 393.1 |
| Ice Point | | | | | |
| 0.0 | 32.0 | 273.15 | Triple Point of Benzoic Acid | | |
| | | | 122.4 | 252.3 | 395.5 |
| Triple Point of Water (d) | | | 124.0 | 255.2 | 397.1 |
| 0.01 | 32.02 | 273.16 | 128.0 | 262.4 | 401.1 |
| | | | 132.0 | 269.6 | 405.1 |
| 4.0 | 39.2 | 277.1 | 136.0 | 276.8 | 409.1 |
| 4.4 | 40.0 | 277.6 | 140.0 | 284.0 | 413.1 |
| 8.0 | 46.4 | 281.1 | | | |
| 8.3 | 47.0 | 281.5 | 144.0 | 291.2 | 417.1 |
| 12.0 | 53.6 | 285.1 | 148.0 | 298.4 | 421.1 |
| 12.2 | 54.0 | 285.4 | 152.0 | 305.6 | 425.1 |
| 16.0 | 60.8 | 289.1 | 156.0 | 312.8 | 429.1 |
| 16.1 | 61.0 | 289.3 | | | |

**Table A–1.** (*continued*)

| °C | °F | K | °C | °F | K |
|---|---|---|---|---|---|
| Freezing Point of Indium | | | 308.0 | 586.4 | 581.1 |
| 156.6 | 313.9 | 429.8 | 312.0 | 593.6 | 585.1 |
| | | | 316.0 | 600.8 | 589.1 |
| 160.0 | 320.0 | 433.1 | 320.0 | 608.0 | 593.1 |
| 164.0 | 327.2 | 437.1 | | | |
| 168.0 | 334.4 | 441.1 | Freezing Point of Cadmium | | |
| 172.0 | 341.6 | 445.1 | 320.9 | 609.6 | 594.1 |
| 176.0 | 348.8 | 449.1 | | | |
| 180.0 | 356.0 | 453.1 | 324.0 | 615.2 | 597.1 |
| 184.0 | 363.2 | 457.1 | | | |
| 188.0 | 370.4 | 461.1 | Freezing Point of Lead | | |
| 192.0 | 377.6 | 465.1 | 327.5 | 621.5 | 600.7 |
| 196.0 | 384.8 | 469.1 | | | |
| 200.0 | 392.0 | 473.1 | 328.0 | 622.4 | 601.1 |
| 204.0 | 399.2 | 477.1 | 332.0 | 629.6 | 605.1 |
| 208.0 | 406.4 | 481.1 | 336.0 | 636.8 | 609.1 |
| 212.0 | 413.6 | 485.1 | 340.0 | 644.0 | 613.1 |
| 216.0 | 420.8 | 489.1 | 344.0 | 651.2 | 617.1 |
| 220.0 | 428.0 | 493.1 | 348.0 | 658.4 | 621.1 |
| 224.0 | 435.2 | 497.1 | 352.0 | 665.6 | 625.1 |
| 228.0 | 442.4 | 501.1 | 356.0 | 672.8 | 629.1 |
| Freezing Point of Tin (*d*) | | | Boiling Point of Mercury | | |
| 231.97 | 449.55 | 505.12 | 356.6 | 673.9 | 629.7 |
| 232.0 | 449.6 | 505.1 | 360.0 | 680.0 | 633.1 |
| 236.0 | 456.8 | 509.1 | 364.0 | 687.2 | 637.1 |
| 240.0 | 464.0 | 513.1 | 368.0 | 694.4 | 641.1 |
| 244.0 | 471.2 | 517.1 | 372.0 | 701.6 | 645.1 |
| 248.0 | 478.4 | 521.1 | 376.0 | 708.8 | 649.1 |
| | | | 380.0 | 716.0 | 653.1 |
| 252.0 | 485.6 | 525.1 | | | |
| 256.0 | 492.8 | 529.1 | 384.0 | 723.2 | 657.1 |
| 260.0 | 500.0 | 533.1 | 388.0 | 730.4 | 661.1 |
| 264.0 | 507.2 | 537.1 | 392.0 | 737.6 | 665.1 |
| 268.0 | 514.4 | 541.1 | 396.0 | 744.8 | 669.1 |
| | | | 400.0 | 752.0 | 673.1 |
| Freezing Point of Bismuth | | | 404.0 | 759.2 | 677.1 |
| 271.3 | 520.3 | 544.5 | 408.0 | 766.4 | 681.1 |
| | | | 412.0 | 773.6 | 685.1 |
| 272.2 | 522.0 | 545.4 | 416.0 | 780.8 | 689.1 |
| 276.0 | 528.8 | 549.1 | | | |
| 280.0 | 536.0 | 553.1 | Freezing Point of Zinc (*d*) | | |
| 284.0 | 543.2 | 557.1 | 419.58 | 787.24 | 692.73 |
| 288.0 | 550.4 | 561.1 | | | |
| 292.0 | 557.6 | 565.1 | 420.0 | 788.0 | 693.1 |
| 296.0 | 564.8 | 569.1 | 424.0 | 795.2 | 697.1 |
| 300.0 | 572.0 | 573.1 | 428.0 | 802.4 | 701.1 |
| 304.0 | 579.2 | 577.1 | 432.0 | 809.6 | 705.1 |

**Table A-1.** (*continued*)

| °C | °F | K | °C | °F | K |
|---|---|---|---|---|---|
| 436.0 | 816.8 | 709.1 | 600.0 | 1112. | 873.1 |
| 440.0 | 824.0 | 713.1 | 604.0 | 1119. | 877.1 |
| 444.0 | 831.2 | 717.1 | 608.0 | 1126. | 881.1 |
| | | | 612.0 | 1134. | 885.1 |

Boiling Point of Sulfur

| °C | °F | K | °C | °F | K |
|---|---|---|---|---|---|
| 444.7 | 832.5 | 717.8 | 616.0 | 1141. | 889.1 |
| | | | 620.0 | 1148. | 893.1 |
| 445.0 | 833.0 | 718.1 | 624.0 | 1155. | 897.1 |
| 448.0 | 838.4 | 721.1 | 628.0 | 1162. | 901.1 |

| 452.0 | 845.6 | 725.1 |
|---|---|---|
| 456.0 | 852.8 | 729.1 |

Freezing Point of Antimony

| °C | °F | K |
|---|---|---|
| 630.7 | 1167.3 | 903.9 |

| °C | °F | K | °C | °F | K |
|---|---|---|---|---|---|
| 460.0 | 860.0 | 733.1 | | | |
| 464.0 | 867.2 | 737.1 | 632.2 | 1170. | 905.4 |
| | | | 636.0 | 1177. | 909.1 |
| 468.0 | 874.4 | 741.1 | 640.0 | 1184. | 913.1 |
| 472.0 | 881.6 | 745.1 | 644.0 | 1191. | 917.1 |
| 476.0 | 888.8 | 749.1 | 648.0 | 1198. | 921.1 |
| 480.0 | 896.0 | 753.1 | | | |

| 484.0 | 903.2 | 757.1 |
|---|---|---|
| 488.0 | 910.4 | 761.1 |

Freezing Point of Magnesium

| °C | °F | K |
|---|---|---|
| 648.8 | 1199.8 | 922.0 |

| °C | °F | K | °C | °F | K |
|---|---|---|---|---|---|
| 492.0 | 917.6 | 765.1 | | | |
| 496.0 | 924.8 | 769.1 | 652.0 | 1206. | 925.1 |
| | | | 656.0 | 1213. | 929.1 |
| 500.0 | 932.0 | 773.1 | 660.0 | 1220. | 933.1 |
| 504.0 | 939.2 | 777.1 | | | |

| 508.0 | 946.4 | 781.1 |
|---|---|---|
| 512.0 | 953.6 | 785.1 |

Freezing Point of Aluminum

| °C | °F | K |
|---|---|---|
| 660.4 | 1220.7 | 933.6 |

| °C | °F | K | °C | °F | K |
|---|---|---|---|---|---|
| 516.0 | 960.8 | 789.1 | | | |
| 520.0 | 968.0 | 793.1 | 664.0 | 1227. | 937.1 |
| | | | 668.0 | 1234. | 941.1 |
| 524.0 | 975.2 | 797.1 | 672.0 | 1242. | 945.1 |
| 528.0 | 982.4 | 801.1 | 676.0 | 1249. | 949.1 |
| 532.0 | 989.6 | 805.1 | | | |
| 536.0 | 996.8 | 809.1 | 680.0 | 1256. | 953.1 |
| 540.0 | 1004. | 813.1 | 684.0 | 1263. | 957.1 |
| 544.0 | 1011. | 817.1 | 688.0 | 1270. | 961.1 |
| | | | 692.0 | 1278. | 965.1 |
| 548.0 | 1018. | 821.1 | 696.0 | 1285. | 969.1 |
| 552.0 | 1026. | 825.1 | 700.0 | 1292. | 973.1 |
| 556.0 | 1033. | 829.1 | 704.0 | 1299. | 977.1 |
| 560.0 | 1040. | 833.1 | 708.0 | 1306. | 981.1 |
| 564.0 | 1047. | 837.1 | 712.0 | 1314. | 985.1 |
| | | | 716.0 | 1321. | 989.1 |
| 568.0 | 1054. | 841.1 | 720.0 | 1328. | 993.1 |
| 572.2 | 1062. | 845.1 | | | |
| 576.0 | 1063. | 849.1 | 724.0 | 1335. | 997.1 |
| 580.0 | 1076. | 853.1 | 728.0 | 1342. | 1001. |
| 584.0 | 1083. | 857.1 | 732.0 | 1350. | 1005. |
| 588.0 | 1090. | 861.1 | 736.0 | 1357. | 1009. |
| | | | 740.0 | 1364. | 1013. |
| 592.0 | 1098. | 865.1 | 744.0 | 1371. | 1017. |
| 596.0 | 1105. | 869.1 | | | |

**Table A–1.** (*continued*)

| °C | °F | K | °C | °F | K |
|----|----|----|----|----|----|
| 748.0 | 1378. | 1021. | 924.0 | 1695. | 1197. |
| 752.0 | 1386. | 1025. | 928.0 | 1702. | 1201. |
| 756.0 | 1393. | 1029. | 932.0 | 1710. | 1205. |
| 760.0 | 1400. | 1033. | 936.0 | 1717. | 1209. |
| 764.0 | 1407. | 1037. | | | |

**Melting Point of Germanium**

| °C | °F | K |
|----|----|----|
| 937.4 | 1719.3 | 1210.5 |

| °C | °F | K |
|----|----|----|
| 768.0 | 1414. | 1041. |
| 772.0 | 1422. | 1045. |
| 776.0 | 1429. | 1049. |
| 780.0 | 1436. | 1053. |

| °C | °F | K |
|----|----|----|
| 940.0 | 1724. | 1213. |
| 944.0 | 1731. | 1217. |
| 948.0 | 1738. | 1221. |
| 952.0 | 1746. | 1225. |
| 956.0 | 1753. | 1229. |
| 960.0 | 1760. | 1233. |

| °C | °F | K |
|----|----|----|
| 784.0 | 1443. | 1057. |
| 784.4 | 1444. | 1058. |
| 788.0 | 1450. | 1061. |
| 788.3 | 1451. | 1062. |
| 792.0 | 1458. | 1065. |
| 796.0 | 1465. | 1069. |
| 800.0 | 1472. | 1073. |
| 804.0 | 1479. | 1077. |

**Freezing Point of Silver** (*d*)

| °C | °F | K |
|----|----|----|
| 961.93 | 1763.47 | 1235.08 |

| °C | °F | K |
|----|----|----|
| 964.0 | 1767. | 1237. |
| 968.0 | 1774. | 1241. |
| 972.2 | 1782. | 1245. |
| 976.0 | 1789. | 1249. |
| 980.0 | 1796. | 1253. |
| 984.0 | 1803. | 1257. |
| 988.0 | 1810. | 1261. |
| 992.0 | 1818. | 1265. |
| 996.0 | 1825. | 1269. |
| 1000. | 1832. | 1273. |
| 1004. | 1839. | 1277. |
| 1008. | 1846. | 1281. |
| 1012. | 1854. | 1285. |

| °C | °F | K |
|----|----|----|
| 808.0 | 1486. | 1081. |
| 812.0 | 1494. | 1085. |
| 816.0 | 1501. | 1089. |
| 820.0 | 1508. | 1093. |
| 824.0 | 1515. | 1097. |
| 828.0 | 1522. | 1101. |
| 832.0 | 1530. | 1105. |
| 836.0 | 1537. | 1109. |
| 840.0 | 1544. | 1113. |
| 844.0 | 1551. | 1117. |
| 848.0 | 1558. | 1121. |
| 852.0 | 1566. | 1125. |
| 856.0 | 1573. | 1129. |
| 860.0 | 1580. | 1133. |
| 864.0 | 1587. | 1137. |
| 868.0 | 1594. | 1141. |
| 872.0 | 1602. | 1145. |
| 876.0 | 1609. | 1149. |
| 880.0 | 1616. | 1153. |
| 884.0 | 1623. | 1157. |
| 888.0 | 1630. | 1161. |

| °C | °F | K |
|----|----|----|
| 1016. | 1861. | 1289. |
| 1020. | 1868. | 1293. |
| 1024. | 1875. | 1297. |
| 1028. | 1882. | 1301. |
| 1032. | 1890. | 1305. |
| 1036. | 1897. | 1309. |
| 1040. | 1904. | 1313. |
| 1044. | 1911. | 1317. |
| 1048. | 1918. | 1321. |
| 1052. | 1926. | 1325. |
| 1056. | 1933. | 1329. |
| 1060. | 1940. | 1333. |

| °C | °F | K |
|----|----|----|
| 892.0 | 1638. | 1165. |
| 896.0 | 1645. | 1169. |
| 900.0 | 1652. | 1173. |
| 904.0 | 1659. | 1177. |
| 908.0 | 1666. | 1181. |
| 912.0 | 1674. | 1185. |
| 916.0 | 1681. | 1189. |
| 920.0 | 1688. | 1193. |

**Freezing Point of Gold** (*d*)

| °C | °F | K |
|----|----|----|
| 1064.43 | 1947.97 | 1337.58 |

| °C | °F | K |
|----|----|----|
| 1064. | 1947. | 1337. |
| 1068. | 1954. | 1341. |

**Table A–1.** (*continued*)

| °C | °F | K | °C | °F | K |
|---|---|---|---|---|---|
| 1072. | 1962. | 1345. | [b]1440. | 2624. | 1713. |
| 1076. | 1969. | 1349. | 1450. | 2642. | 1723. |
| 1080. | 1976. | 1353. | | | |
| 1084. | 1983. | 1357. | | | |

Melting Point of Nickel

| °C | °F | K |
|---|---|---|
| 1453. | 2647. | 1726. |

Freezing Point of Copper

| °C | °F | K | °C | °F | K |
|---|---|---|---|---|---|
| 1083.4 | 1982.1 | 1356.5 | 1460. | 2660. | 1733. |
| | | | 1470. | 2678. | 1743. |
| 1088. | 1990. | 1361. | 1480. | 2696. | 1753. |
| 1092. | 1998. | 1365. | 1490. | 2714. | 1763. |
| 1096. | 2005. | 1369. | 1500. | 2732. | 1773. |
| 1100. | 2012. | 1373. | 1510. | 2750. | 1783. |
| 1110. | 2030. | 1383. | 1520. | 2768. | 1793. |
| 1120. | 2048. | 1393. | 1530. | 2786. | 1803. |
| 1130. | 2066. | 1403. | | | |

Melting Point of Iron

| °C | °F | K |
|---|---|---|
| 1535. | 2795. | 1808. |

| °C | °F | K | °C | °F | K |
|---|---|---|---|---|---|
| 1140. | 2084. | 1413. | 1540. | 2804. | 1813. |
| 1150. | 2102. | 1423. | 1550. | 2822. | 1823. |
| 1160. | 2120. | 1433. | | | |
| 1170. | 2138. | 1443. | | | |
| 1180. | 2156. | 1453. | | | |
| 1190. | 2174. | 1463. | | | |
| 1200. | 2192. | 1473. | | | |

Freezing Point of Palladium

| °C | °F | K |
|---|---|---|
| 1554. | 2829. | 1827. |

| °C | °F | K | °C | °F | K |
|---|---|---|---|---|---|
| 1210. | 2210. | 1483. | 1560. | 2840. | 1833. |
| 1220. | 2228. | 1493. | 1570. | 2858. | 1843. |
| 1230. | 2246. | 1503. | 1580. | 2876. | 1853. |
| 1240. | 2264. | 1513. | 1590. | 2894. | 1863. |
| 1250. | 2282. | 1523. | 1600. | 2912. | 1873. |
| 1260. | 2300. | 1533. | 1610. | 2930. | 1883. |
| 1270. | 2318. | 1543. | 1620. | 2948. | 1893. |
| 1280. | 2336. | 1553. | 1630. | 2966. | 1903. |
| 1290. | 2354. | 1563. | | | |
| 1300. | 2372. | 1573. | 1640. | 2984. | 1913. |
| 1310. | 2390. | 1583. | 1650. | 3002. | 1923. |
| 1320. | 2408. | 1593. | | | |

Melting Point of Titanium

| °C | °F | K |
|---|---|---|
| 1660. | 3020. | 1933. |

| °C | °F | K | °C | °F | K |
|---|---|---|---|---|---|
| 1330. | 2426. | 1603. | 1670. | 3038. | 1943. |
| 1340. | 2444. | 1613. | 1680. | 3056. | 1953. |
| 1350. | 2462. | 1623. | 1690. | 3074. | 1963. |
| 1360. | 2480. | 1633. | 1700. | 3092. | 1973. |
| [b]1370. | 2498. | 1643. | 1710. | 3110. | 1983. |
| 1380. | 2516. | 1653. | 1720. | 3128. | 1993. |
| 1390. | 2534. | 1663. | 1730. | 3146. | 2003. |
| 1400. | 2552. | 1673. | 1740. | 3164. | 2013. |

Melting Point of Silicon

| °C | °F | K |
|---|---|---|
| 1410. | 2570. | 1683. |

| °C | °F | K | °C | °F | K |
|---|---|---|---|---|---|
| 1420. | 2588. | 1693. | 1750. | 3182. | 2023. |
| 1430. | 2606. | 1703. | 1760. | 3200. | 2033. |

**Table A–1.** (*continued*)

| °C | °F | K | °C | °F | K |
|---|---|---|---|---|---|
| Freezing Point of Platinum | | | 2200. | 3992. | 2473. |
| 1772. | 3222. | 2045. | 2210. | 4010. | 2483. |
| | | | 2220. | 4028. | 2493. |
| 1780. | 3236. | 2053. | 2230. | 4046. | 2503. |
| 1790. | 3254. | 2063. | 2240. | 4064. | 2513. |
| 1800. | 3272. | 2073. | 2250. | 4082. | 2523. |
| 1810. | 3290. | 2083. | 2260. | 4100. | 2533. |
| 1820. | 3308. | 2093. | 2270. | 4118. | 2543. |
| 1830. | 3326. | 2103. | 2280. | 4136. | 2553. |
| 1840. | 3344. | 2113. | | | |
| 1850. | 3362. | 2123. | 2290. | 4154. | 2563. |
| 1860. | 3380. | 2133. | 2300. | 4172. | 2573. |
| 1870. | 3398. | 2143. | 2310. | 4190. | 2583. |
| 1880. | 3416. | 2153. | 2320. | 4208. | 2593. |
| 1890. | 3434. | 2163. | 2330. | 4226. | 2603. |
| 1900. | 3452. | 2173. | 2340. | 4244. | 2613. |
| 1910. | 3470. | 2183. | 2350. | 4262. | 2623. |
| 1920. | 3488. | 2193. | 2360. | 4280. | 2633. |
| 1930. | 3506. | 2203. | 2370. | 4298. | 2643. |
| 1940. | 3524. | 2213. | 2380. | 4316. | 2653. |
| 1950. | 3542. | 2223. | | | |
| 1960. | 3560. | 2233. | 2390. | 4334. | 2663. |
| | | | 2400. | 4352. | 2673. |
| Freezing Point of Rhodium | | | Freezing Point of Iridium | | |
| 1966. | 3571. | 2239. | 2410. | 4370. | 2683. |
| 1970. | 3578. | 2243. | 2420. | 4388. | 2693. |
| 1980. | 3596. | 2253. | 2430. | 4406. | 2703. |
| 1990. | 3614. | 2263. | 2440. | 4424. | 2713. |
| 2000. | 3632. | 2273. | 2450. | 4442. | 2723. |
| 2010. | 3650. | 2283. | 2460. | 4460. | 2733. |
| 2020. | 3668. | 2293. | | | |
| 2030. | 3686. | 2303. | Melting Point of Niobium (Columbium) | | |
| 2040. | 3704. | 2313. | 2468. | 4474. | 2740. |
| 2050. | 3722. | 2323. | 2470. | 4478. | 2743. |
| 2060. | 3740. | 2333. | 2480. | 4496. | 2753. |
| 2070. | 3758. | 2343. | 2490. | 4514. | 2763. |
| 2080. | 3776. | 2353. | 2500. | 4532. | 2773. |
| 2090. | 3794. | 2363. | 2510. | 4550. | 2783. |
| 2100. | 3812. | 2373. | 2520. | 4568. | 2793. |
| 2110. | 3830. | 2383. | 2530. | 4586. | 2803. |
| 2120. | 3848. | 2393. | 2540. | 4604. | 2813. |
| 2130. | 3866. | 2403. | 2550. | 4622. | 2823. |
| 2140. | 3884. | 2413. | 2560. | 4640. | 2833. |
| 2150. | 3902. | 2423. | | | |
| 2160. | 3920. | 2433. | 2570. | 4658. | 2843. |
| 2170. | 3938. | 2443. | 2580. | 4676. | 2853. |
| 2180. | 3956. | 2453. | 2590. | 4694. | 2863. |
| | | | 2600. | 4712. | 2873. |
| 2190. | 3974. | 2463. | 2610. | 4730. | 2883. |

**Table A–1.** (*continued*)

| °C | °F | K | °C | °F | K |
|---|---|---|---|---|---|
| **Melting Point of Molybdenum** | | | 3030. | 5486. | 3303. |
| 2617. | 4743. | 2890. | 3040. | 5504. | 3313. |
| | | | 3050. | 5522. | 3323. |
| 2620. | 4748. | 2893. | 3060. | 5540. | 3333. |
| 2630. | 4766. | 2903. | 3070. | 5558. | 3343. |
| 2640. | 4784. | 2913. | 3080. | 5576. | 3353. |
| 2650. | 4802. | 2923. | 3090. | 5594. | 3363. |
| 2660. | 4820. | 2933. | 3100. | 5612. | 3373. |
| | | | 3110. | 5630. | 3383. |
| 2670. | 4838. | 2943. | 3120. | 5648. | 3393. |
| 2680. | 4856. | 2953. | 3130. | 5666. | 3403. |
| 2690. | 4874. | 2963. | 3140. | 5684. | 3413. |
| 2700. | 4892. | 2973. | | | |
| 2710. | 4910. | 2983. | 3150. | 5702. | 3423. |
| 2720. | 4928. | 2993. | 3160. | 5720. | 3433. |
| 2730. | 4946. | 3003. | 3170. | 5738. | 3443. |
| 2740. | 4964. | 3013. | | | |
| 2750. | 4982. | 3023. | **Melting Point of Rhenium** | | |
| 2760. | 5000. | 3033. | 3180. | 5756. | 3453. |
| 2770. | 5018. | 3043. | 3190. | 5774. | 3463. |
| 2780. | 5036. | 3053. | 3200. | 5792. | 3473. |
| 2790. | 5054. | 3063. | 3210. | 5810. | 3483. |
| 2800. | 5072. | 3073. | 3220. | 5828. | 3493. |
| 2810. | 5090. | 3083. | 3230. | 5846. | 3503. |
| 2820. | 5108. | 3093. | 3240. | 5864. | 3513. |
| 2830. | 5126. | 3103. | | | |
| 2840. | 5144. | 3113. | 3250. | 5882. | 3523. |
| 2850. | 5162. | 3123. | 3260. | 5900. | 3533. |
| 2860. | 5180. | 3133. | 3270. | 5918. | 3543. |
| | | | 3280. | 5936. | 3553. |
| 2870. | 5198. | 3143. | 3290. | 5954. | 3563. |
| 2880. | 5216. | 3153. | 3300. | 5972. | 3573. |
| 2890. | 5234. | 3163. | 3310. | 5990. | 3583. |
| 2900. | 5252. | 3173. | 3320. | 6008. | 3593. |
| 2910. | 5270. | 3183. | 3330. | 6026. | 3603. |
| 2920. | 5288. | 3193. | 3340. | 6044. | 3613. |
| 2930. | 5306. | 3203. | 3350. | 6062. | 3623. |
| 2940. | 5324. | 3213. | 3360. | 6080. | 3633. |
| 2950. | 5342. | 3223. | 3370. | 6098. | 3643. |
| 2960. | 5360. | 3233. | 3380. | 6116. | 3653. |
| | | | 3390. | 6134. | 3663. |
| 2970. | 5378. | 3243. | 3400. | 6152. | 3673. |
| 2980. | 5396. | 3253. | | | |
| 2990. | 5414. | 3263. | **Melting Point of Tungsten** | | |
| | | | 3410. | 6170. | 3683. |
| **Melting Point of Tantalum** | | | | | |
| 2996. | 5425. | 3269. | 3420. | 6188. | 3693. |
| | | | 3430. | 6206. | 3703. |
| 3000. | 5432. | 3273. | 3440. | 6224. | 3713. |
| 3010. | 5450. | 3283. | 3450. | 6242. | 3723. |
| 3020. | 5468. | 3293. | | | |

**Table A–1.** (*continued*)

| °C | °F | K | °C | °F | K |
|------|------|-------|------|------|-------|
| 3460. | 6260. | 3733. | 3490. | 6314. | 3763. |
| 3470. | 6278. | 3743. | 3500. | 6332. | 3773. |
| 3480. | 6296. | 3753. | 3510. | 6350. | 3783. |

[a] (*d*) = defining fixed point of the IPTS-68.
[b] Melting points of commonly used stainless steels are between about 1370 and 1440°C (2500 and 2600°F, 1640 and 1710 K).

**Table A–2. RECOMMENDATIONS FOR METALLIC SENSOR MATERIALS IN CONTACT WITH MEASURED FLUIDS**

| *Substance* | *Conditions* | *Recommended Metal* |
|-------------|--------------|---------------------|
| Acetate solvents | Crude or pure | Monel or nickel |
| Acetic acid | 10%, 70°F | 304 stainless steel |
| | 50%, 70°F | 304 stainless steel |
| | 50%, 212°F | 316 stainless steel |
| | 99%, 70°F | 430 stainless steel |
| | 99%, 212°F | 430 stainless steel |
| Acetic anhydride | | Monel |
| Acetone | 212°F | 304 stainless steel |
| Acetylene | | 304, Monel, nickel |
| Alcohol ethyl | 70°F | 304 stainless steel |
| | 212°F | 304 stainless steel |
| Alcohol methyl | 70°F | 304 stainless steel |
| | 212°F | 304 stainless steel |
| Aluminum | Molten | Cast iron |
| Aluminum acetate | Saturated | 304 stainless steel |
| Aluminum sulfate | 10%, 70°F | 304 stainless steel |
| | Saturated 70° F | 304 stainless steel |
| | 10%, 212°F | 316 stainless steel |
| | Saturated, 212°F | 316 stainless steel |
| Ammonia | All concentrations, 70°F | 304 stainless steel |
| Ammonium chloride | All concentrations, 212°F | 316 stainless steel |
| Ammonium nitrate | All concentrations, 70°F | 304 stainless steel |
| | All concentrations, 212°F | 304 stainless steel |
| Ammonium sulfate | 5%, 70°F | 304 stainless steel |
| | 10%, 212°F | 316 stainless steel |
| | Saturated, 212°F | 316 stainless steel |
| Aniline | All concentrations, 70°F | 304 stainless steel |
| Amylacetate | | Monel |
| Asphalt | | Steel (C1018), phosphor bronze, Monel, nickel |
| Barium carbonate | 70°F | 304 stainless steel |
| Barium chloride | 5%, 70°F | Monel |
| | Saturated, 70°F | Monel |
| | Aqueous, hot | 316 stainless steel |
| Barium hydroxide | | Steel (C1018) |

**Table A-2.**  (*continued*)

| Substance | Conditions | Recommended Metal |
|---|---|---|
| Barium sulfite | | Nichrome |
| Benzaldehyde | | Steel (C1018) |
| Benzene | 70°F | 304 stainless steel |
| Benzine | | Steel (C1018), Monel, Inconel |
| Benzol | Hot | 304 stainless steel |
| Boracic acid | 5%, hot or cold | 304 stainless steel |
| Bromine | 70°F | Tantalum |
| Butadiene | | Brass, 304 stainless steel |
| Butane | 70°F | 304 stainless steel |
| Butylacetate | | Monel |
| Butyl alcohol | | Copper |
| Butylenes | | Steel (C1018), phosphor bronze |
| Butyric acid | 5%, 70°F | 304 stainless steel |
| | 5%, 150°F | 304 stainless steel |
| Calcium bisulfite | 70°F | 316 stainless steel |
| Calcium chlorate | Dilute, 70°F | 304 stainless steel |
| | Dilute, 150°F | 304 stainless steel |
| Calcium hydroxide | 10%, 212°F | 304 stainless steel |
| | 20%, 212°F | 304 stainless steel |
| | 50%, 212°F | 317 stainless steel |
| Carbolic acid | All, 212°F | 316 stainless steel |
| Carbon dioxide | Dry | Steel (C1018), Monel |
| Carbon dioxide | Wet | Aluminum, Monel, nickel |
| Carbon tetrachloride | 10%, 70°F | Monel |
| Chlorex caustic | | 316, 317 stainless steel |
| Chlorine gas | Dry, 70°F | 317 stainless steel |
| | Moist, 70°F | Hastelloy C |
| | Moist, 212°F | Hastelloy C |
| Chromic acid | 5%, 70°F | 304 stainless steel |
| | 10%, 212°F | 316 stainless steel |
| | 50%, 212°F | 316 stainless steel |
| Citric acid | 15%, 70°F | 304 stainless steel |
| | 15%, 212°F | 316 stainless steel |
| | Concentrated, 212°F | 317 stainless steel |
| Coal tar | Hot | 304 stainless steel |
| Coke oven gas | | Aluminum |
| Copper nitrate | | 304, 316 stainless steel |
| Copper sulfate | | 304, 316 stainless steel |
| Core oils | | 316 stainless steel |
| Cottonseed oil | | Steel (C1018), Monel, nickel |
| Creosols | | 304 stainless steel |
| Creosote crude | | Steel (C1018), Monel, nickel |
| Cyanogen gas | | 304 stainless steel |
| Dowtherm | | Steel (C1018) |
| Epsom salt | Hot and cold | 304 stainless steel |

**Table A–2.** (*continued*)

| Substance | Conditions | Recommended Metal |
|---|---|---|
| Ether | 70°F | 304 stainless steel |
| Ethyl acetate | | Monel |
| Ethyl chloride | 70°F | 304 stainless steel |
| Ethylene glycol | | Steel (C1018) |
| Ethyl sulfate | 70°F | Monel |
| Ferric chloride | 1%, 70°F | 316 stainless steel |
| | 5%, 70°F | Tantalum |
| | 5%, boiling | Tantalum |
| Ferric sulfate | 5%, 70°F | 304 stainless steel |
| Ferrous sulfate | Dilute, 70°F | 304 stainless steel |
| Formaldehyde | | 304 stainless steel |
| Freon | | Steel (C1018) |
| Formic acid | 5%, 70°F | 316 stainless steel |
| | 5%, 150°F | 316 stainless steel |
| Gallic acid | 5%, 70°F | Monel |
| | 5%, 150°F | Monel |
| Gasoline | 70°F | 304 stainless steel |
| Glucose | 70°F | 304 stainless steel |
| Glycerine | 70°F | 304 stainless steel |
| Glycerol | | 304 stainless steel |
| Heat treating | | 446 stainless steel |
| Hydrobromic acid | 48%, 212°F | Hastelloy B |
| Hydrochloric acid | 1%, 70°F | Hastelloy C |
| | 1%, 212°F | Hastelloy B |
| | 5%, 70°F | Hastelloy C |
| | 5%, 212°F | Hastelloy B |
| | 25%, 70°F | Hastelloy B |
| | 25%, 212°F | Hastelloy B |
| Hydrocyanic acid | | 316 stainless steel |
| Hydrofluoric acid | | Hastelloy C |
| Hydrogen peroxide | 70°F | 316 stainless steel |
| | 212°F | 316 stainless steel |
| Hydrogen sulfide | Wet and dry | 316 stainless steel |
| Iodine | 70°F | Tantalum |
| Kerosene | 70°F | 304 stainless steel |
| Lactic acid | 5%, 70°F | 304 stainless steel |
| | 5%, 150°F | 316 stainless steel |
| | 10%, 212°F | Tantalum |
| Lacquer | 70°F | 316 stainless steel |
| Latex | | Steel (C1018) |
| Lime sulfur | | Steel (C1018), 304 stainless steel, Monel |
| Linseed oil | 70°F | 304 stainless steel |
| Magnesium chloride | 5%, 70°F | Monel |
| | 5%, 212°F | nickel |
| Magnesium sulfate | Cold and hot | Monel |
| Malic acid | Cold and hot | 316 stainless steel |
| Mercury | | Steel (C1018), 304 stainless steel, Monel |

**Table A–2.** (*continued*)

| Substance | Conditions | Recommended Metal |
|---|---|---|
| Methane | 70°F | Steel (1020) |
| Milk | | 304 stainless steel, nickel |
| Mixed acids (sulfuric and nitric—all temp. and %) | | Carpenter 20 |
| Molasses | | Steel (C1018), 304 stainless steel, Monel, nickel |
| Muriatic acid | 70°F | Tantalum |
| Naphtha | 70°F | 304 stainless steel |
| Natural gas | 70°F | 304 stainless steel |
| Neon | 70°F | 304 stainless steel |
| Nickel chloride | 70°F | 304 stainless steel |
| Nickel sulfate | Hot and cold | 304 stainless steel |
| Nitric acid | 5%, 70°F | 304 stainless steel |
| | 20%, 70°F | 304 stainless steel |
| | 50%, 70°F | 304 stainless steel |
| | 50%, 212°F | 304 stainless steel |
| | 65%, 212°F | 316 stainless steel |
| | Concentrated, 70°F | 304 stainless steel |
| | Concentrated, 212°F | Tantalum |
| Nitrobenzene | 70°F | 304 stainless steel |
| Nitrous acid | | 304 stainless steel |
| Oleic acid | 70°F | 316 stainless steel |
| Oleum | 70°F | 316 stainless steel |
| Oxalic acid | 5%, Hot and cold | 304 stainless steel |
| | 10%, 212°F | Monel |
| Oxygen | 70°F | Steel (C1018) |
| Oxygen | Liquid | 304 stainless steel |
| Palmitic acid | | 316 stainless steel |
| Petroleum ether | | 304 stainless steel |
| Phenol | | 304 stainless steel |
| Pentane | | 304 stainless steel |
| Phosphoric acid | 1%, 70°F | 304 stainless steel |
| | 5%, 70°F | 304 stainless steel |
| | 10%, 70°F | 316 stainless steel |
| | 10%, 212°F | Hastelloy C |
| | 30%, 70°F | Hastelloy B |
| | 30%, 212°F | Hastelloy B |
| | 85%, 70°F | Hastelloy B |
| | 85%, 212°F | Hastelloy B |
| Picric acid | 70°F | 304 stainless steel |
| Potassium bromide | 70°F | 316 stainless steel |
| Potassium carbonate | 1%, 70°F | 304 stainless steel |
| Potassium chlorate | 70°F | 304 stainless steel |
| Potassium chloride | 5%, 70°F | 304 stainless steel |
| | 5%, 212°F | 304 stainless steel |
| Potassium hydroxide | 5%, 70°F | 304 stainless steel |
| | 25%, 212°F | 304 stainless steel |
| | 50%, 212°F | 316 stainless steel |
| Potassium nitrate | 5%, 70°F | 304 stainless steel |

**Table A–2.**  (*continued*)

| Substance | Conditions | Recommended Metal |
|-----------|-----------|-------------------|
| Potassium | 5%, 212°F | 304 stainless steel |
| Permanganate | 5%, 70°F | 304 stainless steel |
| Potassium sulfate | 5%, 70°F | 304 stainless steel |
| | 5%, 212°F | 304 stainless steel |
| Potassium sulfide | 70°F | 304 stainless steel |
| Propane | | 304 stainless steel |
| Pyrogallic acid | | 304 stainless steel |
| Quinine bisulfate | Dry | 316 stainless steel |
| Quinine sulfate | Dry | 304 stainless steel |
| Resin | | 304 stainless steel |
| Rosin | Molten | 304 stainless steel |
| Seawater | | Monel |
| Salammoniac | | Monel |
| Salicylic acid | | Nickel |
| Shellac | | 304 stainless steel |
| Soap | 70°F | 304 stainless steel |
| Sodium bicarbonate | All concentrations, 70°F | 304 stainless steel |
| | 5%, 150°F | 304 stainless steel |
| Sodium bisulfate | | Monel |
| Sodium carbonate | 5%, 70°F | 304 stainless steel |
| | 5%, 150°F | 304 stainless steel |
| Sodium chloride | 5%, 70°F | 316 stainless steel |
| | 5%, 150°F | 316 stainless steel |
| | Saturated, 70°F | 316 stainless steel |
| | Saturated, 212°F | 316 stainless steel |
| Sodium fluoride | 5%, 70°F | Monel |
| Sodium hydroxide | | 304 stainless steel |
| Sodium hypochlorite | 5% still | 316 stainless steel |
| Sodium nitrate | Fused | 317 stainless steel |
| Sodium peroxide | | 304 stainless steel |
| Sodium phosphate | | Steel (C1018) |
| Sodium silicate | | Steel (C1018) |
| Sodium sulfate | 70°F | 304 stainless steel |
| Sodium sulfide | 70°F | 316 stainless steel |
| Sodium sulfite | 150°F | 304 stainless steel |
| Steam | | 304 stainless steel |
| Stearic acid | | 304 stainless steel |
| Sulphur dioxide | Moist gas, 70°F | 316 stainless steel |
| | Gas, 575°F | 304 stainless steel |
| Sulphur | Dry, molten | 304 stainless steel |
| | Wet | 316 stainless steel |
| Sulphuric acid | 5%, 70°F | Carp. 20, Hastelloy B |
| | 5%, 212°F | Carp. 20, Hastelloy B |
| | 10%, 70°F | Carp. 20, Hastelloy B |
| | 10%, 212°F | Carp. 20, Hastelloy B |
| | 50%, 70°F | Carp. 20, Hastelloy B |
| | 50%, 212°F | Carp. 20, Hastelloy B |
| | 90%, 70°F | Carp. 20, Hastelloy B |
| | 90%, 212°F | Hastelloy D |

**Table A–2.** (*continued*)

| Substance | Conditions | Recommended Metal |
|---|---|---|
| Tannic acid | 70°F | 304 stainless steel |
| Tar | | Steel (C1018), 304 stainless steel, Monel, nickel |
| Tartaric acid | 70°F | 304 stainless steel |
| | 150°F | 316 stainless steel |
| Tin | Molten | Cast iron |
| Toluene | | Aluminum, phosphor bronze, Monel |
| Trichloroethylene | | Steel (C1018) |
| Turpentine | | 304 stainless steel |
| Varnish | | 304 stainless steel |
| Vegetable oils | | Steel (C1018), 304 stainless steel, Monel |
| Vinegar | | 304 stainless steel |
| Water | Fresh | Copper, steel (C1018) Monel |
| Water | Salt | Aluminum, brass, Monel |
| Whiskey, wine | | 304 stainless steel, nickel |
| Xylene | | Copper |
| Zinc | Molten | Cast iron |
| Zinc chloride | | Monel |
| Zinc sulfate | 5%, 70°F | 304 stainless steel |
| | Saturated, 70°F | 304 stainless steel |
| | 25%, 212°F | 304 stainless steel |

Courtesy of Omega Engineering, Inc., an Omega Group Co.
*Notes:* 1. Considerations of compatibility include contamination, electrolysis, and catalytic reactions.
2. Many of the recommended metals are proprietary alloys used in the U.S.A.; equivalent alloys exist, in most cases, in other countries.

# Bibliography

The following publications are potentially useful in providing additional background information and more details in the areas covered by the chapters indicated.

**For Chapters 1–4:**

Benedict, R. P. *Fundamentals of Temperature, Pressure and Flow Measurements.* 3rd ed. Research Triangle Park, NC: Instrument Society of America, 1984.

Chesmond, C. J. *Control System Technology.* Research Triangle Park, NC: Instrument Society of America, 1984.

Christiansen, D., ed. *Electronic Engineers' Handbook.* 3rd ed. New York: McGraw-Hill, 1988.

Considine, D. M., ed. *Process Instruments and Controls Handbook.* New York: McGraw-Hill.

Institute of Environmental Sciences. *Glossary: Digital Analysis of Dynamic Data Terminology.* Mount Prospect, IL: 1983.

Instrument Society of America. "Dynamic Response Testing of Process Control Instrumentation." *ANSI MC 4.1/ISA-S26.* Research Triangle Park, NC: 1975.

Instrument Society of America. "Electrical Transducer Nomenclature and Terminology." *ANSI MC 6.1/ISA-S37.1* Research Triangle Park, NC: 1982.

Instrument Society of America. "Instrumentation Grounding and Noise Minimization Handbook." *Technical Report No. AFRPL-TR-65-1.* Research Triangle Park, NC: 1973.

Instrument Society of America. "Instrumentation Symbols and Identification." *ANSI/ISA Standard S5.1.* Research Triangle Park, NC: 1984.

Instrument Society of America. *ISA Electrical Safety Standards.* Research Triangle Park, NC: 1982.

Instrument Society of America. "Process Instrumentation Terminology." *ANSI/ISA S51.1.* Research Triangle Park, NC: 1979.

Jones, B. E. *Instrumentation, Measurement and Feedback.* Maidenhead, England: McGraw-Hill, 1977.

Kuo, B. C. *Automatic Control Systems.* 3rd ed. Englewood Cliffs, NJ: Prentice-Hall, 1975.

McCaw, L. *Industrial Measurements—A Laboratory Manual.* Research Triangle Park, NC: Instrument Society of America, 1987.

Morrison, R. *Grounding and Shielding Techniques in Instrumentation.* 2d ed. New York: Wiley, 1977.

Morrison, R. *Instrumentation Fundamentals and Applications.* Research Triangle Park, NC: Instrument Society of America, 1984.

Murrill, P. W. *Fundamentals of Process Control Theory.* Research Triangle Park, NC: Instrument Society of America, 1981.

Norton, H. N., ed. *Sensor and Transducer Selection Guide.* Oxford: Elsevier Advanced Technology Publications, 1989.

Strock, O. J. *Introduction to Telemetry.* Research Triangle Park, NC: Instrument Society of America, 1986.

Superintendent of Documents. "Environmental Test Methods for Aerospace and Ground Equipment." *MIL-STD-810.* Washington, DC.

Sydenham, P. H. *Basic Electronics for Instrumentation.* Research Triangle Park, NC: Instrument Society of America, 1982.

Szymkowiak, E. *Optimized Vibration Testing and Analysis.* Mount Prospect, IL: Institute of Environmental Sciences, 1983.

Travers, D. *Precision Signal Handling and Converter-Microprocessor Interface Techniques.* Research Triangle Park, NC: Instrument Society of America, 1984.

**For Chapters 5–11:**

Harris, C. M., and Crede, C. E., eds. *Shock and Vibration Handbook.* 2d ed. New York: McGraw-Hill, 1977.

McConnell, K. G. *Dynamic Force, Motion and Pressure Measurements.* Brookfield Center, CT: SEM Publications, 1984.

McConnell, K. G. *Notes on Vibration Frequency Analysis.* Brookfield Center, CT: SEM Publications, 1978.

Meirovitch, L. *Elements of Vibration Analysis.* New York: McGraw-Hill, 1975.

Roark, R. J., and Young, W. C. *Formulas for Stress and Strain.* 5th ed. Brookfield Center, CT: SEM Publications, 1975.

Standards for Transducers, available from Instrument Society of America, Research Triangle Park, NC:

"Guide for Specifications and Tests for Piezoelectric Acceleration Transducers for Aerospace Testing." *ISA RP37.2* (Reaffirmed 1982).

"Specifications and Tests for Potentiometric Displacement Transducers." *ANSI/ISA S37.12-1977* (Reaffirmed 1982).

"Specifications and Tests for Strain Gage Force Transducers." *ANSI/ISA S37.8-1977* (Reaffirmed 1982).

"Specifications and Tests for Strain Gage Linear Acceleration Transducers." *ANSI/ISA S37.5-1975* (Reaffirmed 1982).

Window, A. L., and Holister, G. S. *Strain Gauge Technology*. Brookfield Center, CT: SEM Publications, 1982.


**For Chapters 12–14, and 18:**

American Society for Testing and Materials. "Relative Humidity by Wet- and Dry-Bulb Psychrometer, Method of Test for." *ANSI Z110.3/ASTM E337*. Philadelphia, PA: Apr. 26, 1974.

Cho, C. H. *Measurement and Control of Liquid Level*. Research Triangle Park, NC: Instrument Society of America, 1982.

DeCarlo, J. P. *Fundamentals of Flow Measurement*. Research Triangle Park, NC: Instrument Society of America, 1984.

Durst, F., Melling, A., and Whitelaw, J. H. *Principles and Practice of Laser Doppler Anemometry*. New York: Academic Press, 1976.

Hinze, I. O. *Turbulence*. New York: McGraw-Hill, 1975.

Instrument Society of America. *Moisture and Humidity—Measurement and Control in Science and Industry* (symposium papers). Research Triangle Park, NC: 1985.

Instrument Society of America. "Specification, Installation, and Calibration of Turbine Flowmeters." *ANSI/ISA RP31.1-1977*. Research Triangle Park, NC: 1977.

Miller, R. W. *Flow Measurement Engineering Handbook*. New York: McGraw-Hill, 1983.


**For Chapters 15–17:**

Examples of ISO Standards Related to Sound Measurement, available from Secretariat, International Organization for Standardization, Geneva, Switzerland:

"Assessment of Occupational Noise Exposure for Hearing Conservation Purposes." *ISO S 1999*.

*"Measurement of Noise Emitted by Vessels on Inland Waterways and Harbors." ISO S 2922*.

"Measurement of Noise inside Railbound Vehicles." *ISO S 3381*.

"Measurement of Reverberation Time in Auditoria." *ISO S 3382*.

"Standard for Sound Level Meters." *ISO S 651*.

Gillum, D. R. *Industrial Pressure Measurement* (Student Text, Instructor's Guide, Slides). Research Triangle Park, NC: Instrument Society of America, 1982.

Ryans, J. L., and Roper, D. L. *Process Vacuum System Design and Operation*. New York: McGraw-Hill, 1986.

Standards for Pressure Transducers, available from Instrument Society of America, Research Triangle Park, NC:

"Specifications and Tests for Piezoelectric Pressure and Sound-Pressure Transducers." *ANSI MC6.4/ISA S37.10-1975* (Reaffirmed 1982).

"Specifications and Tests for Strain Gage Pressure Transducers." *ANSI/ISA S37.3-1975* (Reaffirmed 1982).

"Specifications and Tests of Potentiometric Pressure Transducers." *ANSI/ISA S37.6-1976* (Reaffirmed 1982).

Standards for Sound Measurements, available from American National Standards Institute (ANSI), New York:

"Methods for Sound Power Determination." *ANSI S1.31 through S1.36.*

"Methods for the Measurement of Sound Pressure Levels." *ANSI S1.13.*

"Preferred Frequencies for Acoustical Measurements." *ANSI S1.6.*

"Preferred Reference Quantities for Acoustical Levels." *ANSI S1.8.*

"Procedures for Calibration of Underwater Electroacoustic Transducers." *ANSI S1.20.*

"Specification for Sound Level Meters." *ANSI S1.4.*

Yerges, L. F. *Sound, Noise and Vibration Control.* New York: Van Nostrand Reinhold, 1978.

**For Chapters 19–21:**

American Institute of Physics. *Temperature: Its Measurement and Control in Science and Industry* (symposium papers, two parts). New York: 1982.

Drummond, A. J., and Thekaekara, M. P., eds. *The Extraterrestrial Solar Spectrum.* Mount Prospect, IL: Institute of Environmental Sciences, 1973.

Instrument Society of America. "American National Standard for Temperature Measurement Thermocouples." *ANSI MC96.1.* Research Triangle Park, NC: 1982.

Kerlin, T. W., and Shepard, R. L. *Industrial Temperature Measurement* (Student Text, Instructor's Guide, Slides). Research Triangle Park, NC: Instrument Society of America, 1982.

**For Chapters 22–24:**

Blair, B. E., ed. "Time and Frequency: Theory and Fundamentals." *NBS Monograph 140.* Washington, DC: U.S. Government Printing Office, 1974.

Boyd, R. W. *Radiometry and the Detection of Optical Radiation.* New York: Wiley, 1983.

Dereniak, E. L. *Optical Radiation Detectors.* New York: Wiley, 1984.

Engstrom, R. W. *Photomultiplier Handbook.* Lancaster, PA: Burle Industries, Inc., 1980.

Fox, R. W. *Optoelectronics Guidebook.* Blue Ridge Summit, PA: TAB Books, 1977.

Institute of Electrical and Electronic Engineers (IEEE). *Nuclear IEEE Standards.* Vols. 1 and 2. New York: Wiley, 1979.

Instrument Society of America. "American National Standard for Voltage or Current Reference Devices: Solid State Devices." *ANSI C100.6-3.* Research Triangle Park, NC: 1984.

Instrument Society of America. "Transducer and Transmitter Installation for Nuclear Safety Applications." *ANSI/ISA Standard S67.01.* Research Triangle Park, NC: 1979 (Reaffirmed 1986).

International Society for Optical Engineering. *Materials Technology for Infrared Detectors.* Bellingham, WA: 1986.

Kaelble, E. F. *Handbook of X-rays.* New York: Wiley, 1974.

Kaminow, I. P. *An Introduction to Electrooptic Devices.* New York: Academic Press, 1974.

Keyes, R. J. *Optical and Infrared Detectors.* 2d ed. New York: Springer-Verlag, 1980.

Knoll, G. F. *Radiation Detection and Measurement.* New York: Wiley, 1979.

Melen, R., and Buss, D., eds. *Charge-Coupled Devices: Technology and Applications.* New York: IEEE Press, 1977.

National Council on Radiation Protection and Measurement. "A Handbook of Radioactivity Measurement Procedures." *NCRP Report No. 58.* Washington, DC: 1979.

Stimson, A. *Photometry and Radiometry for Engineers.* New York: Wiley, 1974.

Wolfe, W. L., and Zissis, G. J., eds. *The Infrared Handbook.* ed. Ann Arbor, MI: Environmental Research Institute of Michigan, 1985.

# Index